Applied Mathematics
of Science & Engineering

Introductory Mathematics for Scientists and Engineers

A Wiley Series, edited by

D. S. JONES, M.B.E., F.R.S.E., F.R.S.

Ivory Professor of Mathematics in the University of Dundee

Applied Mathematics
of Science & Engineering

CHARLES DIXON

Department of Mathematics
University of Dundee

JOHN WILEY & SONS   LONDON NEW YORK SYDNEY TORONTO

Library of Congress Catalog Card No. 77-161692

ISBN 0 471 21626 7

Printed in Great Britain by Universities Press, Belfast, Northern Ireland

To my mother

Introductory Mathematics for Scientists and Engineers

Foreword to the Series

The increasing use of high speed digital computers and the growing desire for numerical answers in many disciplines have led to a steady expansion in the numbers of mathematics courses in the first year or two of college and university studies. Many of these courses are intended for students of physics, chemistry, engineering, biology and economics who will regard mathematics as a tool, but a tool with which they must develop some proficiency. This series is designed for such students. However, that does not prevent prospective mathematicians from reading these books with profit.

The authors have, in general, avoided the strict axiomatic approach which is favoured by some pure mathematicians, but there has been no dilution of the standard of mathematical argument. Learning to follow and construct a logical sequence of ideas is one of the important attributes of a course in mathematics.

While the authors' purpose has been to stress mathematical ideas which are central to applications and necessary for subsequent studies, they have attempted, when appropriate, to convey some notion of the connection between a mathematical model and the real world. Exercises have been included which take account of the ready availability of electronic digital computers.

The careful explanation of difficult points and the provision of large numbers of worked examples and exercises should ensure the popularity of the books in this series with students and teachers alike.

D. S. JONES
Department of Mathematics
University of Dundee

Preface

The text covers some of the material covered in most university undergraduate courses in applied mathematics. I have tried to steer a middle course between side-stepping and glossing over difficult points in the theory on the one hand and including too much rigour, to the detriment of the general reader's understanding, on the other. The text includes numerous examples worked out in detail and it is hoped that these will both illustrate some of the applications of the theory and help towards a better understanding of the theory itself.

On the whole new mathematical techniques have been kept separate from their applications which appear later. This has been thought desirable as I feel that students in general have great difficulty trying to assimilate both the new techniques and their applications, simultaneously. This means however that the reader is, in places, asked to learn new techniques while taking it largely on trust that applications will in fact appear later. This applies in particular to some of the material included in Chapters 4 and 5. A reader who requires a little more motivation might perhaps obtain this by glancing at some of the examples considered in Chapters 6, 7 and 8 before starting on Chapter 4. Alternatively, any reader who must be strongly motivated before he learns any new mathematics might like to jump from Chapter 3 to Chapter 6. He would then soon find it necessary to refer back to the material of Chapters 4 and 5. However I do not feel that this method would be using the text to the best advantage.

My thanks are due to several of my colleagues in the Mathematics Department of the University of Dundee for helpful comments, but in particular my thanks are due to my former teacher and present colleague Mr. Henry Jack who read the first draft of my manuscript and made many constructive suggestions.

C. DIXON

Contents

1

Vector operators

1.1 Line integrals

We begin by recalling the definition of the Riemann integral. Suppose the function f of a single variable x is defined and continuous for all values of x such that $a \leqslant x \leqslant b$ for some finite constants a and b. Divide the interval from a to b into n (not necessarily equal) parts by the points $x_0, x_1, \ldots, x_n$, where $x_0 = a$, $x_n = b$ and $x_0 < x_1 < \ldots < x_n$. Denote by m the maximum value of $(x_{i+1} - x_i)$ for $i = 0, 1, \ldots, n - 1$ (figure 1).

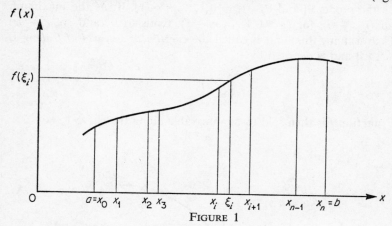

FIGURE 1

Now form the sum

$$S = \sum_{i=0}^{n-1} f(\xi_i)(x_{i+1} - x_i)$$

where, for each value of i, ξ_i is chosen so that $x_i \leqslant \xi_i \leqslant x_{i+1}$. If as m tends to 0 (and so $n \to \infty$), S tends to a limit, this limit is called the Riemann integral of $f(x)$ with respect to x from a to b and is denoted by

$$\int_a^b f(x)\, dx.$$

1

In this case, the function $f(x)$ is said to be integrable (in the Riemann sense) in the interval $a \leqslant x \leqslant b$. If the above is looked upon as the integral of $f(x)$ along the x-axis, then the following generalization of the definition to integration along any simple curve L comes quite naturally.

Suppose a function f is uniquely defined, and continuous for all points along a given simple continuous curve L between the points A and B (figure 2). With L is associated the direction from A to B.

Divide the curve L into n not necessarily equal parts by the points $P_0, P_1, P_2, \ldots, P_n$ where P_0 coincides with A and P_n coincides with B (figure 2). Let $l(P)$ denote the length of the curve from some fixed point O, on the curve, to any other point P also on the curve and let l increase as P moves in the direction from A to B. Let $l_i = l(P_i)$ denote the length of the curve from the point O to the point $P_i (i = 0,1,\ldots,n)$. Now form the sum

$$S = \sum_{i=0}^{n-1} (l_{i+1} - l_i)f_i$$

where f_i denotes the value of the function f at some point on the curve between the points P_i and P_{i+1} $(i = 0,1,\ldots,n-1)$. If, as the maximum value of $(l_{i+1} - l_i)$ for $(i = 0,1,\ldots,n-1)$, tends to 0 (and so $n \to \infty$) S tends to a limit, this limit is called the *curvilinear integral of f along* L from A to B and is denoted by

$$\int_L f \, dl.$$

The function f is then said to be integrable along the curve L.

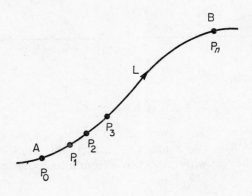

FIGURE 2

If the function f has the constant value k for all points along the curve L then

$$\int_L f \, \mathrm{d}l = \int_L k \, \mathrm{d}l = k \int_L \mathrm{d}l = k(l(B) - l(A))$$

where $(l(B) - l(A))$ is the length of the curve between the points A and B.

If L′ has associated with it the direction from B to A, that is the opposite direction to L, but is otherwise identical with L, then $\int_{L'} f \, \mathrm{d}l$ is obtained by taking the limit, as the maximum value of $(l_{i+1} - l_i)$ for $i = 0, 1, \ldots, n - 1$ tends to zero, of the sum

$$S' = \sum_{i=0}^{n-1} (l_i - l_{i+1}) f_i = -\sum_{i=0}^{n-1} (l_{i+1} - l_i) f_i = -S.$$

Hence

$$\int_{L'} f \, \mathrm{d}l = -\int_L f \, \mathrm{d}l.$$

The following results (already known for integrals along the x-axis) also hold.

1.
$$\int_L (f + g) \, \mathrm{d}l = \int_L f \, \mathrm{d}l + \int_L g \, \mathrm{d}l$$

where f and g are integrable along L.

2.
$$\int_L f \, \mathrm{d}l = \int_{L_1} f \, \mathrm{d}l + \int_{L_2} f \, \mathrm{d}l$$

where the curve L is composed of the two parts L_1, from A to C, and L_2, from C to B. The gradient of the curve L need not be continuous at the point C (figure 3).

3.
$$m(l(B) - l(A)) \leqslant \int_L f \, \mathrm{d}l \leqslant M(l(B) - l(A))$$

where $(l(B) - l(A))$ is the length of the curve L and $m \leqslant f \leqslant M$ for all points along the curve.

4. There exists at least one point P on L such that

$$\int_L f \, \mathrm{d}l = f(P)(l(B) - l(A))$$

where $f(P)$ is the value of the function f at the point P. (This result is called the *mean value theorem* for line integrals).

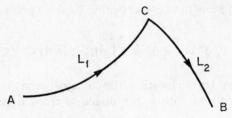

FIGURE 3

If the curve is represented parametrically, using l as parameter, by

$$\mathbf{r}(l) = (x(l), y(l), z(l)),$$

then

$$\int_L f(x, y, z)\, dl = \int_a^b f(x(l), y(l), z(l))\, dl.$$

where a and b are the particular parameter values corresponding to the points A and B respectively.

However, curves are not always represented parametrically in this way using the length l as parameter, but may be represented in terms of some other parameter t, by

$$\mathbf{r}(t) = (x(t), y(t), z(t)).$$

Then

$$\int_L f(x, y, z)\, dl = \int_{l=a}^{l=b} f(x(t), y(t), z(t))\, dl$$
$$= \int_{t=t_A}^{t=t_B} f(x(t), y(t), z(t)) \frac{dl}{dt}\, dt$$

where t_A and t_B, are the parameter values corresponding to the points A and B respectively. But, denoting dx/dt, dy/dt, dz/dt by $\dot{x}$, $\dot{y}$, $\dot{z}$ respectively, we have

$$\frac{dl}{dt} = \sqrt{\dot{x}^2 + \dot{y}^2 + \dot{z}^2}$$

when the sense of l is chosen so that l increases with increasing t. Hence we obtain

$$\int_L f(x, y, z)\, dl = \int_{t_A}^{t_B} f(x(t), y(t), z(t))(\dot{x}^2 + \dot{y}^2 + \dot{z}^2)^{\frac{1}{2}}\, dt$$

if l increases with increasing t. That is

$$\int_L f(x, y, z)\, dl = \int_{t_A}^{t_B} f(x(t), y(t), z(t)) g(t)\, dt$$

where $g(t) = (\dot{x}^2 + \dot{y}^2 + \dot{z}^2)^{\frac{1}{2}}$ is a function dependent on the form of the curve L.

Example   Evaluate

$$\int_L (x^2 y + 3xyz)\, \mathrm{d}l$$

where L is the smaller arc of the circle with centre the origin and radius a from the point $(a,0,0)$ to the point $(0,a,0)$.

The path of integration (figure 4) lies entirely in the x,y-plane and may be represented parametrically by

$$x = a \cos t, \qquad y = a \sin t, \qquad z = 0.$$

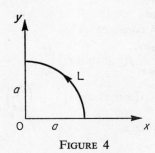

FIGURE 4

Therefore

$$\dot{x}^2 + \dot{y}^2 + \dot{z}^2 = a^2 (\sin^2 t + \cos^2 t) = a^2.$$

Then

$$\int_L (x^2 y + 3xyz)\, \mathrm{d}l = \int_0^{\pi/2} (a^2 \cos^2 t)(a \sin t) a\, \mathrm{d}t$$

$$= a^4 \left[-\frac{\cos^3 t}{3} \right]_0^{\pi/2} = \tfrac{1}{3}a^4.$$

Example   Evaluate

$$\int_L (3y + xz)\, \mathrm{d}l$$

where L is the curve given parametrically by $\mathbf{r}(t) = (2t, \tfrac{1}{3}t^3, t^2)$ from the point $(0,0,0)$ to the point $(2,\tfrac{1}{3},1)$.

Here

$$x(t) = 2t, \qquad y(t) = \tfrac{1}{3}t^3, \qquad z(t) = t^2$$

$$\dot{\mathbf{r}} = (\dot{x},\dot{y},\dot{z}) = (2,t^2,2t)$$

Therefore

$$\mathbf{\dot{r}} \cdot \mathbf{\dot{r}} = \dot{x}^2 + \dot{y}^2 + \dot{z}^2 = 4 + t^4 + 4t^2 = (t^2 + 2)^2.$$

Hence

$$\int_L (3y + xz)\, \mathrm{d}l = \int_0^1 (t^3 + 2t^3)(t^2 + 2)\, \mathrm{d}t$$

$$= \int_0^1 (3t^5 + 6t^3)\, \mathrm{d}t$$

$$= \left[\tfrac{1}{2}t^6 + \tfrac{3}{2}t^4 \right]_0^1 = 2.$$

Suppose now that the path is not given parametrically, but instead is represented in the form

$$y = y(x), \qquad z = z(x).$$

(This implies that no finite length of the path lies in a plane having an equation of the form $x = $ constant.) Then

$$\int_L f(x,y,z)\, \mathrm{d}l = \int_{x=x_0}^{x=x_1} f\big(x,y(x),z(x)\big)\frac{\mathrm{d}l}{\mathrm{d}x}\, \mathrm{d}x$$

$$= \int_{x_0}^{x_1} f\big(x,y(x),z(x)\big)\left(1 + \left(\frac{\mathrm{d}y}{\mathrm{d}x}\right)^2 + \left(\frac{\mathrm{d}z}{\mathrm{d}x}\right)^2\right)^{\frac{1}{2}}\, \mathrm{d}x$$

if l increases with increasing x.

Example   Consider again $\int_L (x^2 y + 3xyz)\, \mathrm{d}l$ *where* L *is the smaller arc of the circle with centre the origin and radius a from the point* $(a,0,0)$ *to the point* $(0,a,0)$ *(figure 4).*

The path of integration may be represented by

$$x^2 + y^2 = a^2, \qquad z = 0$$

or

$$y = \sqrt{a^2 - x^2}, \qquad z = 0$$

(y is positive for the given curve L). Therefore

$$\frac{\mathrm{d}y}{\mathrm{d}x} = \frac{-x}{\sqrt{a^2 - x^2}}, \qquad \frac{\mathrm{d}z}{\mathrm{d}x} = 0$$

and so

$$1 + \left(\frac{\mathrm{d}y}{\mathrm{d}x}\right)^2 + \left(\frac{\mathrm{d}z}{\mathrm{d}x}\right)^2 = 1 + \frac{x^2}{a^2 - x^2} = \frac{a^2}{a^2 - x^2}.$$

Hence

$$\int_L (x^2 y + 3xyz)\, \mathrm{d}l = -\int_a^0 (x^2\sqrt{a^2 - x^2})\left(\frac{a}{\sqrt{a^2 - x^2}}\right)\, \mathrm{d}x$$

The minus sign outside the integral sign is required because, as the curve is described in the given sense and so l increases, x decreases. That is dl/dx is negative and so

$$\frac{dl}{dx} = -\sqrt{1 + \left(\frac{dy}{d}\right)^2 + \left(\frac{dz}{d}\right)^2} = \frac{-a}{\sqrt{a^2 - x^2}}.$$

Therefore

$$\int_L (x^2y + 3xyz)\, dl = -a \int_a^0 x^2\, dx = \tfrac{1}{3}a^4.$$

Now we have seen that

$$\int_L f(x,y,z)\, dl = \int_{x_0}^{x_1} f(x,y(x),z(x))\, \frac{dl}{dx}\, dx$$

provided the curve L can be represented in the form $y = y(x)$, $z = z(x)$. Similarly we can write

$$\int_L f(x,y,z)\, dl = \int_{y_0}^{y_1} f(x(y),y,z(y))\, \frac{dl}{dy}\, dy$$

provided the curve L can be represented in the form $x = x(y)$, $z = z(y)$.
Also

$$\int_L f(x,y,z)\, dl = \int_{z_0}^{z_1} f(x(z),y(z),z)\, \frac{dl}{dz}\, dz$$

provided the curve L can be represented in the form $x = x(z)$, $y = y(z)$. Which form is used will of course depend on the particular problem.

Up to now we have considered only integrals of scalar functions along a curve. We shall now extend the above ideas to integrals of vector functions along a curve. Readers familiar with mechanics will realize that such an integral gives the work done when the point of application of a force moves along a curve. If $\int_{x_0}^{x_1} g_1\, dx$, $\int_{y_0}^{y_1} g_2\, dy$ and $\int_{z_0}^{z_1} g_3\, dz$ represent three line integrals along the same curve L, then

$$\int_{x_0}^{x_1} g_1\, dx + \int_{y_0}^{y_1} g_2\, dy + \int_{z_0}^{z_1} g_3\, dz = \int_L (g_1\, dx + g_2\, dy + g_3\, dz)$$

$$= \int_L \left(g_1\, \frac{dx}{dl} + g_2\, \frac{dy}{dl} + g_3\, \frac{dz}{dl}\right) dl$$

$$= \int_L \left(\mathbf{g} \cdot \frac{d\mathbf{r}}{dl}\right) dl$$

where $\mathbf{g} = (g_1, g_2, g_3)$ and $\mathbf{r} = (x(l), y(l), z(l))$ represents the path of integration L. Now it is shown in books on elementary calculus that

$$\left(\frac{dx}{dl}\right)^2 + \left(\frac{dy}{dl}\right)^2 + \left(\frac{dz}{dl}\right)^2 = 1$$

and so

$$\frac{d\mathbf{r}}{dl} = \left(\frac{dx}{dl}, \frac{dy}{dl}, \frac{dz}{dl}\right)$$

is a unit vector. Its direction is that of the tangent to the curve L at the point having parameter value l.

The line integral

$$\int_L \left(\mathbf{g} \cdot \frac{d\mathbf{r}}{dl}\right) dl$$

is frequently also written as

$$\int_L \mathbf{g} \cdot d\mathbf{r}$$

where $d\mathbf{r}$ is looked upon as the vector (dx, dy, dz).

Example   Evaluate $I \equiv \int_L \{(3x^2 - 6yz)\, dx + (2y + 3xz)\, dy + (1 - 4xyz^2)\, dz\}$
for the following three curves L.
  (i) *The curve given parametrically by* $\mathbf{r} = (t, t^2, t^3)$ *from the point* A(0,0,0) *to the point* B(1,1,1).
  (ii) *The straight lines* AC, CD, DB *where* A *and* B *are as above and* C *and* D *are the points* (0,0,1) *and* (0,1,1) *respectively.*
  (iii) *The straight line* AB (*figure 5*).

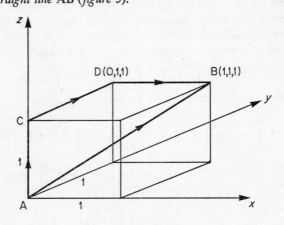

FIGURE 5

(i)

$$x = t, \qquad y = t^2, \qquad z = t^3.$$

Therefore $dx = dt$, $dy = 2t\, dt$ and $dz = 3t^2\, dt$. Hence

$$I = \int_0^1 \{(3t^2 - 6t^5) + (2t^2 + 3t^4)2t + (1 - 4t^9)3t^2\} \, dt$$

$$= \int_0^1 (-12t^{11} + 4t^3 + 6t^2) \, dt = 2.$$

(ii) If we put $\mathbf{F} = (3x^2 - 6yz, 2y + 3xz, 1 - 4xyz^2)$, then

$$I = \int_L \mathbf{F} \cdot d\mathbf{r} = \int_{L_1} \mathbf{F} \cdot d\mathbf{r} + \int_{L_2} \mathbf{F} \cdot d\mathbf{r} + \int_{L_3} \mathbf{F} \cdot d\mathbf{r}$$

where L_1, L_2, L_3 are the straight lines AC, CD, DB respectively. Along AC, $x = 0$ and $y = 0$ and z varies between 0 and 1. Therefore $dx = 0$ and $dy = 0$. Hence

$$\int_{L_1} \mathbf{F} \cdot d\mathbf{r} = \int_0^1 dz = 1.$$

Along CD, $x = 0$, $z = 1$ and y varies between 0 and 1. Therefore $dx = 0$ and $dz = 0$. Hence

$$\int_{L_2} \mathbf{F} \cdot d\mathbf{r} = \int_0^1 2y \, dy = 1.$$

Along DB, $y = 1$, $z = 1$ and x varies between 0 and 1. Therefore $dy = 0$ and $dz = 0$. Hence

$$\int_{L_3} \mathbf{F} \cdot d\mathbf{r} = \int_0^1 (3x^2 - 6) \, dx = -5.$$

Therefore $I = 1 + 1 - 5 = -3$.

(iii) Along AB, $x = y = z$ and so $dx = dy = dz$. Therefore

$$I = \int_0^1 (3x^2 - 6x^2 + 2x + 3x^2 + 1 - 4x^4) \, dx$$

$$= \int_0^1 (-4x^4 + 2x + 1) \, dx$$

$$= 6/5.$$

It is clearly seen from the above example that the line integral of a given function depends not only on the end points of the integration but also on the path of integration itself. In general, different paths between the same end points give different results.

Example   Evaluate

$$\int_L \mathbf{F} \cdot d\mathbf{r}$$

where $\mathbf{F} = (x - y^2, 2xyz, x^2y - z)$ and L is the arc of the parabola $y = x^2$ in the plane $z = \frac{3}{2}$ from the point $(0,0,\frac{3}{2})$ to the point $(1,1,\frac{3}{2})$.

$$\int_L \mathbf{F} \cdot d\mathbf{r} = \int \{(x - y^2)\, dx + 2xyz\, dy + (x^2y - z)\, dz\}$$

with appropriate limits. But on the given curve, $y = x^2$ and so $dy = 2x\, dx$ and $z = \frac{3}{2}$ and so $dz = 0$. Also x varies from 0 to 1. Therefore

$$\int_L \mathbf{F} \cdot d\mathbf{r} = \int_0^1 \{(x - x^4)\, dx + 2x^3 \cdot \tfrac{3}{2} \cdot (2x\, dx)\}$$

$$= \int_0^1 (5x^4 + x)\, dx$$

$$= \tfrac{3}{2}.$$

Exercises

1. Evaluate

$$\int_L (x^2 - y)\, dl$$

where L is the quadrant of the circle $x^2 + y^2 = a^2$, $z = 0$ from the point $(a,0,0)$ to the point $(0,a,0)$.

2. Evaluate

$$\int_L \left(y + \frac{x}{y} - \frac{2xy^2}{z^2}\right) dl$$

where L is the curve given parametrically by $\mathbf{r}(t) = (\frac{1}{4}t^4, t^2, \frac{2}{3}t^3)$ from the point $(\frac{1}{4}, 1, \frac{2}{3})$ to the point $(4, 4, \frac{16}{3})$.

3. Evaluate

$$\int_L (x\, dx - y\, dy + (x + z)\, dz)$$

for the following two curves having the same end points $(0,0,0)$ and $(1,1,1)$.
(i) The curve given parametrically by $\mathbf{r} = (t, t^2, t^3)$.
(ii) The curve given parametrically by $\mathbf{r} = (t^3, t^2, t)$.

4. Evaluate

$$\int_L \{(2x + y)\, dx + (2y - x)\, dy\}$$

for the following curves between the same end points $(\frac{1}{4}, 1, 0)$ and $(1, 2, 0)$.

(i) the parabola $y^2 = 4x$, $z = 0$
(ii) the straight line joining the points
(iii) the straight line from $(\frac{1}{4},1,0)$ to $(1,1,0)$ together with the straight line from $(1,1,0)$ to $(1,2,0)$
(iv) the curve given parametrically by $x = t^2 - 2t + 1$, $y = \frac{2}{3}(t + 1)$, $z = 0$.

5. Evaluate

$$\int_L \mathbf{F} \cdot d\mathbf{r}$$

where $\mathbf{F} = zx\mathbf{i} + zy\mathbf{j} + z^2\mathbf{k}$ and L is
(a) the straight line joining the origin to the point $(1,1,1)$
(b) the curve $x = y$, $z = x^2$ between the origin and the point $(1,1,1)$.

6. Evaluate

$$\int_L \mathbf{F} \cdot d\mathbf{r}$$

where $\mathbf{F} = yz\mathbf{i} + zx\mathbf{j} + xy\mathbf{k}$ and L is
(a) the straight line joining the origin to the point $(1,1,1)$
(b) the straight line from the origin to $(1,0,0)$ together with the straight line from $(1,0,0)$ to $(1,1,0)$ and the straight line from $(1,1,0)$ to $(1,1,1)$
(c) the curve $x = y$, $z = x^2$ between the origin and the point $(1,1,1)$.

7. Find $\int_L \mathbf{F} \cdot d\mathbf{r}$ where $\mathbf{F} = 3x^2\mathbf{i} + (2xz - y)\mathbf{j} + z\mathbf{k}$ and L is
(a) the curve given parametrically by $\mathbf{r} = (2t^2,t,4t^2 - t)$ from $t = 0$ to $t = 1$,
(b) the curve given by $x^2 = 4y$, $3x^3 = 8z$ from $x = 0$ to $x = 2$.

1.2 Surface integrals

Suppose that a function f is uniquely defined, and continuous for all points on a given surface S having area s. Divide the surface into n parts S_1, S_2, $\ldots$, S_n having respective areas s_1, s_2, $\ldots$, s_n where $\sum_{i=1}^{n} s_i = s$. Now form the sum $\sum_{i=1}^{n} f_i s_i$ where f_i denotes the value of the function f at some point on the surface S_i. If, as the maximum value of $s_i(i = 1,2,\ldots,n)$ tends to 0, (and so $n \to \infty$), the above sum tends to a limit, this limit is called the surface integral of f over the surface S and is denoted by

$$\int_S f \, ds.$$

The function f is said to be integrable over the surface S and the following results can be shown to hold.

1. If the function f has the constant value k for all points on the surface S having area s, then

$$\int_S f \, ds = \int_S k \, ds = k \int_S ds = ks,$$

2. $$\int_S f \, ds = \int_{S'} f \, ds + \int_{S''} f \, ds$$

if the surface S consists of both the surfaces S′ and S″ and

3. $$\int_S (f + g) \, ds = \int_S f \, ds + \int_S g \, ds$$

if the function g is also integrable over the surface S.

Furthermore, there exists at least one point P on the surface S such that

$$\int_S f \, ds = f(P)s$$

where $f(P)$ is the value of the function f at the point P. This last result is known as the *mean value theorem* for surface integrals.

If the surface S lies entirely in a plane parallel to the x,y-plane, then it can be shown that $\int_S f \, ds$ may be written either as the double integral $\iint f \, dx \, dy$ or as the double integral $\iint f \, dy \, dx$ between appropriate limits for x and y.

Example   If S is the rectangle in the x,y-plane bounded by the lines of intersection of the planes $x = a, x = b, y = c, y = d$ with the plane $z = 0$ (figure 6), evaluate

(i) $\displaystyle\int_S ds$ *and* (ii) $\displaystyle\int_S (x - x^2 y + 3xz) \, ds$.

(i) $$\int_S ds = \int_c^d \int_a^b dx \, dy$$

since for every value of y between c and d, x varies between a and b for points on S. Hence

$$\int_S ds = \int_c^d \left\{ \left[x \right]_a^b \right\} dy = \int_c^d (b - a) \, dy$$

$$= (b - a)(d - c)$$

$$= \text{the area of the rectangle.}$$

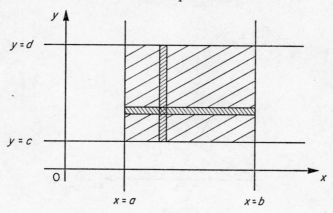

FIGURE 6

Similarly,

$$\int_a^b \int_c^d dy\, dx = (b-a)(d-c).$$

(ii) $\int_S (x - x^2 y + 3xz)\, ds = \int_c^d \int_a^b (x - x^2 y)\, dx\, dy$ (since $z = 0$ on S)

$$= \int_c^d \left\{ \left[\frac{x^2}{2} - \frac{x^3}{3}\, y \right]_a^b \right\} dy$$

$$= \int_c^d \left\{ \tfrac{1}{2}(b^2 - a^2) - \tfrac{1}{3}(b^3 - a^3)y \right\} dy$$

$$= \tfrac{1}{2}(b^2 - a^2)(d - c) - \tfrac{1}{6}(b^3 - a^3)(d^2 - c^2)$$

$$= \tfrac{1}{6}(b - a)(d - c)$$

$$\times \{3(a + b) - (d + c)(a^2 + b^2 + ab)\}.$$

The same result is obtained using $\int_a^b \int_c^d (x - x^2 y)\, dy\, dx$.

Example Evaluate $\int_S ds$ *and* $\int_S (x - x^2 y + 3xz)\, ds$ *when S is the plane triangle having vertices at the points* $(0,0,1)$, *A*$(0,2,1)$ *and B*$(1,0,1)$ *(figure 7).*

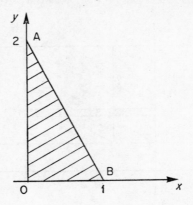

FIGURE 7

The line AB can be represented by $y = 2(1 - x)$, $z = 1$. Thus for every value of x between 0 and 1, y varies between the line $y = 0$, $z = 1$ and the line AB for all points on S, that is y varies between $y = 0$ and $y = 2(1 - x)$. Since the limits for y now depend on x, it is essential that, in evaluating the double integral, the integration with respect to y should be carried out first. That is we must use the form $\iint dy\, dx$ instead of $\iint dx\, dy$. Therefore

$$\int_S ds = \int_0^1 \int_0^{2(1-x)} dy\, dx$$

$$= \int_0^1 \left\{ \left[y \right]_0^{2(1-x)} \right\} dx = \int_0^1 2(1 - x)\, dx = 1.$$

$$= \text{the area of the triangle.}$$

Alternatively, the equations for AB may be written as

$$x = 1 - y/2, \qquad z = 1.$$

Then

$$\int_S ds = \int_0^2 \int_0^{1-y/2} dx\, dy$$

Since for every value of y between 0 and 2, x varies between 0 and $(1 - y/2)$ for points on S. Therefore

$$\int_S ds = \int_0^2 \left\{ \left[x \right]_0^{1-y/2} \right\} dy = \int_0^2 \left(1 - \frac{y}{2} \right) dy$$

$$= 1 \text{ (as before).}$$

Similarly,

$$\int_S (x - x^2y + 3xz)\, ds = \int_0^1 \int_0^{2(1-x)} (x - x^2y + 3x)\, dy\, dx$$

(since $z = 1$ on S)

$$= \int_0^1 \left\{ \left[4xy - \frac{x^2y^2}{2} \right]_0^{2(1-x)} \right\} dx$$

$$= \int_0^1 \{ 8x(1 - x) - 2x^2(1 - x)^2 \}\, dx$$

$$= \int_0^1 (-2x^4 + 4x^3 - 10x^2 + 8x)\, dx$$

$$= \tfrac{19}{15}.$$

The same result is obtained using

$$\int_0^2 \int_0^{1-y/2} dx\, dy.$$

Now let S be an arbitrary plane surface completely contained within the rectangle defined by $a \leqslant x \leqslant b$, $c \leqslant y \leqslant d$, and let this be the smallest such rectangle to contain S (figure 8). Suppose also that the boundary of S is such that the line $x = \alpha$ for $a < \alpha < b$ intersects it in exactly two points and similarly that the line $y = \beta$ for $c < \beta < d$ intersects it in exactly two points.

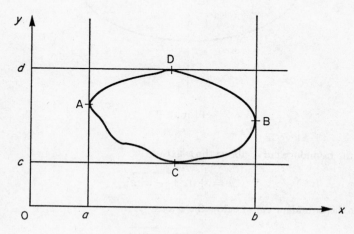

FIGURE 8

Let A and B be points of the surface S having x-coordinates a and b respectively. Then if $y = g_1(x)$ on the bounding curve ACB and $y = g_2(x)$ on the bounding curve ADB,

$$\int_S f \, ds = \int_a^b \int_{g_1(x)}^{g_2(x)} dy \, dx.$$

Alternatively, if $x = h_1(y)$ on the bounding curve CAD and $x = h_2(y)$ on the bounding curve CBD where C and D are points of the surface having ordinates c and d respectively, then

$$\int_S ds = \int_c^d \int_{h_1(y)}^{h_2(y)} dx \, dy.$$

Example   Evaluate $\int_S ds$ *where S is the ellipse given by* $x^2/a^2 + y^2/b^2 = 1$, $z = 0$ *(figure 9). (In other words, determine the area of the given ellipse).*

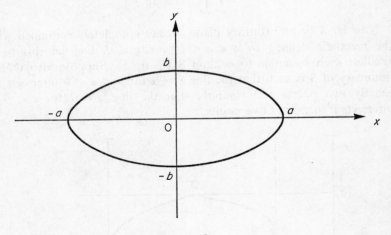

FIGURE 9

For the boundary curve above the x-axis,

$$y = +b\sqrt{1 - x^2/a^2}$$

and for the boundary curve below the x-axis,

$$y = -b\sqrt{1 - x^2/a^2}.$$

Therefore

$$\int_S ds = \int_{-a}^a \int_{-b\sqrt{1-x^2/a^2}}^{+b\sqrt{1-x^2/a^2}} dy \, dx = \int_{-a}^a \left\{ \left[y \right]_{-b\sqrt{1-x^2/a^2}}^{+b\sqrt{1-x^2/a^2}} \right\} dx$$

$$= \int_{-a}^a 2b\sqrt{1 - \frac{x^2}{a^2}} \, dx.$$

In order to carry out this last integration, put $x = a \sin t$ $(-\pi < t \leqslant \pi)$. Therefore

$$\int_S ds = \int_{-\frac{\pi}{2}}^{\frac{\pi}{2}} 2b \cos t (a \cos t \, dt)$$

$$= ab \int_{-\frac{\pi}{2}}^{\frac{\pi}{2}} (1 + \cos 2t) \, dt$$

$$= \pi ab.$$

Now we know that in order to evaluate a definite integral $\int_{x_0}^{x_1} f(x) \, dx$ it is sometimes helpful to change the variable of integration to u say. Then, if $x = x(u)$ where $x(u)$ is a continuous function of u with a continuous derivative in the interval $u_0 \leqslant u \leqslant u_1$ such that $x(u_0) = x_0$ and $x(u_1) = x_1$ (or $x(u_0) = x_1$ and $x(u_1) = x_0$) and if $x(u)$ varies between x_0 and x_1 when u varies between u_0 and u_1

$$\int_{x_0}^{x_1} f(x) \, dx = \int_{u_0}^{u_1} f\left(x(u)\right) \left| \frac{dx}{du} \right| du.$$

For a double integral $\iint_S f(x,y) \, dx \, dy$ we can introduce new variables u and v such that $x = x(u,v)$, $y = y(u,v)$ where $x(u,v)$ and $y(u,v)$ are continuous and have continuous partial derivatives in the region S*, in the u,v-plane, corresponding to S. (That is each point (α,β) in S* corresponds to a point $(x(\alpha,\beta),y(\alpha,\beta))$ in S and conversely). Then

$$\iint_S f(x,y) \, dx \, dy = \iint_{S^*} f\left(x(u,v),y(u,v)\right) |J(u,v)| \, du \, dv$$

where $J(u,v) = (\partial x/\partial u)(\partial y/\partial v) - (\partial x/\partial v)(\partial y/\partial u)$. The function J is known as the Jacobian function of the transformation from variables x and y to variables u and v and is often denoted by

$$\frac{\partial(x,y)}{\partial(u,v)}.$$

Example   Evaluate $\int_S f(x,y)\,\mathrm{d}s$ *where* $f(x,y) \equiv (x+y)^2$ *and S is the region between the circles* $x^2 + y^2 = 1$ *and* $x^2 + y^2 = 4$ *in the plane* $z = 0$.

This example is most easily done by changing to plane polar coordinates r, θ. Then

$$x = r\cos\theta,\ y = r\sin\theta.$$

Therefore

$$J(r,\theta) = \frac{\partial x}{\partial r}\frac{\partial y}{\partial\theta} - \frac{\partial x}{\partial\theta}\frac{\partial y}{\partial r}$$

$$= \cos\theta(r\cos\theta) - (-r\sin\theta)\sin\theta$$

$$= r.$$

Hence

$$\int_S f(x,y)\,\mathrm{d}s = \int_0^{2\pi}\int_1^2 (r\cos\theta + r\sin\theta)^2 r\,\mathrm{d}r\,\mathrm{d}\theta$$

$$= \int_0^{2\pi}\int_1^2 r^3(1 + \sin 2\theta)\,\mathrm{d}r\,\mathrm{d}\theta$$

$$= \frac{15}{4}\int_0^{2\pi}(1 + \sin 2\theta)\,\mathrm{d}\theta = \frac{15\pi}{2}.$$

Now it can be shown for the general surface S (not necessarily plane) given in terms of the parameters u and v by

$$x = x(u,v), \qquad y = y(u,v), \qquad z = z(u,v),$$

that

$$\int_S f(x,y,z)\,\mathrm{d}s = \int_{S'} f\big(x(u,v),y(u,v),z(u,v)\big)g(u,v)\,\mathrm{d}s'$$

where S' is the surface in the u,v-plane corresponding to S in the xyz space and the function g depends on the form of the surface S.* For a

* If $\mathbf{r} = (x(u,v),\, y(u,v),\, z(u,v))$ is the position vector of the general point on the surface S then it can be shown that

$$g(u,v) = \left(\left(\frac{\partial\mathbf{r}}{\partial u}\right)^2\left(\frac{\partial\mathbf{r}}{\partial v}\right)^2 - \left(\frac{\partial^2\mathbf{r}}{\partial u\,\partial v}\right)^2\right)^{\frac{1}{2}} = \left|\frac{\partial\mathbf{r}}{\partial u} \times \frac{\partial\mathbf{r}}{\partial v}\right|$$

surface S lying entirely in the u,v-plane $g(u,v) = 1$. Thus the surface integral has been reduced to an integral over a *plane* surface S' and hence can be interpreted as a double integral.

In particular, if the parameters are the coordinates x and y (and so the surface is given by $z = z(x,y)$), the surface S' will lie wholly in the x,y plane. Then

$$\int_S f(x,y,z)\,\mathrm{d}s = \iint f\big(x,y,z(x,y)\big)g(x,y)\,\mathrm{d}x\,\mathrm{d}y$$

with appropriate limits for x and y. In this case it can be shown that $g(x,y) = \sec\gamma$ where γ is the acute angle between the normal to S at the point (x,y) on it and the z-axis.† Similar expressions are of course obtained using x, z and y, z as parameters.

If $\iint_S u_1\,\mathrm{d}y\,\mathrm{d}z$, $\iint_S u_2\,\mathrm{d}z\,\mathrm{d}x$ and $\iint_S u_3\,\mathrm{d}x\,\mathrm{d}y$ represent surface integrals over the same surface S then

$$\iint_S u_1\,\mathrm{d}y\,\mathrm{d}z + \iint_S u_2\,\mathrm{d}z\,\mathrm{d}x + \iint_S u_3\,\mathrm{d}x\,\mathrm{d}y$$

$$= \int_S (u_1\,\mathrm{d}y\,\mathrm{d}z + u_2\,\mathrm{d}z\,\mathrm{d}x + u_3\,\mathrm{d}x\,\mathrm{d}y)$$

and this is frequently written as

$$\int_S \mathbf{u}\,.\,\mathrm{d}s \quad \text{or as} \quad \int_S (\mathbf{u}\,.\,\hat{\mathbf{n}})\,\mathrm{d}s$$

where $\mathbf{u} = (u_1,u_2,u_3)$ and ds is looked upon as the vector

$$(\mathrm{d}y\,\mathrm{d}z, \mathrm{d}z\,\mathrm{d}x, \mathrm{d}x\,\mathrm{d}y).$$

$\hat{\mathbf{n}}$ is unit vector in the direction of ds. The vector $\hat{\mathbf{n}}$ is of course a function of position, that is it varies from point to point on the surface S, and it gives a unit vector normal to the surface at any point on the surface.

Similarly the expressions

$$\int_S (\mathbf{u}\,.\,\hat{\mathbf{n}}^*)\,\mathrm{d}s \quad \text{and} \quad \int_S \mathbf{u}\,.\,\mathrm{d}s^*,$$

where $\mathrm{d}s^* = \hat{\mathbf{n}}^*\,\mathrm{d}s$ and $\hat{\mathbf{n}}^* = -\hat{\mathbf{n}}$ is also a unit vector normal to the surface at any point on the surface (figure 10), are now meaningful.

---

† If the small element of area δs of the surface S is projected orthogonally on to the x,y-plane the corresponding area $\delta s'$ in the x,y-plane is $\delta s \cos\gamma$. Therefore $\delta s = \delta s' \sec\gamma$.

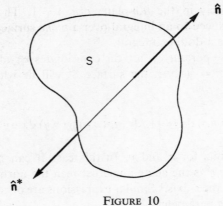

FIGURE 10

Thus when writing a surface integral in this vector notation, care must be taken to specify which unit normal is to be used, otherwise a difference in sign will occur. In other words, we must specify from which side of the surface the unit normal vector is to be drawn.† If S is a closed surface, that is if it completely encloses a given volume, then it is customary to choose the positive sense of the unit normal vector outwards from the enclosed volume.

Suppose now that

$$\hat{n} = (\cos\alpha, \cos\beta, \cos\gamma)$$

where $\cos^2\alpha + \cos^2\beta + \cos^2\gamma = 1$ since $\hat{n}$ is a unit vector. That is α, β and γ are the angles that the direction of $\hat{n}$ makes with the directions of the x, y and z-axes respectively.

Then

$$\int_S (\mathbf{u} \cdot \mathbf{n})\, ds = \int_S (u_1 \cos\alpha + u_2 \cos\beta + u_3 \cos\gamma)\, ds$$

$$= \iint u_1 g_1 \cos\alpha \, dy\, dz + \iint u_2 g_2 \cos\beta \, dz\, dx$$

$$+ \iint u_3 g_3 \cos\gamma \, dx\, dy.$$

But

$$\int_S (\mathbf{u} \cdot \mathbf{n})\, ds = \int_S \mathbf{u} \cdot ds$$

$$= \iint u_1 \, dy\, dz + \iint u_2 \, dz\, dx + \iint u_3 \, dx\, dy.$$

† All the surfaces considered will have two distinct sides that is they will be what is called orientable. A Moebius strip is an example of a one-sided or non-orientable surface.

Thus the functions g_1, g_2, g_3 may be taken so that

$$g_1 \cos \alpha = g_2 \cos \beta = g_3 \cos \gamma = 1.†$$

Example   Find the surface area of a hemisphere of radius a.

Consider the sphere with centre the origin and radius a. Let S denote that part of the surface of this sphere which lies above the x,y-plane.

Then S is given by

$$x^2 + y^2 + z^2 = a^2, \qquad z \geqslant 0$$

or alternatively by

$$z = \sqrt{a^2 - x^2 - y^2}.$$

It is required to evaluate $\displaystyle\int_S \mathrm{d}s$. This is equal to

$$\iint \frac{1}{\cos \gamma} \, \mathrm{d}x \, \mathrm{d}y$$

with appropriate limits for x and y, where γ is the angle that the normal to the surface S at the point (x,y,z) on it makes with the positive z-axis, and where the sense of the normal is chosen so that $\cos \gamma \geqslant 0$.

Now the normal to a sphere, with centre the origin, at a point P(x,y,z) on it has the same direction as the radius vector to the point P. Thus the vector (x,y,z) gives a vector in the direction of the normal to the surface at the point P on it. Hence a unit normal vector is given by

$$\left(\frac{x}{\sqrt{x^2 + y^2 + z^2}}, \frac{y}{\sqrt{x^2 + y^2 + z^2}}, \frac{z}{\sqrt{x^2 + y^2 + z^2}} \right) = \left(\frac{x}{a}, \frac{y}{a}, \frac{z}{a} \right)$$

since P(x,y,z) lies on the surface $x^2 + y^2 + z^2 = a^2$. Therefore for this unit vector

$$\cos \gamma = \frac{z}{a} \geqslant 0$$

for all points on the surface S since $z \geqslant 0$ on the surface S. Hence the above unit normal vector has the desired sense. Therefore

$$\int_S \mathrm{d}s = \int_{S'} \frac{a}{z} \, \mathrm{d}s$$

where S$'$ is the circle with centre the origin and radius a lying in the x,y-plane (figure 11).

Now, if plane polar coordinates r, θ are used,

$$z = \sqrt{a^2 - x^2 - y^2} = \sqrt{a^2 - r^2}$$

† Note that $\mathrm{d}s \cos \alpha$, $\mathrm{d}s \cos \beta$, $\mathrm{d}s \cos \gamma$ are the orthogonal projections of $\mathrm{d}s$ onto the yz-, zx- and xy-planes respectively.

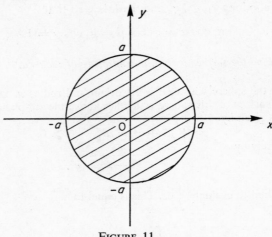

FIGURE 11

and the integral becomes

$$\int_0^{2\pi} \int_0^a \frac{a}{\sqrt{a^2 - r^2}} |J(r,\theta)|\, dr\, d\theta = \int_0^{2\pi} \int_0^a \frac{ar}{\sqrt{a^2 - r^2}}\, dr\, d\theta$$

$$= \int_0^{2\pi} \left[-a\sqrt{a^2 - r^2} \right]_0^a d\theta = 2\pi a^2.$$

Example   Evaluate $\int_S \mathbf{u} \cdot d\mathbf{s}$ *where* $\mathbf{u} = (y^2 - 3xy,\ x^2(y + 2z),\ x + z)$, S *is the surface of the plane triangle with vertices* A(1,0,0), B(0,2,0), C(0,0,1) *and the unit normal to* S *is positive in the sense outwards from the origin (figure* 12).

S is the plane surface $2x + y + 2z = 2$ bounded by the lines AB (given by $2x + y = 2$, $z = 0$), BC (given by $y + 2z = 2$, $x = 0$) and CA (given by $x + z = 1$, $y = 0$).

The projection of S onto the y,z-plane is the triangle OBC, and as z varies between 0 and 1, y will vary between the z-axis and the line CB, that is between 0 and $2 - 2z$, for points in this triangle.

Similarly, for the triangle OCA, x varies between 0 and 1 while z varies between 0 and $1 - x$, and for the triangle OAB, y varies between 0 and 2 while x varies between 0 and $1 - y/2$. Therefore

$$\int_S \mathbf{u} \cdot d\mathbf{s} = \int_0^1 \int_0^{2-2z} (y^2 - 3xy)\, dy\, dz + \int_1^1 \int_0^{1-x} x^2(y + 2z)\, dz\, dx$$

$$+ \int_1^2 \int_0^{1-y/2} (x + z)\, dx\, dy.$$

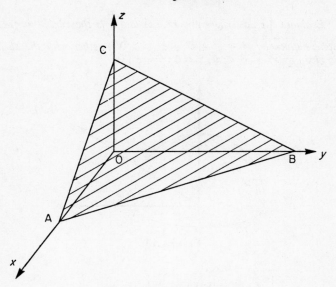

FIGURE 12

But for points on the given surface S,

$$x = 1 - z - y/2, \qquad y = 2(1 - x - 2) \quad \text{and} \quad z = 1 - x - y/2.$$

Hence

$$\int_S \mathbf{u} \cdot \mathrm{d}\mathbf{s} = \int_0^1 \int_0^{2(1-z)} \left(y^2 - 3y\left(1 - z - \frac{y}{2}\right)\right) \mathrm{d}y \, \mathrm{d}z$$

$$+ \int_0^1 \int_0^{1-x} 2x^2(1 - x) \, \mathrm{d}z \, \mathrm{d}x + \int_0^2 \int_0^{1-y/2} \left(1 - \frac{y}{2}\right) \mathrm{d}x \, \mathrm{d}y$$

$$= \int_0^1 \left\{ \left[\frac{y^3}{3} - \frac{3y^2}{2} (1 - z) + \frac{y^3}{2} \right]_0^{2(1-z)} \right\} \mathrm{d}z$$

$$+ \int_0^1 \left\{ \left[2x^2(1 - x)z \right]_0^{1-x} \right\} \mathrm{d}x + \int_0^2 \left\{ \left[\left(1 - \frac{y}{2}\right)x \right]_0^{1-y/2} \right\} \mathrm{d}y$$

$$= \int_0^1 \tfrac{2}{3}(1 - z)^3 \, \mathrm{d}z + \int_0^1 2x^2(1 - x)^2 \, \mathrm{d}x + \int_0^2 \left(1 - \frac{y}{2}\right)^2 \mathrm{d}y = \tfrac{9}{10}.$$

Example   Evaluate $\int_S \mathbf{r} \cdot d\mathbf{s}$ *where* $\mathbf{r} = (x,y,z)$ *and* S *is the curved surface* S′ *of the hemisphere given by* $x^2 + y^2 + z^2 = a^2$, $z \geqslant 0$ *together with the surface* S″ *of the disc given by* $x^2 + y^2 \leqslant a^2$, $z = 0$ *(figure 13).*

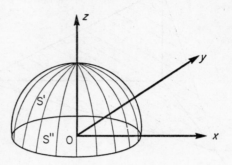

FIGURE 13

$$\int_S \mathbf{r} \cdot d\mathbf{s} = \int_{S'} \mathbf{r} \cdot d\mathbf{s} + \int_{S''} \mathbf{r} \cdot d\mathbf{s}$$

$$= \int_{S'} (\mathbf{r} \cdot \hat{\mathbf{n}}) \, ds + \int_{S''} (\mathbf{r} \cdot \hat{\mathbf{n}}) \, ds.$$

Now for S′, $\mathbf{r}$ and $\hat{\mathbf{n}}$ are parallel and have the same sense and so

$$\mathbf{r} \cdot \hat{\mathbf{n}} = |\mathbf{r}| \, |\hat{\mathbf{n}}| = |\mathbf{r}|$$

since $\hat{\mathbf{n}}$ is unit vector. But $|\mathbf{r}| = a$ since the point (x,y,z) lies on the hemisphere and so $x^2 + y^2 + z^2 = a^2$. Hence

$$\mathbf{r} \cdot \hat{\mathbf{n}} = a.$$

Therefore

$$\int_{S'} (\mathbf{r} \cdot \hat{\mathbf{n}}) \, ds = \int_{S'} a \, ds = a \int_{S'} ds = a(2\pi a^2)$$

since the surface area of the hemisphere is $2\pi a^2$. Now for S″, $\hat{\mathbf{n}} = -\mathbf{k}$ where $\mathbf{k}$ is unit vector in the positive direction of the z-axis. Also, for points on S″, $\mathbf{r} = (x,y,0)$ and so $\mathbf{r} \cdot \hat{\mathbf{n}} = 0$. Therefore

$$\int_{S''} (\mathbf{r} \cdot \hat{\mathbf{n}}) \, ds = \int_{S''} 0 \, ds = 0.$$

Hence

$$\int_S \mathbf{r} \cdot d\mathbf{s} = 2\pi a^3.$$

Example   Show that $\int_S \mathbf{c} \cdot d\mathbf{s} = 0$ *where* $\mathbf{c}$ *is a constant vector and* S *is any closed surface.*

Let $\mathbf{c} = (c_1, c_2, c_3)$ where c_1, c_2, c_3 are constants. Therefore

$$\int_S \mathbf{c} \cdot d\mathbf{s} = \int_{S_1} c_1 \, ds + \int_{S_2} c_2 \, ds + \int_{S_3} c_3 \, ds$$

where S_1, is the surface in the y,z-plane corresponding to S, that is S_1 is the orthogonal projection of S onto the y,z-plane, S_2 and S_3 are similarly defined in the x,z- and x,y-planes respectively.

Now since S is closed, it can be divided into two parts such that the surfaces in the y,z-plane corresponding to each of these parts are identical. The positive side of this common surface is however different for the two parts of S. Therefore

$$\int_{S_1} c_1 \, ds = \iint c_1 \, dy \, dz - \iint c_1 \, dy \, dz$$

where the corresponding limits for the two double integrals are equal since each represents an integral over the same area in the y,z-plane. Therefore

$$\int_{S_1} c_1 \, ds = 0.$$

Similarly it can be shown that $\int_{S_2} c_2 \, ds = 0$ and $\int_{S_3} c_3 \, ds = 0$. Hence

$$\int_S \mathbf{c} \cdot d\mathbf{s} = 0.$$

Exercises

1. Evaluate $\int_S (xy^2 - yz^2 + xyz) \, ds$ where S is the rectangle in the plane $x = \frac{1}{2}$ bounded by the lines of intersection of the planes $y = 1$, $y = 2$, $z = -1$, $z = 1$ with the plane $x = \frac{1}{2}$.

2. Evaluate $\int_S (x^2 - 3xyz + y^2z^2) \, ds$ where S is the surface of the plane triangle with vertices $(0,1,0)$, $(2,1,0)$ and $(2,1,3)$.

3. Determine the area of a circle of radius a by evaluating $\int_S ds$ where S is the circle of radius a, centre the origin lying in the x,y-plane.

4. Evaluate $\int_S z \, ds$ where S is the surface of the hemisphere $x^2 + y^2 + z^2 = a^2$, $z \geqslant 0$.

5. Evaluate $\int_S \mathbf{u} \cdot d\mathbf{s}$ where $\mathbf{u} = (x, z^2 - xz, -xy)$ and S is the surface of the triangle with vertices $(2,0,0)$, $(0,2,0)$ and $(0,0,4)$.

6. Evaluate $\int_S \mathbf{c} \cdot d\mathbf{s}$ where $\mathbf{c}$ is a constant vector and S is that part of the surface of the sphere with centre the origin and radius $2a$ which lies in the first octant. The unit normal to S at any point on it is taken to be positive in the sense outwards from the origin.

1.3 Volume integrals

Suppose that a function f is uniquely defined and continuous for all points within a given three-dimensional region V of volume v. Divide the region V into n parts $V_1, V_2, \ldots , V_n$ having respective volumes $v_1, v_2, \ldots ,$ v_n where $\sum_{i=1}^{n} v_i = v$.

Now form the sum $\sum_{i=1}^{n} f_i v_i$ where f_i denotes the value of the function f at some point in the region V_i. If as the maximum value of v_i $(i = 1,2,\ldots,n)$ tends to zero (and $n \to \infty$) the above sum tends to a limit, this limit is called the volume integral of f throughout the region V and is denoted by

$$\int_V f \, dv.$$

The function f is said to be integrable within the region V and the following results can be shown to hold.

1. If the function f has the constant value k for all points within the region V, then

$$\int_V f \, dv = \int_V k \, dv = k \int_V dv = kv.$$

2.
$$\int_V f \, dv = \int_{V'} f \, dv + \int_{V''} f \, dv$$

if the region V consists of both the regions V' and V'', and

3.
$$\int_V (f + g) \, dv = \int_V f \, dv + \int_V g \, dv$$

if the function g is also integrable within the region V.

If rectangular cartesian coordinates x, y, z are used it can be shown that $\int_V f \, dv$ may be written as the triple integral $\iiint f \, dx \, dy \, dz$ between appropriate limits for x, y and z. The order of integration may be changed with the corresponding modification of the limits as for double integrals.

Example   Evaluate $\int_V \mathrm{d}v$ *where* V *is the region bounded by the planes* $x = 0$, $x = a$, $y = 0$, $y = b$, $z = 0$, $z = c$ (*figure 14*).

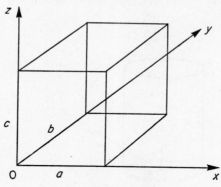

FIGURE 14

For each value z_0 of z between 0 and c, the point (x,y,z_0) of V lies in the rectangle $0 \leqslant x \leqslant a$, $0 \leqslant y \leqslant b$, $z = z_0$ (figure 15). Thus for each value of y

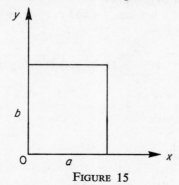

FIGURE 15

between 0 and b, x varies between 0 and a. Therefore

$$\int_V \mathrm{d}v = \int_0^c \int_0^b \int_0^a \mathrm{d}x \, \mathrm{d}y \, \mathrm{d}z = \int_0^c \int_0^b \left\{ \left[x \right]_0^a \right\} \mathrm{d}y \, \mathrm{d}z$$

$$= \int_0^c \left\{ \left[ay \right]_0^b \right\} \mathrm{d}z$$

$$= \left[abz \right]_0^c = abc$$

$$= \text{the volume of the region V.}$$

Example   Evaluate $\displaystyle\int_V (x+y+z)\,dv$ *where* V *is the region bounded by the surfaces* $x^2+y^2=a^2$, $z=0$, $z=h$. *That is* V *is the right circular cylinder with axis the z-axis, radius a, base in the x,y-plane and height h.* (*figure* 16).

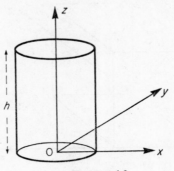

FIGURE 16

For each value z_0 of z between 0 and h, the point (x,y,z_0) of V lies inside the circle $x^2+y^2=a^2$, $z=z_0$ (figure 17).

Thus, for each value of y between $-a$ and a, x varies between $-\sqrt{a^2-y^2}$ and $+\sqrt{a^2-y^2}$. Therefore

$$\int_V (x+y+z)\,dv = \int_0^h\int_{-a}^a\int_{-\sqrt{a^2-y^2}}^{\sqrt{a^2-y^2}}(x+y+z)\,dx\,dy\,dz$$

$$= \int_0^h\int_{-a}^a\left\{\left[\frac{x^2}{2}+xy+zx\right]_{-\sqrt{a^2-y^2}}^{\sqrt{a^2-y^2}}\right\}\,dy\,dz$$

$$= \int_0^h\int_{-a}^a\{2y\sqrt{a^2-y^2}+2z\sqrt{a^2-y^2}\}\,dy\,dz$$

$$= \int_0^h\int_{-a}^a 2y\sqrt{a^2-y^2}\,dy\,dz + \int_0^h\int_{-\pi/2}^{\pi/2}2za\cos\theta a\cos\theta\,d\theta\,dz$$

where in the second integral, the substitution $y=a\sin\theta$ has been used. Hence

$$\int_V (x+y+z)\,dv = \int_0^h\left\{\left[-\tfrac{2}{3}(a^2-y^2)^{3/2}\right]_{-a}^a\right\}\,dz$$

$$+ \int_0^h\left\{a^2z\left[\theta+\tfrac{1}{2}\sin 2\theta\right]_{-\pi/2}^{\pi/2}\right\}\,dz$$

$$= \int_0^h \pi a^2 z\,dz$$

$$= \tfrac{1}{2}\pi a^2 h^2.$$

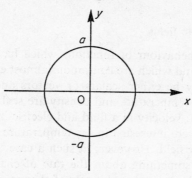

FIGURE 17

Exercises

1. Evaluate $\int_V xyz\,dv$ where V is the region bounded by the planes $x = 0$, $x = 2$, $y = 0$, $y = 1$, $z = 0$, $z = 1$.

2. By evaluating $\int_V dv$ where V is the right circular cylinder with axis the z-axis, radius a, base in the x,y-plane and height h, show that the volume of the cylinder is $\pi a^2 h$.

3. Evaluate $\int_V z\,dv$ where V is the region in the first octant bounded by the coordinate planes and the plane $x + y + z = 1$.

4. Evaluate

$$\int_V \frac{x^2 + y^2}{\sqrt{z^2 - y^2}}\,dv$$

where V is the conical region bounded by the surfaces $x^2 + y^2 = z^2$, $z = 0$ and $z = h$.

5. If V is the hemispherical region enclosed between the surface

$$x^2 + y^2 + z^2 = a^2, z \geqslant 0$$

and the plane $z = 0$, show that

$$\int_V (x + y + z)\,dv = \frac{\pi a^4}{4}.$$

1.4 Scalar and vector fields

We consider the behaviour of functions which have unique values at each point in space and which are continuous almost everywhere in space. These functions may be either scalars or vectors and are said to form fields. For example, temperature and density are scalar fields (having no associated direction); velocity in a fluid and electric intensity of force are vector fields. If we were interested in say temperature in a solid body, we should have a scalar field. However, in such a case, we would probably also want to know something about the rate of change of temperature with distance in the body, and as this rate of change generally depends on the direction chosen in the body, we thus introduce a vector field. We shall show that given *any* scalar field we can derive a corresponding vector field.

If φ is a continuous scalar field, then through any point P of the region considered we can construct a surface such that at each point on it the function φ has the same value as at P. No two of these surfaces corresponding to different values of the function can intersect since this would imply that the function did not have a unique value there. These surfaces are called equipotential surfaces. (Surfaces of constant temperature, that is isothermal surfaces, provide a particular example).

1.5 The gradient of the scalar function φ

φ is a function of position, that is φ at a point depends on the coordinates of that point. Using rectangular cartesian coordinates $\varphi \equiv \varphi(x,y,z)$. (Note however that since φ is *uniquely* defined at any point then its value at any point must be independent of the coordinate *system* used to define the point.)

Let $P(x,y,z)$ and $P'(x + p, y + q, z + r)$ be such that the length of the line joining P and P' is s. The potentials of the equipotential surfaces through P and P' will be

$$\varphi(P) \equiv \varphi(x,y,z) \quad \text{and} \quad \varphi(P') \equiv \varphi(x + p, y + q, z + r)$$

respectively (figure 18). Now

$$\varphi(x + p, y + q, z + r) - \varphi(x,y,z)$$
$$= \{\varphi(x + p, y + q, z + r) - \varphi(x, y + q, z + r)\}$$
$$+ \{\varphi(x, y + q, z + r) - \varphi(x, y, z + r)\}$$
$$+ \{(x, y, z + r) - \varphi(x, y, z)\}$$

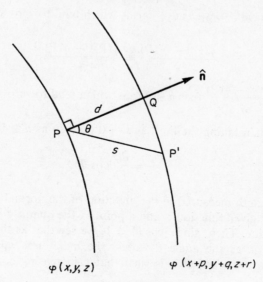

FIGURE 18

Therefore

$$\frac{\varphi(x + p, y + q, z + r) - \varphi(x,y,z)}{s}$$

$$= \frac{\varphi(x + p, y + q, z + r) - \varphi(x, y + q, z + r)}{p} \frac{p}{s}$$

$$+ \frac{\varphi(x, y + q, z + r) - \varphi(x, y, z + r)}{q} \frac{q}{s} + \frac{\varphi(x, y, z + r) - \varphi(x,y,z)}{r} \frac{r}{s}.$$

Now taking the limit as $s \to 0$ (and so as p, q, r also tend to zero) we obtain

$$\frac{\partial \varphi}{\partial l} = \frac{\partial \varphi}{\partial x} \frac{dx}{dl} + \frac{\partial \varphi}{\partial y} \frac{dy}{dl} + \frac{\partial \varphi}{\partial z} \frac{dz}{dl}$$

where l is length measured along the direction of PP'. Thus we associate $\partial \varphi / \partial l$ with the direction PP'.

Now let PQ be the normal at P to the equipotential surface through P, and let d be the length of PQ where Q is the intersection of this normal with the equipotential surface through P' (figure 18). Suppose also that **n** is a unit vector in the direction of the normal PQ and that θ is the angle between PP' and PQ.

Then we have $d = s(\cos \theta + $ terms which tend to zero as $s \rightarrow 0)$.
Therefore

$$\frac{\varphi(P') - \varphi(P)}{s} = \frac{\varphi(P') - \varphi(P)}{d} \frac{d}{s}$$

$$= \frac{\varphi(P') - \varphi(P)}{d} (\cos \theta + \text{ terms which tend to zero as } s \rightarrow 0).$$

Therefore again taking the limit as $s \rightarrow 0$ (and so as $d \rightarrow 0$) we obtain

$$\frac{\partial \varphi}{\partial l} = \frac{\partial \varphi}{\partial n} \cos \theta$$

where n is length measured in the direction of the normal at P.

Now for a given function φ and a point P, the quantity $\partial \varphi / \partial n$ will have a specific value. Then, since $|\cos \theta| \leqslant 1$, we see that as the direction PP′ varies (that is as the angle θ varies) the maximum value of $\partial \varphi / \partial l$ is $\partial \varphi / \partial n$ where the sense of n is such that n increasing corresponds to φ increasing.

Thus the maximum value of $\partial \varphi / \partial l$ is $\partial \varphi / \partial n$ and occurs when $\theta = 0$, that is when PP′ lies along the normal PQ.

The vector quantity $(\partial \varphi / \partial n)\hat{n}$ is called the *gradient of* φ and is written grad φ or $\nabla \varphi$ (pronounced 'del φ').

Thus grad φ, at any point P in the region for which φ is defined, is the vector having magnitude equal to the magnitude of the derivative of φ normal to the equipotential surface at P, direction along the normal at P and sense towards increasing φ.

Now grad φ is independent of any choice of coordinate system, but if rectangular cartesians are again introduced and $\cos \alpha$, $\cos \beta$, $\cos \gamma$ are taken as the direction cosines of the unit vector $\hat{n}$ (figure 19), we have,

$$\hat{n} = (\cos \alpha, \cos \beta, \cos \gamma).$$

Therefore

$$\text{grad } \varphi = \frac{\partial \varphi}{\partial n} \hat{n} = \left(\frac{\partial \varphi}{\partial n} \cos \alpha, \frac{\partial \varphi}{\partial n} \cos \beta, \frac{\partial \varphi}{\partial n} \cos \gamma \right).$$

But from the equation

$$\frac{\partial \varphi}{\partial l} = \frac{\partial \varphi}{\partial n} \cos \theta$$

we obtain

$$\frac{\partial \varphi}{\partial n} \cos \alpha = \frac{\partial \varphi}{\partial x}, \frac{\partial \varphi}{\partial n} \cos \beta = \frac{\partial \varphi}{\partial y} \quad \text{and} \quad \frac{\partial \varphi}{\partial n} \cos \gamma = \frac{\partial \varphi}{\partial z}.$$

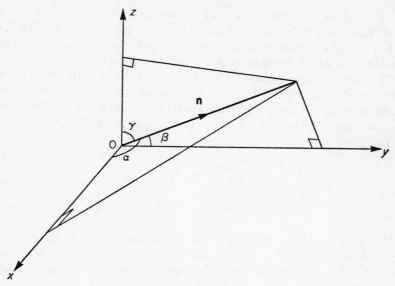

FIGURE 19

Therefore in rectangular cartesian coordinates,

$$\text{grad } \varphi \equiv \left(\frac{\partial \varphi}{\partial x}, \frac{\partial \varphi}{\partial y}, \frac{\partial \varphi}{\partial z} \right)$$

$$\equiv \frac{\partial \varphi}{\partial x} \mathbf{i} + \frac{\partial \varphi}{\partial y} \mathbf{j} + \frac{\partial \varphi}{\partial z} \mathbf{k}$$

where $\mathbf{i}, \mathbf{j}, \mathbf{k}$ are unit vectors in the directions of the respective coordinate axes.

This is often written as

$$\left(\mathbf{i} \frac{\partial}{\partial x} + \mathbf{j} \frac{\partial}{\partial y} + \mathbf{k} \frac{\partial}{\partial z} \right) \varphi$$

and the differential operator $\left(\mathbf{i} \dfrac{\partial}{\partial x} + \mathbf{j} \dfrac{\partial}{\partial y} + \mathbf{k} \dfrac{\partial}{\partial z} \right)$

is denoted by the symbol ∇, thus leading to the notation $\nabla \varphi$ for the gradient.

It is clear, from the elementary properties of partial derivatives, that the operator ∇ is linear, that is that

$$\nabla(\alpha \varphi + \beta \psi) = \alpha \nabla \varphi + \beta \nabla \psi$$

if α and β are constants.

The *component* of the vector grad φ in any given direction gives the rate of increase of φ with respect to distance in that particular direction.

For example, consider the function $\varphi = r^m$ where m is a real number and r is the distance of the variable point P from a fixed origin O.

The surfaces $\varphi = $ constant are concentric spheres with centre O and therefore the unit normal $\hat{n}$ to such a surface at any point is parallel to the position vector $\mathbf{r}$ of that point relative to O. Therefore $\hat{n} = \hat{r}$ and r measures distance along the normal. ($\hat{r}$ denotes unit vector parallel to $\mathbf{r}$).

Therefore

$$\text{grad } \varphi = \frac{\partial \varphi}{\partial n} \hat{n} = \frac{\partial \varphi}{\partial r} \hat{r}$$

$$= \frac{\partial r^m}{\partial r} \hat{r} = m r^{m-1} \hat{r} = m r^{m-2} \mathbf{r}.$$

We shall see later that in the case when $m = -1$, $\varphi = 1/r$ gives the gravitational potential at a point due to a unit point mass at the origin and grad $\varphi = (-1/r^2)\hat{r} = (-1/r^3)\mathbf{r}$ gives the force field associated with that mass.

Corresponding results will also be obtained for point charges in electrostatics and point sources in hydrodynamics.

Example   Find grad φ when $\varphi \equiv 3x^2 + z^2 - 2yz + 2zx$ and hence find a unit vector normal to the surface given implicitly by $3x^2 + z^2 - 2yz + 2zx = 0$ at the point $(0,\frac{1}{2},1)$ on it.

$$\text{grad } \varphi = \left(\frac{\partial \varphi}{\partial x}, \frac{\partial \varphi}{\partial y}, \frac{\partial \varphi}{\partial z}\right) = (6x + 2z, -2z, 2z - 2y + 2x)$$

and this gives a vector normal to surfaces of the form $\varphi = $ constant. Therefore putting $x = 0$, $y = \frac{1}{2}$ and $z = 1$ in the above expression for grad φ, we obtain a vector normal to the given surface at the point $(0,\frac{1}{2},1)$ namely $(2,-2,1)$. Now the length of this vector is $(4 + 4 + 1)^{\frac{1}{2}} = 3$. Hence the required unit vector is $(\frac{2}{3},-\frac{2}{3},\frac{1}{3})$.

Example   Find the rate of increase of $\varphi = x^3 - 3xyz^2 + y^2z + 2x + y + 4z$ in the direction of the vector (yz,zx,xy) at the point $(1,2,3)$.

We require the component of grad φ at the point $(1,2,3)$ in the direction of the given vector at this point. This is obtained by taking the scalar product of grad φ at the point with unit vector in the given direction at the point.

$$\text{grad } \varphi = (3x^2 - 3yz^2 + 2, -3xz^2 + 2yz + 1, -6xyz + y^2 + 4)$$

$$= (-49,-14,-28) \text{ at the point } (1,2,3).$$

Also, at the point (1,2,3) the vector (yz,zx,xy) is (6,3,2) which has length 7. Therefore unit vector in this direction is $(\frac{6}{7},\frac{3}{7},\frac{2}{7})$.

Hence the required rate of increase is

$$(-49,-14,-28) \cdot (\tfrac{6}{7},\tfrac{3}{7},\tfrac{2}{7}) = -42 - 6 - 8$$
$$= -56$$

The negative sign implies that φ is in fact decreasing in the given direction.

Exercises

1. Determine grad φ at the point $(-1,3,-2)$ when $\varphi = xyz^2 - 2x + x^2y$.

2. If $\mathbf{a} = (y^2, 2xy, -xz^2)$ and $\varphi = z^3 - x^2y$, find $\mathbf{a} \cdot \nabla\varphi$ and $\mathbf{a} \times \nabla\varphi$ at the point $(-1,-1,1)$.

3. Find unit vector normal to the surface $x^3 - 3xy^2 + yz^2 = 1$ at the point (1,0,2) on it.

4. Find the rate of increase of $\varphi = y^2 + 2yz - x^2$ in the direction of the vector $(1/x, 1/y, 1/z)$ at the point (1,2,1).

5. Using rectangular cartesian coordinates prove that

$$\text{grad } (\varphi_1\varphi_2) = \varphi_1 \text{ grad } \varphi_2 + \varphi_2 \text{ grad } \varphi_1$$

and that

$$\text{grad } \left(\frac{\varphi_1}{\varphi_2}\right) = \frac{\varphi_2 \text{ grad } \varphi_1 - \varphi_1 \text{ grad } \varphi_2}{\varphi_2^2} \quad \text{if} \quad \varphi_2 \neq 0.$$

6. Using rectangular cartesian coordinates, show that if $\mathbf{r}$ is the position vector of a general point with respect to a fixed origin, then grad $(1/r) = -\mathbf{r}/r^3$ where $r = |\mathbf{r}|$.

1.6 The divergence of a vector

The divergence of a vector $\mathbf{a}$ (denoted by div $\mathbf{a}$) at a point P is defined by

$$\text{div } \mathbf{a} = \lim_{v \to 0} \frac{\displaystyle\int_S \mathbf{a} \cdot d\mathbf{s}}{v}$$

where S is a closed surface surrounding the point P and enclosing a volume v. Div $\mathbf{a}$ is a scalar quantity and its value at any point is independent of the coordinate system used to define that point. Thus, given any vector field we have now defined an associated scalar field.

We now proceed to determine an expression for div **a** using rectangular cartesian coordinates x, y and z. Let $\mathbf{a} \equiv \mathbf{a}(x,y,z) = (a_1, a_2, a_3)$ where a_1, a_2, a_3 are themselves functions of x, y, z. To find div **a** at any point $P(x_0, y_0, z_0)$, construct the rectangular box ABCDEFGH having sides of length $2h$, $2k$, $2l$ and with P at its centre (figure 20). Denote its surface by S.

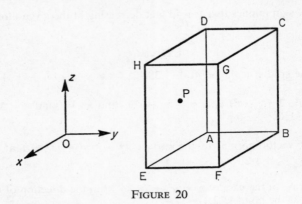

FIGURE 20

With the convention that for a closed surface the positive sense of the normal is outwards from the volume enclosed, we see that in calculating the contribution to $\int_S \mathbf{a} \cdot \mathbf{ds}$ due to the face ABCD, the positive sense of the normal is in the negative x direction. Therefore

$$\int_{ABCD} \mathbf{a} \cdot \mathbf{ds} = -\int_{z_0-l}^{z_0+l} \int_{y_0-k}^{y_0+k} a_1(x_0 - h, y, z)\, \mathrm{d}y\, \mathrm{d}z$$

since $x = x_0 - h$ on ABCD. Hence, applying the mean value theorem for surface integrals to this we see that

$$\int_{ABCD} \mathbf{a} \cdot \mathbf{ds} = -a_1(x_0 - h, y_0 + \varepsilon, z_0 + \eta)4kl$$

where $(x_0 - h, y_0 + \varepsilon, z_0 + \eta)$ is some point in the rectangle ABCD at which the function a_1 has its 'average' value for the face. Since $y_0 - k \leqslant y_0 + \varepsilon \leqslant y_0 + k$ and $z_0 - l \leqslant z_0 + \eta \leqslant z_0 + l$ for the above point to lie in the rectangle ABCD, we see that $\varepsilon \to 0$ as $k \to 0$ and $\eta \to 0$ as $l \to 0$.

Similarly, $\int_{EFGH} \mathbf{a} \cdot \mathbf{ds} = a_1(x_0 + h, y_0 + \varepsilon', z_0 + \eta')4kl$ where $\varepsilon' \to 0$ as $k \to 0$ and $\eta^1 \to 0$ as $l \to 0$.

Now the volume v enclosed by the surface S is easily seen to be $8hkl$. Hence the contribution to div **a** from the pair of opposite faces ABCD and EFGH is

$$\lim_{v \to 0} \frac{1}{v} \left\{ \int_{ABCD} \mathbf{a} \cdot \mathbf{ds} + \int_{EFGH} \mathbf{a} \cdot \mathbf{ds} \right\}$$

$$= \lim_{\substack{h \to 0 \\ k \to 0 \\ l \to 0}} \left\{ \frac{a_1(x_0 + h, y_0 + \varepsilon', z_0 - \eta') - a_1(x_0 - h, y_0 + \varepsilon, z_0 + \eta)}{2h} \right\}$$

$$= \lim_{h \to 0} \left\{ \frac{a_1(x_0 + h, y_0, z_0) - a_1(x_0 - h, y_0, z_0)}{2h} \right\}$$

$$= \frac{\partial a_1}{\partial x} \quad \text{at} \quad P(x_0, y_0, z_0).$$

Similarly the other pairs of opposite faces contribute $\partial a_2/\partial y$ and $\partial a_3/\partial z$ both evaluated at P. Therefore div **a** at a point P is given in terms of cartesian coordinates by

$$\frac{\partial a_1}{\partial x} + \frac{\partial a_2}{\partial y} + \frac{\partial a_3}{\partial z}$$

where the partial derivatives are calculated at the point P. It can be shown that this result is *independent* of the closed surface S originally chosen to surround the point P.

Thus we now have, associated with any vector field **a**, a scalar field div **a** given in cartesian coordinates by

$$\text{div } \mathbf{a} = \frac{\partial a_1}{\partial x} + \frac{\partial a_2}{\partial y} + \frac{\partial a_3}{\partial z}.$$

This is also written as $\nabla \cdot \mathbf{a}$ where again ∇ is to be regarded as the operator

$$\left(\mathbf{i} \frac{\partial}{\partial x} + \mathbf{j} \frac{\partial}{\partial y} + \mathbf{k} \frac{\partial}{\partial z} \right).$$

Clearly

$$\text{div } (\alpha \mathbf{a} + \beta \mathbf{b}) = \alpha \text{ div } \mathbf{a} + \beta \text{ div } \mathbf{b}$$

if α and β are scalar constants.

If div $\mathbf{a} \equiv 0$, then **a** is said to be *solenoidal*.

For example, consider the vector field given by **r** the position vector of a variable point with respect to a fixed origin.

$$\mathbf{r} = (x, y, z).$$

Therefore

$$\operatorname{div} \mathbf{r} = \frac{\partial x}{\partial x} + \frac{\partial y}{\partial y} + \frac{\partial z}{\partial z} = 3.$$

Now consider the derivation of this same result starting from the fundamental definition of div $\mathbf{r}$. To determine div $\mathbf{r}$ at any point P having position vector $\mathbf{r}$, surround this point by a sphere of radius c and centre at P (figure 21). Call the surface of this sphere S. The volume v enclosed is $\frac{4}{3}\pi c^3$.

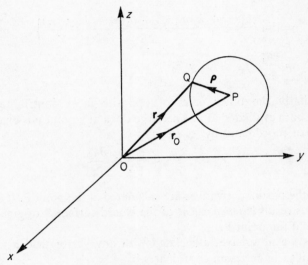

FIGURE 21

Then

$$\operatorname{div} \mathbf{r} = \lim_{v \to 0} \frac{\displaystyle\int_S \mathbf{r} \cdot \mathbf{ds}}{v} = \lim_{c \to 0} \frac{\displaystyle\int_S \mathbf{r} \cdot \mathbf{ds}}{\frac{4}{3}\pi c^3}.$$

Now at Q, any point on S, $\mathbf{r}$ may be written in the form $\mathbf{r} = \mathbf{r}_0 + \boldsymbol{\rho}$ where $\boldsymbol{\rho} = \overline{PQ}$ and is such that $|\boldsymbol{\rho}| = c$.

Therefore

$$\int_S \mathbf{r} \cdot \mathbf{ds} = \int_S (\mathbf{r}_0 + \boldsymbol{\rho}) \cdot \mathbf{ds} = \int_S \mathbf{r}_0 \cdot \mathbf{ds} + \int_S \boldsymbol{\rho} \cdot \mathbf{ds}.$$

But since $\mathbf{r}_0$ is a constant vector and S is closed, we have seen (in a previous example on page 24) that

$$\int_S \mathbf{r}_0 \cdot \mathbf{ds} = 0.$$

Also

$$\int_S \boldsymbol{\rho} \cdot d\mathbf{s} = \int_S (\boldsymbol{\rho} \cdot \hat{\mathbf{n}}) \, ds = \int_S |\boldsymbol{\rho}| \, ds$$

since $\boldsymbol{\rho}$ and $\hat{\mathbf{n}}$ are parallel and in the same sense and $|\hat{\mathbf{n}}| = 1$.
Hence

$$\int_S \boldsymbol{\rho} \cdot d\mathbf{s} = c \int_S ds = c4\pi c^2$$

since the area of S $= 4\pi c^2$.
Hence

$$\text{div } \mathbf{r} = \lim_{r \to 0} \frac{4\pi c^3}{\frac{4}{3}\pi c^2} = 3.$$

Example   Find div F *for* $\mathbf{F} = (3x^2 - 2yz, \; y^3 + yz^2, \; xyz - 3xz^2)$.

$$\text{div } \mathbf{F} = \frac{\partial F_1}{\partial x} + \frac{\partial F_2}{\partial y} + \frac{\partial F_3}{\partial z}$$

where $\mathbf{F} = (F_1, F_1, F_3)$
Therefore

$$\text{div } \mathbf{F} = \frac{\partial}{\partial x}(3x^2 - 2yz) + \frac{\partial}{\partial y}(y^3 + yz^2) + \frac{\partial}{\partial z}(xyz - 3xz^2)$$

$$= 6x + 3y^2 + z^2 + xy - 6xz.$$

Exercises

1. Find div F when $\mathbf{F} = (x^3 - y^2/x, \; y + 3xyz, \; z - x^2 + y)$.

2. If $\mathbf{F} = (x^2 + y^2 + z^2, \; 3xyz, \; xy^2 - zy^2)$, show that div $\mathbf{F} = 0$ at the point $(2, 1, -\frac{1}{2})$.

3. Find a, b and c such that the vector

$$\mathbf{F} = (axz + x^2, \; by + xy^2, \; z - z^2 + cxz - 2xyz)$$

is solenoidal.

4. If $\varphi = x^3 y^4 z^2$, find the vector a such that a $=$ grad φ and hence find div a.

1.7 The curl or rotation of a vector a

Whereas the divergence of a vector is a scalar, the curl or rotation of a vector is another vector. The component in a direction **n** of the vector curl **a** at a point P is defined as

$$\lim_{s \to 0} \frac{\displaystyle\int_L \mathbf{a} \cdot d\mathbf{l}}{s}$$

where L is a closed curve surrounding P and lying in the plane through P which is perpendicular to n̂, and s is the area enclosed by the curve L. The integration round L is performed in the same sense as a rotation represented by a vector having the same direction and sense as n̂. For example, if L lies in the plane of the paper and n̂ points into the paper then the integration round L is performed in a clockwise sense and vice versa if n̂ points out of the paper. It can be shown that the above definition is *independent* of the shape of the area s.

Before proceeding to determine a cartesian expression for the vector curl **a**, consider the following physical example of a fluid rotating like a rigid body with angular speed ω about the z-axis.

The velocity vector **q** at any point $P(x,y,z)$ in the fluid may be written

$$\mathbf{q} = -\omega y\mathbf{i} + \omega x\mathbf{j} + O\mathbf{k}$$

where **i**, **j**, **k** are unit vectors in the directions of the x, y, z axes respectively (figure 22).

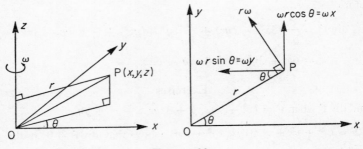

FIGURE 22

To obtain the z-component (that is the component in the direction **k**) of curl **q**, at a point $P_0(x_0,y_0,z_0)$ we must first integrate **q** round a closed curve L enclosing P_0 and lying in the plane through P_0 parallel to the x,y-plane. Take as the closed curve the rectangle ABCD with centre at P_0 and sides of length 2α and 2β (figure 23). The area s of this rectangle is $4\alpha\beta$.

$$\int_L \mathbf{q} \cdot d\mathbf{l} = \int_{AB} \mathbf{q} \cdot d\mathbf{l} + \int_{BC} \mathbf{q} \cdot d\mathbf{l} + \int_{CD} \mathbf{q} \cdot d\mathbf{l} + \int_{DA} \mathbf{q} \cdot d\mathbf{l}.$$

But on AB, $y = y_0 - \beta$ and so $dy = 0$ and $z = z_0$ and so $dz = 0$ and x varies from $(x_0 - \alpha)$ to $(x_0 + \alpha)$. Similarly on BC, $x = x_0 + \alpha$, $dx = 0$,

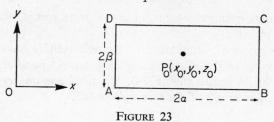

FIGURE 23

$dz = 0$ and y varies from $(y_0 - \beta)$ to $(y_0 + \beta)$, on CD, $y = y_0 + \beta$, $dy = 0$, $dz = 0$ and x varies from $(x_0 + \alpha)$ to $(x_0 - \alpha)$ and on DA, $x = x_0 - \alpha$, $dx = 0$, $dz = 0$ and y varies from $(y_0 + \beta)$ to $(y_0 - \beta)$. Therefore

$$\int_L \mathbf{q} \cdot d\mathbf{l} = \int_{x_0-\alpha}^{x_0+\alpha} - \omega(y_0 - \beta)\, dx + \int_{y_0-\beta}^{y_0+\beta} \omega(x_0 + \alpha)\, dy$$

$$+ \int_{x_0+\alpha}^{x_0-\alpha} - \omega(y_0 + \beta)\, dx + \int_{y_0+\beta}^{y_0-\beta} \omega(x_0 - \alpha)\, dy$$

$$= -\omega(y_0 - \beta)\Big[x\Big]_{x_0-\alpha}^{x_0+\alpha} + \omega(x_0 + \alpha)\Big[y\Big]_{y_0-\beta}^{y_0+\beta}$$

$$- \omega(y_0 + \beta)\Big[x\Big]_{x_0+\alpha}^{x_0-\alpha} + \omega(x_0 - \alpha)\Big[y\Big]_{y_0+\beta}^{y_0-\beta}$$

$$= -2\alpha\omega(y_0 - \beta) + 2\beta\omega(x_0 + \alpha) + 2\alpha\omega(y_0 + \beta) - 2\beta\omega(x_0 - \alpha)$$

$$= 8\omega\alpha\beta.$$

Therefore

$$\lim_{s \to 0} \frac{\int_L \mathbf{a} \cdot d\mathbf{l}}{s} = \lim_{\substack{\alpha \to 0 \\ \beta \to 0}} \frac{8\omega\alpha\beta}{4\alpha\beta} = 2\omega.$$

Hence the $\mathbf{k}$ component of curl $\mathbf{q}$ is 2ω.

Similarly, by considering closed curves in planes through P_0 and parallel to the y,z- and z,x-planes respectively it can be shown that the $\mathbf{i}$ and $\mathbf{j}$ components of curl $\mathbf{q}$ are 0.

Thus we have shown that curl $\mathbf{q} = (0,0,2\omega)$.

In hydrodynamics the curl of the velocity vector is called the *vorticity* vector.

Let us now obtain an expression for curl **a** using rectangular cartesian coordinates x, y and z.

For the x-component, consider the rectangle ABCD having sides of length 2α, 2β lying in the plane through $P_0(x_0,y_0,z_0)$ parallel to the y,z-plane with P_0 at its centre (figure 24). Let the bounding curve of this rectangle be L. The area s of the rectangle is $4\alpha\beta$.

$$\int_L \mathbf{a}\cdot d\mathbf{l} = \int_{AB} \mathbf{a}\cdot d\mathbf{l} + \int_{BC} \mathbf{a}\cdot d\mathbf{l} + \int_{CD} \mathbf{a}\cdot d\mathbf{l} + \int_{DA} \mathbf{a}\cdot d\mathbf{l}.$$

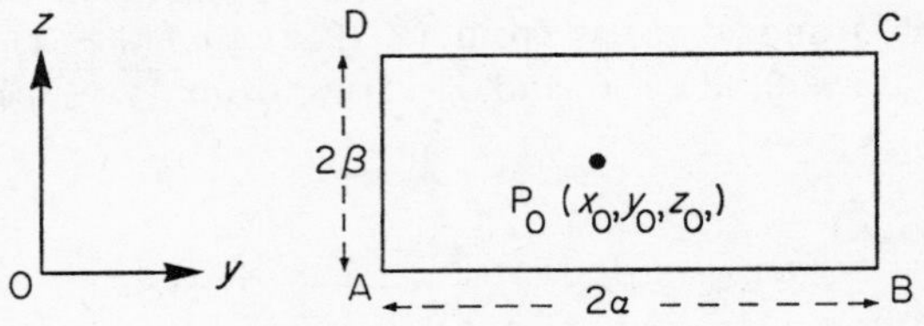

FIGURE 24

Now on AB, $z = z_0 - \beta$ and so $dz = 0$, $x = x_0$ and so $dx = 0$ and y varies from $(y_0 - \alpha)$ to $(y_0 + \alpha)$. Similarly, on CD, $z = z_0 + \beta$, $dz = 0$, $x = x_0$, $dx = 0$, and y varies from $(y_0 + \alpha)$ to $(y_0 - \alpha)$. Hence

$$\int_{AB} \mathbf{a}\cdot d\mathbf{l} + \int_{CD} \mathbf{a}\cdot d\mathbf{l} = \int_{y_0-\alpha}^{y_0+\alpha} a_2(x_0, y, z_0 - \beta)\, dy + \int_{y_0+\alpha}^{y_0-\alpha} a_2(x_0, y, z_0 + \beta)\, dy.$$

Then, applying the mean value theorem for integrals to the above, we see that

$$\int_{AB} \mathbf{a}\cdot d\mathbf{l} + \int_{CD} \mathbf{a}\cdot d\mathbf{l} = 2\alpha a_2(x_0, y_0 + \varepsilon, z_0 - \beta) - 2\alpha a_2(x_0, y_0 + \eta, z_0 + \beta)$$

where $y_0 + \varepsilon$ is the y-coordinate of the point in AB at which a_2 has its mean value for AB. Since this point lies in AB, $-\alpha \leqslant \varepsilon \leqslant \alpha$ and so $\varepsilon \to 0$ as $\alpha \to 0$. η is similarly defined for CD and so also satisfies the condition $\eta \to 0$ as $\alpha \to 0$. Therefore

$$\lim_{s\to 0} \frac{\left\{ \int_{AB} \mathbf{a}\cdot d\mathbf{l} + \int_{CD} \mathbf{a}\cdot d\mathbf{l} \right\}}{s}$$

$$= \lim_{\substack{\alpha\to 0 \\ \beta\to 0}} \frac{2\alpha\{a_2(x_0, y_0 + \varepsilon, z_0 - \beta) - a_2(x_0, y_0 + \eta, z_0 + \beta)\}}{4\alpha\beta}$$

$$= \lim_{\beta \to 0} \frac{a_2(x_0, y_0, z_0 - \beta) - a_2(x_0, y_0, z_0 + \beta)}{2\beta}$$

$$= -\frac{\partial a_2}{\partial z} \quad \text{at} \quad P_0.$$

Similarly,

$$\lim_{s \to 0} \frac{\left\{ \int_{BC} \mathbf{a} \cdot d\mathbf{l} + \int_{DA} \mathbf{a} \cdot d\mathbf{l} \right\}}{s} = \frac{\partial a_3}{\partial y}$$

at P_0 so that the x-component of curl $\mathbf{a}$ at P_0 is

$$\left(\frac{\partial a_3}{\partial y} - \frac{\partial a_2}{\partial z} \right)$$

evaluated at P_0. By cyclic permutation the y- and z-components are then

$$\left(\frac{\partial a_1}{\partial z} - \frac{\partial a_3}{\partial x} \right)$$

and

$$\left(\frac{\partial a_2}{\partial x} - \frac{\partial a_1}{\partial y} \right)$$

respectively. Thus curl $\mathbf{a}$ at a point is given in terms of cartesian coordinates by

$$\text{curl } \mathbf{a} = \left(\frac{\partial a_3}{\partial y} - \frac{\partial a_2}{\partial z} \right)\mathbf{i} + \left(\frac{\partial a_1}{\partial z} - \frac{\partial a_3}{\partial x} \right)\mathbf{j} + \left(\frac{\partial a_2}{\partial x} - \frac{\partial a_1}{\partial y} \right)\mathbf{k}$$

where the partial derivatives are calculated at that point.

Thus we now have associated with any vector field $\mathbf{a}$ another vector field curl $\mathbf{a}$.

It is easily seen from the above equation that curl $\mathbf{a}$, in cartesian coordinates, can also be written as $\nabla \times \mathbf{a}$ where, as before, ∇ is to be regarded as the operator $[\mathbf{i}(\partial/\partial x) + \mathbf{j}(\partial/\partial y) + \mathbf{k}(\partial/\partial z)]$. Then, from the determinant rule for writing down a vector product, we see that curl $\mathbf{a}$ in cartesian coordinates may be written symbolically as

$$\text{curl } \mathbf{a} = \begin{vmatrix} \mathbf{i} & \mathbf{j} & \mathbf{k} \\ \dfrac{\partial}{\partial x} & \dfrac{\partial}{\partial y} & \dfrac{\partial}{\partial z} \\ a_1 & a_2 & a_3 \end{vmatrix}.$$

If curl $\mathbf{a} \equiv \mathbf{0}$ then $\mathbf{a}$ is said to be *irrotational*.

For example consider the vector field given by $\mathbf{r}$ the position vector of a variable point with respect to a fixed origin.

$$\mathbf{r} = (x, y, z).$$

Therefore

$$\text{curl } \mathbf{r} = \left(\frac{\partial z}{\partial y} - \frac{\partial y}{\partial z}\right)\mathbf{i} + \left(\frac{\partial x}{\partial z} - \frac{\partial z}{\partial x}\right)\mathbf{j} + \left(\frac{\partial y}{\partial x} - \frac{\partial x}{\partial y}\right)\mathbf{k}$$

$$= \mathbf{0}.$$

That is the vector field defined by $\mathbf{r}$ is irrotational.

Now consider the derivation of this same result starting from the fundamental definition of curl $\mathbf{r}$. To determine the x-component of curl $\mathbf{r}$ at any point P_0 having position vector $\mathbf{r}_0 = (x_0, y_0, z_0)$, surround this point by a circle of radius c, centre at P_0 and lying wholly in the plane through P_0 and parallel to the y,z-plane (figure 25). Let the bounding circumference of this circle be L. The area enclosed is πc^2.

$$\int_L \mathbf{r} \cdot d\mathbf{l} = \int_L (\mathbf{r}_0 + \mathbf{\rho}) \cdot d\mathbf{l}$$

where Q is any point on L, $\mathbf{\rho} = \overline{P_0 Q}$ and is such that $|\mathbf{\rho}| = c$. Now $\int_L \mathbf{\rho} \cdot d\mathbf{l} = \int_L (\mathbf{\rho} \cdot \mathbf{t}) \, dl$ where $\mathbf{t}$ is a unit vector in the direction of the tangent to the curve L at the general point Q on it.

Hence $\int_L \mathbf{\rho} \cdot d\mathbf{l} = 0$ since $\mathbf{\rho}$ and $\mathbf{t}$ are always perpendicular so that $\mathbf{\rho} \cdot \mathbf{t} \equiv 0$. Also any point on L may be expressed (figure 26) in terms of the parameter θ by the vector $(x_0, y_0 + c \cos \theta, z_0 + c \sin \theta)$ for $0 \leqslant \theta < 2\pi$.

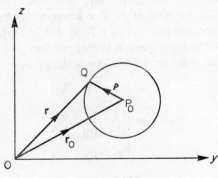

FIGURE 25

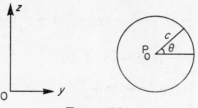

FIGURE 26

Therefore

$$\int_L \mathbf{r}_0 \cdot \mathbf{dl} = \int_L (x_0 \, dx + y_0 \, dy + z_0 \, dz)$$

$$= \int_0^{2\pi} y_0(-c \sin \theta \, d\theta) + \int_0^{2\pi} z_0 c \cos \theta \, d\theta \quad (dx = 0 \text{ on L})$$

$$= cy_0 \Big[\cos \theta \Big]_0^{2\pi} + cz_0 \Big[\sin \theta \Big]_0^{2\pi} = 0.$$

(This is a special case of the general result that $\int_L \mathbf{b} \cdot \mathbf{dl} = 0$ for any constant vector $\mathbf{b}$ and closed curve L.) Therefore the x-component of curl $\mathbf{r} = 0$.

Similarly, it can be shown that the y- and z-components are also zero. Therefore curl $\mathbf{r} = \mathbf{0}$.

Example If

$$\mathbf{F}_1 = (y^2 - z^2 + 3yz - 2x, \ 3xz + 2xy, \ 3xy - 2xz + 2z)$$

and

$$\mathbf{F}_2 = (z^2 + 2x + 3y, \ 3x + 2y + z, \ y + 2xz),$$

show that $\mathbf{F}_1$, is both irrotational and solenoidal whereas $\mathbf{F}_2$ is irrotational but not solenoidal.

$$\text{curl } \mathbf{F}_1 = \left(\frac{\partial}{\partial y} (3xy - 2xz + 2z) - \frac{\partial}{\partial z} (3xz + 2xy), \right.$$

$$\frac{\partial}{\partial z} (y^2 - z^2 + 3yz - 2x) - \frac{\partial}{\partial x} (3xy - 2xz + 2z),$$

$$\left. \frac{\partial}{\partial x} (3xz + 2xy) - \frac{\partial}{\partial y} (y^2 - z^2 + 3yz - 2x) \right)$$

$$= (3x - 3x, \ -2z + 3y - 3y + 2z, \ 3z + 2y - 2y - 3z)$$

$$= \mathbf{0}.$$

Therefore $\mathbf{F}_1$ is irrotational.

$$\text{div } \mathbf{F}_1 = \frac{\partial}{\partial x}(y^2 - z^2 + 3yz - 2x) + \frac{\partial}{\partial y}(3xz + 2xy)$$

$$+ \frac{\partial}{\partial z}(3xy - 2xz + 2z)$$

$$= -2 + 2x - 2x + 2$$

$$= 0.$$

Therefore $\mathbf{F}_1$ is solenoidal.

$$\text{curl } \mathbf{F}_2 = \left(\frac{\partial}{\partial y}(y + 2xz) - \frac{\partial}{\partial z}(3x + 2y + z), \frac{\partial}{\partial z}(z^2 + 2x + 3y)\right.$$

$$\left. - \frac{\partial}{\partial x}(y + 2xz), \frac{\partial}{\partial x}(3x + 2y + z) - \frac{\partial}{\partial y}(z^2 + 2x + 3y)\right)$$

$$= (1 - 1, 2z - 2z, 3 - 3)$$

$$= \mathbf{0}.$$

Therefore $\mathbf{F}_2$ is irrotational.

$$\text{div } \mathbf{F}_2 = \frac{\partial}{\partial x}(z^2 + 2x + 3y) + \frac{\partial}{\partial y}(3x + 2y + z) + \frac{\partial}{\partial z}(y + 2xz)$$

$$= 2 + 2 + 2x$$

$$\neq 0.$$

Therefore $\mathbf{F}_2$ is not solenoidal.

Example *Show that* div $(\varphi \mathbf{a}) = \varphi$ div $\mathbf{a} + \mathbf{a} \cdot$ grad φ *where* φ *and* $\mathbf{a}$ *define arbitrary scalar and vector fields respectively.*

$$\text{div } \varphi \mathbf{a} = \text{div } (\varphi a_1, \varphi a_2, \varphi a_3) \quad \text{where} \quad \mathbf{a} = (a_1, a_2, a_3).$$

Therefore

$$\text{div } \varphi \mathbf{a} = \frac{\partial}{\partial x}(\varphi a_1) + \frac{\partial}{\partial y}(\varphi a_2) + \frac{\partial}{\partial z}(\varphi a_3)$$

$$= \varphi \frac{\partial a_1}{\partial x} + a_1 \frac{\partial \varphi}{\partial x} + \varphi \frac{\partial a_2}{\partial y} + a_2 \frac{\partial \varphi}{\partial y} + \varphi \frac{\partial a_3}{\partial z} + a_3 \frac{\partial \varphi}{\partial z}$$

$$= \varphi \left(\frac{\partial a_1}{\partial x} + \frac{\partial a_2}{\partial y} + \frac{\partial a_3}{\partial z}\right) + \left(a_1 \frac{\partial \varphi}{\partial x} + a_2 \frac{\partial \varphi}{\partial y} + a_3 \frac{\partial \varphi}{\partial z}\right)$$

$$= \varphi \text{ div } \mathbf{a} + \mathbf{a} \cdot \text{ grad } \varphi.$$

The following results (the proofs of which are left to the reader) are also easily obtained.

$$\text{curl } (\varphi \mathbf{a}) = \text{grad } \varphi \times \mathbf{a} + \varphi \text{ curl } \mathbf{a}$$

and
$$\text{div } (\mathbf{a} \times \mathbf{b}) = \mathbf{b} \cdot \text{curl } \mathbf{a} - \mathbf{a} \cdot \text{curl } \mathbf{b}$$

where **b** also defines an arbitrary vector field.

Example *Show that if* **c** *is any constant vector and* **r** *is the position vector of the general point, then* grad $(\mathbf{c} \cdot \mathbf{r}) = \mathbf{c}$, div $(\mathbf{c} \times \mathbf{r}) = 0$ *and* curl $(\mathbf{c} \times \mathbf{r}) = 2\mathbf{c}$.

Since **c** is any constant vector we can choose one of the coordinate axes, say the x-axis, to be in the direction of **c**. Then $\mathbf{c} = (c,0,0)$.
Therefore $\mathbf{c} \cdot \mathbf{r} = (c,0,0) \cdot (x,y,z) = cx$.
Then

$$\text{grad } (\mathbf{c} \cdot \mathbf{r}) = \frac{\partial}{\partial x}(cx)\mathbf{i} + \frac{\partial}{\partial y}(cx)\mathbf{j} + \frac{\partial}{\partial z}(cx)\mathbf{k}$$

$$= c\mathbf{i} \quad \text{since } c \text{ is constant.}$$

Hence grad $(\mathbf{c} \cdot \mathbf{r}) = \mathbf{c}$.

$$\mathbf{c} \times \mathbf{r} = (c,0,0) \times (x,y,z) = (0,-cz,cy)$$

Therefore

$$\text{div } (\mathbf{c} \times \mathbf{r}) = \frac{\partial}{\partial y}(-cz) + \frac{\partial}{\partial z}(cy) = 0.$$

Alternatively div$(\mathbf{c} \times \mathbf{r}) = \mathbf{r} \cdot \text{curl } \mathbf{c} - \mathbf{c} \cdot \text{curl } \mathbf{r} = 0$ since curl $\mathbf{c} = 0$ as **c** is a constant vector and it has already been seen that curl $\mathbf{r} = 0$.

$$\text{curl } (\mathbf{c} \times \mathbf{r}) = \begin{vmatrix} \mathbf{i} & \mathbf{j} & \mathbf{k} \\ \dfrac{\partial}{\partial x} & \dfrac{\partial}{\partial y} & \dfrac{\partial}{\partial z} \\ 0 & -cz & cy \end{vmatrix}$$

$$= (c + c,0,0).$$

$$= 2\mathbf{c}.$$

Example *If* **c** *is a constant vector and* **r** *the position vector of a general point, show that*

$$\text{div } [(\mathbf{c} \cdot \mathbf{r})r^n]\mathbf{r} = (n + 4)(\mathbf{c} \cdot \mathbf{r})r^n$$

and

$$\text{curl } [(\mathbf{c} \cdot \mathbf{r})r^n]\mathbf{r} = (\mathbf{c} \times \mathbf{r})r^n.$$

$$\text{div } [(\mathbf{c} \cdot \mathbf{r})r^n]\mathbf{r} = (\mathbf{c} \cdot \mathbf{r})r^n \text{ div } \mathbf{r} + \mathbf{r} \cdot \text{grad } [(\mathbf{c} \cdot \mathbf{r})r^n]$$

$$= (\mathbf{c} \cdot \mathbf{r})r^n \text{ div } \mathbf{r} + \mathbf{r} \cdot [(\mathbf{c} \cdot \mathbf{r}) \text{ grad } r^n + r^n \text{ grad } (\mathbf{c} \cdot \mathbf{r})].$$

But on page 34 we see that grad $r^n = nr^{n-2}\mathbf{r}$, on page 38 we see that div $\mathbf{r} = 3$

and from the last example we have grad $(\mathbf{c} \cdot \mathbf{r}) = \mathbf{c}$. Therefore

$$\text{div } [(\mathbf{c} \cdot \mathbf{r})r^n]\mathbf{r} = 3(\mathbf{c} \cdot \mathbf{r})r^n + \mathbf{r} \cdot [(\mathbf{c} \cdot \mathbf{r})nr^{n-2}\mathbf{r} + r^n\mathbf{c}]$$
$$= 3(\mathbf{c} \cdot \mathbf{r})r^n + (\mathbf{c} \cdot \mathbf{r})nr^{n-2}\mathbf{r} \cdot \mathbf{r} + r^n(\mathbf{c} \cdot \mathbf{r})$$
$$= (n + 4)(\mathbf{c} \cdot \mathbf{r})r^n.$$

$$\text{curl } [(\mathbf{c} \cdot \mathbf{r})r^n]\mathbf{r} = \text{grad } [(\mathbf{c} \cdot \mathbf{r})r^n] \times \mathbf{r} + (\mathbf{c} \cdot \mathbf{r})r^n \text{ curl } \mathbf{r}.$$

But on page 44 we see that curl $\mathbf{r} = \mathbf{0}$. Therefore

$$\text{curl } [(\mathbf{c} \cdot \mathbf{r})r^n]\mathbf{r} = ((\mathbf{c} \cdot \mathbf{r})nr^{n-2}\mathbf{r} + r^n\mathbf{c}) \times \mathbf{r}$$
$$= r^n(\mathbf{c} \times \mathbf{r}).$$

Exercises

1. If $\mathbf{F} = (x^2 + y^2 + z^2, 3xyz, xy^2 - zy^2)$, find curl $\mathbf{F}$.

2. If $\mathbf{F} = (x^4 - y^2z^2, 2xyz^2 - y^3z, 4x^2z^2 - 4y^2z^2)$, show that curl $\mathbf{F} = \mathbf{0}$ at the point $(-1,2,1)$.

3. Show that the vector $\mathbf{F} = (2x + 3yz - z^2, 2y + 3xz, 3xy - 2xz)$ is irrotational.

4. Verify that when $\mathbf{a} = (3x, yz, z^2x)$ and $\mathbf{b} = (y, 3x^2yz, -z^3)$

$$\text{div } (\mathbf{a} \times \mathbf{b}) = \mathbf{b} \cdot \text{curl } \mathbf{a} - \mathbf{a} \cdot \text{curl } \mathbf{b}.$$

5. Verify that div $(\varphi\mathbf{a}) = \varphi \text{ div } \mathbf{a} + \mathbf{a} \cdot \text{grad } \varphi$ when $\varphi = x^2y^2z$ and

$$\mathbf{a} = (xz, y^2, xyz^2).$$

6. If $\varphi = x^2y^2z$ and $\mathbf{a} = (xz, y^2, xyz^2)$ verify that

$$\text{curl } (\varphi\mathbf{a}) = \text{grad } \varphi \times \mathbf{a} + \varphi \text{ curl } \mathbf{a}.$$

1.8 Second-order expressions

We have defined the first-order expressions 'div', 'grad' and 'curl'. Second-order expressions are any meaningful double combination of these, for example curl grad φ, div curl $\mathbf{a}$, grad div $\mathbf{a}$. Note however that since grad φ has only been defined for scalar functions φ, the expressions grad grad φ and grad curl $\mathbf{a}$ are meaningless, as is curl div $\mathbf{a}$ since we have only defined the curl of a vector.

1. $$\text{div grad } \varphi = \text{div } \left(\frac{\partial\varphi}{\partial x}, \frac{\partial\varphi}{\partial y}, \frac{\partial\varphi}{\partial z}\right) = \frac{\partial^2\varphi}{\partial x^2} + \frac{\partial^2\varphi}{\partial y^2} + \frac{\partial^2\varphi}{\partial z^2}$$

$$= \left(\frac{\partial^2}{\partial x^2} + \frac{\partial^2}{\partial y^2} + \frac{\partial^2}{\partial z^2}\right)\varphi$$

and is denoted by $\nabla^2\varphi$. The equation $\nabla^2\varphi = 0$ is called Laplace's equation and we shall be concerned later with its solutions.

2. $$\text{curl grad } \varphi = \text{curl} \left(\frac{\partial \varphi}{\partial x}, \frac{\partial \varphi}{\partial y}, \frac{\partial \varphi}{\partial z} \right)$$

$$= \left(\frac{\partial^2 \varphi}{\partial y \, \partial z} - \frac{\partial^2 \varphi}{\partial z \, \partial y} \right) \mathbf{i} + \left(\frac{\partial^2 \varphi}{\partial z \, \partial x} - \frac{\partial^2 \varphi}{\partial x \, \partial z} \right) \mathbf{j}$$

$$+ \left(\frac{\partial^2 \varphi}{\partial x \, \partial y} - \frac{\partial^2 \varphi}{\partial y \, \partial x} \right) \mathbf{k}$$

$$= \mathbf{0}.$$

Therefore curl grad $\varphi \equiv \mathbf{0}$.

We shall see later that if curl $\mathbf{a} \equiv \mathbf{0}$ then $\mathbf{a}$ may be expressed as the gradient of a scalar function, (that is $\mathbf{a} = \text{grad } \varphi$ for some scalar function φ).

3. $$\text{div curl } \mathbf{a} = \text{div} \left(\frac{\partial a_3}{\partial y} - \frac{\partial a_2}{\partial z}, \frac{\partial a_1}{\partial z} - \frac{\partial a_3}{\partial x}, \frac{\partial a_2}{\partial x} - \frac{\partial a_1}{\partial y} \right)$$

$$\text{if} \quad \mathbf{a} = (a_1, a_2, a_3)$$

$$= \frac{\partial}{\partial x} \left(\frac{\partial a_3}{\partial y} - \frac{\partial a_2}{\partial z} \right) + \frac{\partial}{\partial y} \left(\frac{\partial a_1}{\partial z} - \frac{\partial a_3}{\partial x} \right) + \frac{\partial}{\partial z} \left(\frac{\partial a_2}{\partial x} - \frac{\partial a_1}{\partial y} \right)$$

$$= \frac{\partial^2 a_3}{\partial x \, \partial y} - \frac{\partial^2 a_2}{\partial x \, \partial z} + \frac{\partial^2 a_1}{\partial y \, \partial z} - \frac{\partial^2 a_3}{\partial y \, \partial x} + \frac{\partial^2 a_2}{\partial z \, \partial x} - \frac{\partial^2 a_1}{\partial z \, \partial y}$$

$$= 0.$$

Therefore div curl $\mathbf{a} \equiv 0$.

We shall see later that if div $\mathbf{a} \equiv 0$ then $\mathbf{a}$ may be expressed as the curl of some vector function (that is $\mathbf{a} = \text{curl } \mathbf{b}$ for some vector function $\mathbf{b}$).

4. $$\text{curl curl } \mathbf{a} = \text{curl} \left(\frac{\partial a_3}{\partial y} - \frac{\partial a_2}{\partial z}, \frac{\partial a_1}{\partial z} - \frac{\partial a_3}{\partial x}, \frac{\partial a_2}{\partial x} - \frac{\partial a_1}{\partial y} \right)$$

Therefore the x-component is

$$\frac{\partial}{\partial y} \left(\frac{\partial a_2}{\partial x} - \frac{\partial a_1}{\partial y} \right) - \frac{\partial}{\partial z} \left(\frac{\partial a_1}{\partial z} - \frac{\partial a_3}{\partial x} \right)$$

$$= \frac{\partial^2 a_2}{\partial x \, \partial y} - \frac{\partial^2 a_1}{\partial y^2} - \frac{\partial^2 a_1}{\partial z^2} + \frac{\partial^2 a_3}{\partial x \, \partial z}$$

$$= \frac{\partial^2 a_1}{\partial x^2} + \frac{\partial^2 a_2}{\partial x \, \partial y} + \frac{\partial^2 a_3}{\partial x \, \partial z} - \left(\frac{\partial^2 a_1}{\partial x^2} + \frac{\partial^2 a_1}{\partial y^2} + \frac{\partial^2 a_1}{\partial z^2} \right)$$

$$= \frac{\partial}{\partial x} \left(\frac{\partial a_1}{\partial x} + \frac{\partial a_2}{\partial y} + \frac{\partial a_3}{\partial z} \right) - \left(\frac{\partial^2}{\partial x^2} + \frac{\partial^2}{\partial y^2} + \frac{\partial^2}{\partial z^2} \right) a_1$$

$$= \frac{\partial}{\partial x} (\text{div } \mathbf{a}) - \nabla^2 a_1.$$

Therefore

$$\text{curl curl } \mathbf{a} = \left(\frac{\partial}{\partial x}(\text{div } \mathbf{a}), \frac{\partial}{\partial y}(\text{div } \mathbf{a}), \frac{\partial}{\partial z}(\text{div } \mathbf{a})\right) - (\nabla^2 a_1, \nabla^2 a_2, \nabla^2 a_3)$$

$$= \text{grad div } \mathbf{a} - \nabla^2 \mathbf{a}$$

where $\nabla^2 \mathbf{a}$ means in *cartesian coordinates* the vector $(\nabla^2 a_1, \nabla^2 a_2, \nabla^2 a_3)$. It must be emphasized that this expression for $\nabla^2 \mathbf{a}$ does not apply in other coordinate systems.

Higher order expressions such as curl curl curl $\mathbf{a}$, grad div grad φ etc. can be determined in the same way as above.

Example *If $\varphi = x^2 yz$ and $\psi = x^2 + y^2 - z^2$ find*
(*i*) grad (grad φ . grad ψ)
(*ii*) div (grad $\varphi \times$ grad ψ)
(*iii*) curl (grad $\varphi \times$ grad ψ).

$$\text{grad } \varphi = (2xyz, x^2 z, x^2 y),$$

$$\text{grad } \psi = (2x, 2yz, -2z),$$

Therefore

$$\text{grad } \varphi . \text{grad } \psi = 4x^2 yz + 2x^2 yz - 2x^2 yz$$

$$= 4x^2 yz.$$

$$\text{grad } \varphi \times \text{grad } \psi = \begin{vmatrix} \mathbf{i} & \mathbf{j} & \mathbf{k} \\ 2xyz & x^2 z & x^2 y \\ 2x & 2y & -2z \end{vmatrix}$$

$$= (-2x^2(y^2 + z^2), 2xy(2z^2 + x^2), 2xz(2y^2 - x^2)).$$

(i)   grad (grad φ . grad ψ) = $(8xyz, 4x^2 z, 4x^2 y)$.

(ii)   div (grad $\varphi \times$ grad ψ)

$$= -4x(y^2 + z^2) + 2x(2z^2 + x^2) + 2x(2y^2 - x^2) = 0.$$

(iii)   curl (grad $\varphi \times$ grad ψ)

$$= \begin{vmatrix} \mathbf{i} & \mathbf{j} & \mathbf{k} \\ \dfrac{\partial}{\partial x} & \dfrac{\partial}{\partial y} & \dfrac{\partial}{\partial z} \\ -2x^2(y^2 + z^2) & 2xy(2z^2 + x^2) & 2xz(2y^2 - x^2) \end{vmatrix}$$

$$= (8xyz - 8xyz, -4x^2 z - 4y^2 z + 6x^2 z, 4yz^2 + 6x^2 y - 4x^2 y)$$

$$= (0, 2z(x^2 - 2y^2), 2y(x^2 + 2z^2)).$$

Exercises

1. Find $\nabla^2\varphi$ when $\varphi = x^3y + x^2y^2z + z^3y^3 + 3xyz - 2y + 4$.

2. Show that $\nabla^2(r^n) = n(n + 1)r^{n-2}$ where $r^2 = x^2 + y^2 + z^2$ and hence deduce that $\nabla^2(1/r) = 0$ except at $r = 0$.

3. If $\mathbf{b} = -\text{grad } \varphi$ where $\nabla^2\varphi = 0$, $\mathbf{c}$ is a constant vector and $\mathbf{r}$ the position vector, show that
   (i) div $[\varphi\mathbf{c} + (\mathbf{c}\cdot\mathbf{r})\mathbf{b}] = 0$,
   (ii) div $[\varphi\mathbf{b} + (\mathbf{c}\cdot\mathbf{r})\mathbf{c}] = a^2 - b^2$,
   (iii) curl $[\varphi\mathbf{c} + (\mathbf{c}\cdot\mathbf{r})\mathbf{b}] = 2\mathbf{c}\times\mathbf{b}$,
   (iv) curl $[\varphi\mathbf{b} + (\mathbf{c}\cdot\mathbf{r})\mathbf{c}] = 0$.

4. Verify that for $\mathbf{a} = (y^2, 2xyz^2, x^3z)$, curl curl $\mathbf{a} = \text{grad }(\text{div } \mathbf{a}) - \nabla^2\mathbf{a}$.

5. Prove that $\nabla^2(\varphi_1\varphi_2) = \varphi_1\nabla^2\varphi_2 + 2\nabla\varphi_1\cdot\nabla\varphi_2 + \varphi_2\nabla^2\varphi_1$.

6. If $\varphi_1 = 2x^2yz^2$ and $\varphi_2 = x^2y^2 - y^2z^2$ find
   (i) grad (grad $\varphi_1\cdot$ grad φ_2),
   (ii) div (grad $\varphi_1\times$ grad φ_2).

1.9 Orthogonal curvilinear coordinates

If we can find functions $u_1(x,y,z)$, $u_2(x,y,z)$ and $u_3(x,y,z)$ such that to each point (x,y,z) there corresponds one point (u_1,u_2,u_3) and conversely, if the tangent planes to the surfaces given by $u_1 = $ constant, $u_2 = $ constant, $u_3 = $ constant are mutually orthogonal at each point, then u_1, u_2, u_3 form a system of orthogonal curvilinear coordinates. The axes of u_1, u_2, u_3 are the intersections of the mutually orthogonal tangent planes. The orientation of the axes changes from point to point. For example (1) *spherical polar coordinates*

$u_1 \equiv r = \sqrt{x^2 + y^2 + z^2} = $ the distance of a point P from a fixed point (taken as the origin)

$u_2 \equiv \theta = $ the value of $\tan^{-1}(\sqrt{x^2 + y^2}/z)$ which lies in the range 0 to π.
   $= $ the angle (between 0 and π) which the radius vector from the origin to the point makes with a fixed direction (taken as the positive z-axis).

$u_3 \equiv \psi = $ the principal value of $\tan^{-1}(y/x)$
   $= $ the angle (between π and $-\pi$) which the projection in the x,y-plane of the radius vector to the point makes with a fixed direction (taken as the positive x-axis) (figure 27).

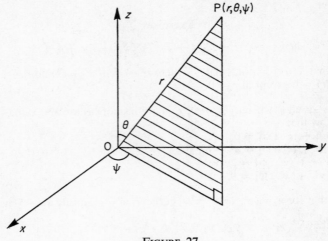

FIGURE 27

The surfaces $r =$ constant are spheres with centre at the origin O.
The surfaces $\theta =$ constant are circular cones with axis Oz and vertex at O.
The surfaces $\psi =$ constant are planes containing Oz.
The surface $\theta = 0$ does not conform with the definitions in that it degenerates into a line (the z-axis) and does not have a unique tangent plane at any point on it. Points on this line are thus usually given special consideration.

(2) *cylindrical polar coordinates*

$u_1 \equiv \rho = \sqrt{x^2 + y^2} =$ the distance of a point P from a fixed line (taken as the z-axis).

$u_2 \equiv \psi =$ the principal value of $\tan^{-1}(y/x)$
     = the angle (between π and $-\pi$) which the projection in the x,y-plane of the radius vector to the point makes with a fixed direction (taken as the positive x-axis).

$u_3 \equiv z =$ the height of the point P above the x,y-plane (figure 28).

The surfaces $\rho =$ constant are circular cylinders with axis the z-axis.
The surfaces $\psi =$ constant are planes containing the z-axis.
The surfaces $z =$ constant are planes perpendicular to the z-axis.
Once again, the z-axis (corresponding here to the surface $\rho = 0$) does not conform with the definitions and must be given special consideration.

There are also other examples of orthogonal curvilinear coordinate systems which are frequently used in practice, but these will not be considered here. Which coordinate system used will of course depend on

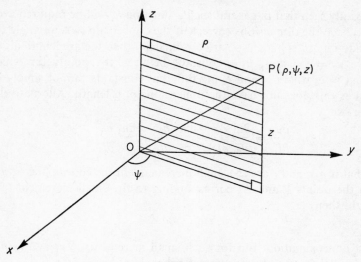

FIGURE 28

the particular physical problem being dealt with, and we shall see later that much tedious work can be avoided by making the correct choice.

Now, returning to the general argument, let P be the point $(u_1(P), u_2(P), u_3(P))$. The arc length l_1 along the curve given by $u_2 = $ constant $(= u_2(P))$, $u_3 = $ constant $(= u_3(P))$ from P to the point A having coordinates $(u_1(P) + \alpha, u_2(P), u_3(P))$ is to a first approximation $l_1 = h_1\alpha$ say where h_1 is a function of the coordinates of P* (figure 29).

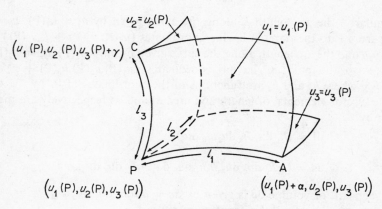

FIGURE 29

* For the particular coordinate systems considered here, this relation is in fact exact.

It is readily seen that in general such a function h_1 will be required even if only to keep the dimensions correct. If the coordinate u_1 is an angle (as in the 2nd and 3rd spherical polar coordinates), then the arc length along the curve $u_2 = $ constant, $u_3 = $ constant between the points for which $u_1 = c_1, u_1 = c_2$ where c_1 and c_2 are different constants cannot simply be $(c_2 - c_1)$ as this does not have the dimensions of a length. Alternatively, since

$$\frac{\partial u_1}{\partial l_1} = \frac{\partial u_1}{\partial x}\frac{\partial x}{\partial l_1} + \frac{\partial u_1}{\partial y}\frac{\partial y}{\partial l_1} + \frac{\partial u_1}{\partial z}\frac{\partial z}{\partial l_1}$$

we see that if p, q and r are the small increments in the coordinates x, y, z between the points P and A, corresponding to the small increment l_1 in arc length, then

$$\alpha = \nabla u_1 \cdot (p,q,r)$$

to a first approximation. But for such small increments, ∇u_1 and (p,q,r) will be parallel (to the same degree of approximation) and so

$$\alpha = |\nabla u_1|\,|(p,q,r)| = |\nabla u_1|\sqrt{p^2 + q^2 + r^2} = l_1|\nabla u_1|$$

or

$$l_1 = h_1\alpha$$

where

$$h_1 = \frac{1}{|\nabla u_1|} = \frac{1}{\sqrt{\left(\dfrac{\partial u_1}{\partial x}\right)^2 + \left(\dfrac{\partial u_1}{\partial y}\right)^2 + \left(\dfrac{\partial u_1}{\partial z}\right)^2}}.$$

Similarly, the arc length l_2 along the curve given by $u_1 = u_1(P), u_3 = u_3(P)$ from P to the point B having coordinates $(u_1(P), u_2(P) + \beta, u_3(P))$ is $l_2 = h_2\beta$ and the arc length along the curve given by $u_1 = u_1(P), u_2 = u_2(P)$ from P to the point C having coordinates $(u_1(P), u_2(P), u_3(P) + \gamma)$ is $l_3 = h_3\gamma$ where h_2 and h_3 are functions of the coordinates of P.

The 'vector elements' of length and area dl and ds respectively are then given by

$$\mathbf{dl} = (h_1\,du_1, h_2\,du_2, h_3\,du_3)$$

and

$$\mathbf{ds} = (h_2 h_3\,du_2\,du_3, h_3 h_1\,du_3\,du_1, h_1 h_2\,du_1\,du_2).$$

The 'element' of volume dv is given by d$v = h_1 h_2 h_3\,du_1\,du_2\,du_3$. For example (1) *spherical polars* (figure 30).

$$P \equiv (r_0, \theta_0, \psi_0), \ A \equiv (r_0 + \alpha, \theta_0, \psi_0), \ B \equiv (r_0, \theta_0 + \beta, \psi_0), \ C \equiv (r_0, \theta_0, \psi_0 + \gamma).$$

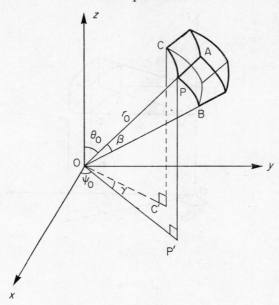

FIGURE 30

The curve $\theta = \theta_0$, $\psi = \psi_0$ is the straight line OP and the arc length PA is then α. Thus in this case $h_1 = 1$.

The curve $r = r_0$, $\psi = \psi_0$ is the circle centre 0 and radius OP and the arc length PB is $r_0\beta$. Hence h_2 at P is r_0 and so in general $h_2 = r$.

The curve $r = r_0$, $\theta = \theta_0$ is the circle in the plane through P parallel to the x,y-plane which has its centre on the z-axis and which passes through P. The radius of this circle is $r_0 \sin \theta_0$ and so the arc length PC is $(r_0 \sin \theta_0)\gamma$. Hence h_3 at P is $r_0 \sin \theta_0$ and in general, we have $h_3 = r \sin \theta$.

(2) *cylindrical polars* (figure 31).

$$P \equiv (\rho_0,\psi_0,z_0), \quad A \equiv (\rho_0 + \alpha,\psi_0,z_0), \quad B \equiv (\rho_0,\psi_0 + \beta,z_0), \quad C \equiv (\rho_0,\psi_0,z_0 + \gamma).$$

The curve $\psi = \psi_0$, $z = z_0$ is the straight line PN perpendicular to Oz and the arc length PA is then α. Therefore $h_1 = 1$.

The curve $\rho = \rho_0$, $z = z_0$ is the circle with centre N on the z-axis and radius NP and the arc length PB is thus $\rho_0\beta$. Therefore h_2 at P is ρ_0 and so in general $h_2 = r$.

The curve $\rho = \rho_0$, $\psi = \psi_0$ is the straight line through P parallel to Oz and so the arc length PC is γ. Therefore $h_3 = 1$.

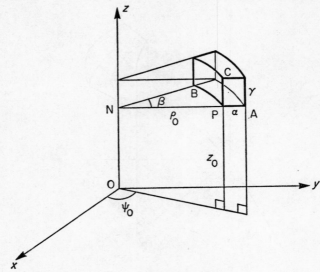

FIGURE 31

Example   Evaluate $\int_S z^2 \, ds$ *where S is the surface of the hemisphere represented by* $x^2 + y^2 + z^2 = a^2, z \geqslant 0$.

Use spherical polar coordinates. The hemispherical surface S is then defined by $r = a; 0 \leqslant \theta \leqslant \pi/2; 0 \leqslant \psi < 2\pi$. Now, in the notation of the above section, the 'element of area' on the surface $u_1 =$ constant is $h_2 h_3 \, du_2 \, du_3$. Hence in spherical polar coordinates, the 'element of area' on the surface $r = a$ is $a(a \sin \theta) \, d\theta \, d\psi$. Hence

$$\int_S z^2 \, ds = \int_0^{2\pi} \int_0^{\frac{1}{2}\pi} (a \cos \theta)^2 a^2 \sin \theta \, d\theta \, d\psi$$

since $z = a \cos \theta$. Therefore

$$\int_S z^2 \, ds = a^4 \int_0^{2\pi} \int_0^{\frac{1}{2}\pi} \cos^2 \theta \sin \theta \, d\theta \, d\psi$$

$$= a^4 \int_0^{2\pi} \left[-\frac{\cos^3 \theta}{3} \right]_0^{\frac{1}{2}\pi} d\psi = \frac{a^4}{3} \left[\psi \right]_0^{2\pi} = \tfrac{2}{3}\pi a^4.$$

Example   Evaluate $\int_S (x^2 + y^2) \, ds$ *where S is the surface of the cone* $z^2 = 3(x^2 + y^2)$ *between the planes* $z = 0$ *and* $z = 3$.

Use spherical polar coordinates. Since $r^2 = x^2 + y^2 + z^2$, we see that for points on S,

$$r^2 = \frac{z^2}{3} + z^2 = \tfrac{4}{3}z^2 \quad \text{for} \quad 0 \leqslant z \leqslant 3.$$

Also, since $\theta = \tan^{-1} [\sqrt{(x^2 + y^2)}/z,]$ we see that for points on S,

$$\theta = \tan^{-1} \frac{1}{\sqrt{3}} = \frac{\pi}{6}$$

Thus S is defined by $\theta = \tfrac{1}{6}\pi$; $0 \leqslant r \leqslant 2\sqrt{3}$; $0 \leqslant \psi < 2\pi$. The 'element of area' on the surface $\theta = \tfrac{1}{6}\pi$ is $r \sin (\tfrac{1}{6}\pi) \, \mathrm{d}\psi \, \mathrm{d}r$. Also, on the surface S, $x^2 + y^2 = r \sin (\tfrac{1}{6}\pi)^2 = r^2/4$. Hence

$$\int_S (x^2 + y^2) \, \mathrm{d}s = \int_0^{2\sqrt{3}} \int_0^{2\pi} \frac{r^2}{4} \frac{r}{2} \, \mathrm{d}\psi \, \mathrm{d}r$$

$$= \frac{\pi}{4} \int_0^{2\sqrt{3}} r^3 \, \mathrm{d}r$$

$$= \frac{\pi}{4} \frac{(2\sqrt{3})^4}{4} = 9\pi.$$

Example   Evaluate $\displaystyle\int_V (x^2 + y^2) \, \mathrm{d}v$ *where V is the region bounded by the planes* $x = 0, y = 0, z = 0, z = h$ *and the cylindrical surface* $x^2 + y^2 = a^2$ *and in which* x *and* y *are both positive.*

Use cylindrical polar coordinates. The region V is then defined by $0 \leqslant \rho \leqslant a$, $0 \leqslant \psi \leqslant \tfrac{1}{2}\pi$, $0 \leqslant z \leqslant h$. The 'element of volume' in cylindrical polar coordinates is $\rho \, \mathrm{d}\rho \, \mathrm{d}\psi \, \mathrm{d}z$. Hence

$$\int_V (x^2 + y^2) \, \mathrm{d}v = \int_0^h \int_0^{\frac{1}{2}\pi} \int_0^a \rho^2 \rho \, \mathrm{d}\rho \, \mathrm{d}\psi \, \mathrm{d}z$$

$$= \frac{a^4}{4} \int_0^h \int_0^{\frac{1}{2}\pi} \mathrm{d}\psi \, \mathrm{d}z$$

$$= \frac{\pi a^4}{8} \int_0^h \mathrm{d}z = \frac{\pi a^4 h}{8}.$$

1.91 Grad φ

Let $\mathbf{k}_1$, $\mathbf{k}_2$, $\mathbf{k}_3$ be unit vectors normal to the surfaces $u_1 = $ constant, $u_2 = $ constant and $u_3 = $ constant respectively. Then $\mathbf{k}_1$, $\mathbf{k}_2$ and $\mathbf{k}_3$ form an orthogonal set of unit vectors whose orientation is a function of position.

It is worth emphasizing that while $\mathbf{k}_1$, $\mathbf{k}_2$, $\mathbf{k}_3$ have constant length (unity) they are not in fact constant vectors as their orientation varies from point to point.

The components of grad φ at any point are

$$\frac{\partial \varphi}{\partial l_1}, \frac{\partial \varphi}{\partial l_2}, \frac{\partial \varphi}{\partial l_3} \quad \text{or} \quad \frac{1}{h_1}\frac{\partial \varphi}{\partial u_1}, \frac{1}{h_2}\frac{\partial \varphi}{\partial u_2}, \frac{1}{h_3}\frac{\partial \varphi}{\partial u_3}$$

in the directions of $\mathbf{k}_1$, $\mathbf{k}_2$, $\mathbf{k}_3$ respectively at that point.

That is

$$\text{grad } \varphi = \frac{1}{h_1}\frac{\partial \varphi}{\partial u_1}\mathbf{k}_1 + \frac{1}{h_2}\frac{\partial \varphi}{\partial u_2}\mathbf{k}_2 + \frac{1}{h_3}\frac{\partial \varphi}{\partial u_3}\mathbf{k}_3.$$

For example in spherical polar coordinates $u_1 = r$, $u_2 = \theta$, $u_3 = \psi$, $h_1 = 1$, $h_2 = r$ and $h_3 = r \sin \theta$ (figure 32).

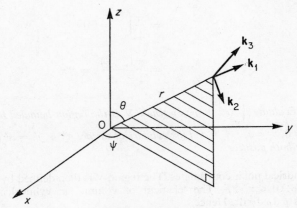

FIGURE 32

Therefore

$$\text{grad } \varphi = \frac{\partial \varphi}{\partial r}\mathbf{k}_1 + \frac{1}{r}\frac{\partial \varphi}{\partial \theta}\mathbf{k}_2 + \frac{1}{r \sin \theta}\frac{\partial \varphi}{\partial \psi}\mathbf{k}_3.$$

In cylindrical polar coordinates $u_1 = \rho$, $u_2 = \psi$, $u_3 = z$, $h_1 = 1$, $h_2 = \rho$, $h_3 = 1$ (figure 33).

Therefore

$$\text{grad } \varphi = \frac{\partial \varphi}{\partial \rho}\mathbf{k}_1 + \frac{1}{\rho}\frac{\partial \varphi}{\partial \psi}\mathbf{k}_2 + \frac{\partial \varphi}{\partial z}\mathbf{k}_3.$$

Example   Obtain the vector grad φ *when* $\varphi = r(\cos \theta + \sin \theta)$ *and r is the distance of a variable point* P *from a fixed origin* O *and* OP *makes an angle* θ

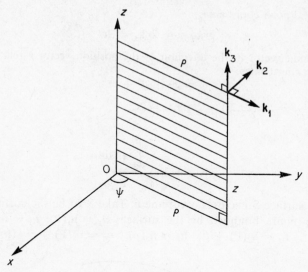

FIGURE 33

with a fixed direction O*z. Hence obtain a unit vector normal to the surface*
$\varphi = $ *constant at any point on it for which* $\theta = \frac{1}{3}\pi.$

r and θ as given can be chosen as the first two spherical polar coordinates.
Therefore

$$\text{grad } \varphi = \frac{\partial \varphi}{\partial r} \mathbf{k}_1 + \frac{1}{r} \frac{\partial \varphi}{\partial \theta} \mathbf{k}_2 + \frac{1}{r \sin \theta} \frac{\partial \varphi}{\partial \psi} \mathbf{k}_3$$

$$= (\cos \theta + \sin \theta)\mathbf{k}_1 + (\cos \theta - \sin \theta)\mathbf{k}_2.$$

The length of this vector is $\sqrt{2}$ and so unit vector in the direction of grad φ is

$$\frac{1}{\sqrt{2}} (\cos \theta + \sin \theta)\mathbf{k}_1 + \frac{1}{\sqrt{2}} (\cos \theta - \sin \theta)\mathbf{k}_2.$$

Hence unit vector normal to the surface $\varphi = $ constant at a point on it for which
$\theta = \pi/3$ is

$$\frac{1}{\sqrt{2}}\left(\frac{1}{2} + \frac{\sqrt{3}}{2}\right)\mathbf{k}_1 + \frac{1}{\sqrt{2}}\left(\frac{1}{2} - \frac{\sqrt{3}}{2}\right)\mathbf{k}_2$$

$$= \frac{1}{2\sqrt{2}} (1 + \sqrt{3})\mathbf{k}_1 + \frac{1}{2\sqrt{2}} (1 - \sqrt{3})\mathbf{k}_2.$$

Example   Obtain the vector grad φ *when* $\varphi = r^2$ *and* r *is distance from a fixed
point* O.

In spherical polar coordinates,

$$\text{grad } \varphi = 2r\mathbf{k}_1 = 2\mathbf{r}$$

since $\mathbf{k}_1$ is unit vector in the direction of the position vector $\mathbf{r}$ relative to O. In cartesian coordinates,

$$\varphi = x^2 + y^2 + z^2$$

and so grad $\varphi = (2x, 2y, 2z) = 2(x, y, z) = 2\mathbf{r}$.

1.92 Div a

$$\text{div } \mathbf{a} = \lim_{v \to 0} \frac{\int_S \mathbf{a} \cdot d\mathbf{s}}{v}$$

where the surface S encloses a volume v. Take S to be the surface of the curvilinear figure bounded by the surfaces $u_1 = u_1(\text{P})$, $u_1 = u_1(\text{P}) + \alpha$, $u_2 = u_2(\text{P})$, $u_2 = u_2(\text{P}) + \beta$, $u_3 = u_3(\rho)$, $u_3 = u_3(\text{P}) + \gamma$ (figure 34).

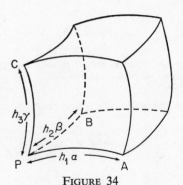

FIGURE 34

If α, β and γ are small, then the volume v enclosed is approximately $l_1 l_2 l_3 = h_1 \alpha h_2 \beta h_3 \gamma$ where h_1, h_2 and h_3 are to be evaluated at P.

Suppose $\mathbf{a} = a_1 \mathbf{k}_1 + a_2 \mathbf{k}_2 + a_3 \mathbf{k}_3$. If S_1 consists of the two faces of the figure which lie in the surfaces $u_1 = u_1(\text{P})$ and $u_1 = u_1(\text{P}) + \alpha$, we have

$$\int_{S_1} \mathbf{a} \cdot d\mathbf{s} = (-a_1 h_3 \gamma h_2 \beta)_\text{P} + (a_1 h_3 \gamma h_2 \beta)_\text{A} + \text{smaller terms}$$

Now using Taylor's theorem we obtain

$$\int_{S_1} \mathbf{a} \cdot d\mathbf{s} = (-a_1 h_3 \gamma h_2 \beta)_\text{P} + \left\{ a_1 h_3 \gamma h_2 \beta + \frac{\partial}{\partial u_1} (a_1 h_3 \gamma h_2 \beta) \alpha \right\}_\text{P}$$

$$+ \text{smaller terms.}$$

$$= \left\{ \frac{\partial}{\partial u_1} (a_1 h_2 h_3) \alpha \beta \gamma \right\}_\text{P} + \text{smaller terms.}$$

Therefore

$$\frac{\int_{S_1} \mathbf{a} \cdot d\mathbf{s}}{v} = \frac{1}{h_1 h_2 h_3} \frac{\partial}{\partial u_1} (a_1 h_2 h_3) + \text{smaller terms,}$$

where the 'smaller terms' tend to zero as α, β, $\gamma \to 0$.

Similarly expressions of this form are obtained on considering the other pairs of opposite faces, and then on taking the limit as $v \to 0$ (that is as α, β, $\gamma \to 0$) it is obtained that

$$\text{div } \mathbf{a} \text{ at } P = \lim_{v \to 0} \frac{\int_S \mathbf{a} \cdot d\mathbf{s}}{v}$$

$$= \frac{1}{h_1 h_2 h_3} \left\{ \frac{\partial}{\partial u_1} (a_1 h_2 h_3) + \frac{\partial}{\partial u_2} (a_2 h_3 h_1) + \frac{\partial}{\partial u_3} (a_3 h_1 h_2) \right\}$$

evaluated at P. For example in rectangular cartesian coordinates $u_1 = x$, $u_2 = y$, $u_3 = z$, $h_1 = 1$, $h_2 = 1$, $h_3 = 1$.

Therefore

$$\text{div } \mathbf{a} = \frac{\partial a_x}{\partial x} + \frac{\partial a_y}{\partial y} + \frac{\partial a_z}{\partial z}$$

where $\mathbf{a} = (a_x, a_y, a_z)$ in terms of cartesian coordinates.

In spherical polar coordinates $u_1 = r$, $u_2 = \theta$, $u_3 = \psi$, $h_1 = 1$, $h_2 = r$, $h_3 = r \sin \theta$.

Therefore

$$\text{div } \mathbf{a} = \frac{1}{r^2 \sin \theta} \left\{ \frac{\partial}{\partial r} (a_r r^2 \sin \theta) + \frac{\partial}{\partial \theta} (a_\theta r \sin \theta) + \frac{\partial}{\partial \psi} (a_\psi r) \right\}$$

where $\mathbf{a} = (a_r, a_\theta, a_\psi)$ in terms of spherical polar coordinates.

In cylindrical polar coordinates, $u_1 = \rho$, $u_2 = \psi$, $u_3 = z$, $h_1 = 1$, $h_2 = \rho$, $h_3 = 1$. Therefore

$$\text{div } \mathbf{a} = \frac{1}{\rho} \left\{ \frac{\partial}{\partial \rho} (a_\rho \rho) + \frac{\partial}{\partial \psi} (a_\psi) + \frac{\partial}{\partial z} (a_z \rho) \right\}$$

where $\mathbf{a} = (a_\rho, a_\psi, a_z)$ in terms of cylindrical polar coordinates.

Example   Evaluate div **a** *when* **a**, *expressed in terms of spherical polar coordinates, is given by* $(\cos \theta + \sin \theta, \cos \theta - \sin \theta, 0)$.

$$\text{div } \mathbf{a} = \frac{1}{r^2 \sin \theta} \left\{ \frac{\partial}{\partial r} \left(r^2 \sin \theta (\cos \theta + \sin \theta) \right) \right.$$

$$\left. + \frac{\partial}{\partial \theta} \left(r \sin \theta (\cos \theta - \sin \theta) \right) + \frac{\partial}{\partial \psi} (0) \right\}$$

$$= \frac{1}{r^2 \sin \theta} \{ 2r \sin \theta (\cos \theta + \sin \theta)$$

$$+ r[\cos \theta (\cos \theta - \sin \theta) + \sin \theta (-\sin \theta - \cos \theta)] \}$$

$$= \frac{1}{r \sin \theta}.$$

Example   Determine for what values of n the vector $r^n \mathbf{r}$ *is solenoidal where* **r** *is the position of the general point relative to a fixed origin and* $r = |\mathbf{r}|$.

$$r^n \mathbf{r} = r^{n+1} \hat{\mathbf{r}} = (r^{n+1}, 0, 0)$$

in terms of spherical polar coordinates. Therefore

$$\text{div } (r^n \mathbf{r}) = \frac{1}{r^2 \sin \theta} \left\{ \frac{\partial}{\partial r} (r^{n+3} \sin \theta) \right\} = \frac{1}{r^2 \sin \theta} (n + 3) r^{n+2} \sin \theta$$

$$= (n + 3) r^n \equiv 0 \text{ when } n = -3 \text{ (and } r \neq 0).$$

That is the vector $r^n \mathbf{r}$ is solenoidal only when $n = -3$.

1.93 $\nabla^2 \varphi$

Since $\nabla^2 \varphi = \text{div (grad } \varphi)$ the expression for $\nabla^2 \varphi$ in terms of general orthogonal curvilinear coordinates is easily obtained from those for div **a** and grad φ as follows.

Put

$$\mathbf{a} = \text{grad } \varphi = \left(\frac{1}{h_1} \frac{\partial \varphi}{\partial u_1}, \frac{1}{h_2} \frac{\partial \varphi}{\partial u_2}, \frac{1}{h_3} \frac{\partial \varphi}{\partial u_3} \right)$$

in the expression for div **a**.

Therefore

$$\nabla^2 \varphi = \frac{1}{h_1 h_2 h_3} \left\{ \frac{\partial}{\partial u_1} \left(\frac{h_2 h_3}{h_1} \frac{\partial \varphi}{\partial u_1} \right) + \frac{\partial}{\partial u_2} \left(\frac{h_3 h_1}{h_2} \frac{\partial \varphi}{\partial u_2} \right) + \frac{\partial}{\partial u_3} \left(\frac{h_1 h_2}{h_3} \frac{\partial \varphi}{\partial u_3} \right) \right\}.$$

For example in spherical polar coordinates

$$\nabla^2 \varphi = \frac{1}{r^2 \sin \theta} \left\{ \frac{\partial}{\partial r} \left(r^2 \sin \theta \frac{\partial \varphi}{\partial r} \right) + \frac{\partial}{\partial \theta} \left(\sin \theta \frac{\partial \varphi}{\partial \theta} \right) + \frac{\partial}{\partial \psi} \left(\frac{1}{\sin \theta} \frac{\partial \varphi}{\partial \psi} \right) \right\}$$

$$= \frac{\partial^2 \varphi}{\partial r^2} + \frac{2}{r} \frac{\partial \varphi}{\partial r} + \frac{1}{r^2} \frac{\partial^2 \varphi}{\partial \theta^2} + \frac{\cot \theta}{r^2} \frac{\partial \varphi}{\partial \theta} + \frac{1}{r^2 \sin^2 \theta} \frac{\partial^2 \varphi}{\partial \psi^2}.$$

In cylindrical polar coordinates

$$\nabla^2 \varphi = \frac{1}{\rho}\left\{\frac{\partial}{\partial \rho}\left(\rho \frac{\partial \varphi}{\partial \rho}\right) + \frac{\partial}{\partial \psi}\left(\frac{1}{\rho}\frac{\partial \varphi}{\partial \psi}\right) + \frac{\partial}{\partial z}\left(\rho \frac{\partial \psi}{\partial z}\right)\right\}$$

$$= \frac{\partial^2 \varphi}{\partial \rho^2} + \frac{1}{\rho}\frac{\partial \varphi}{\partial \rho} + \frac{1}{\rho^2}\frac{\partial^2 \varphi}{\partial \psi^2} + \frac{\partial^2 \varphi}{\partial z^2}.$$

Example   Evaluate $\nabla^2 \varphi$ when $\varphi = r(\cos\theta + \sin\theta)$ and r and θ are spherical polar coordinates.

$$\nabla^2 \varphi = \frac{2}{r}(\cos\theta + \sin\theta) + \frac{1}{r}(-\cos\theta - \sin\theta) + \frac{\cot\theta}{r}(-\sin\theta + \cos\theta)$$

$$= \frac{1}{r\sin\theta}.$$

Example   Obtain an expression for $\nabla^2 \varphi$ when $\varphi = r^2$ and r is distance from a fixed point.

In terms of spherical polar coordinates,

$$\nabla^2 \varphi = \frac{\partial}{\partial r^2}(r^2) + \frac{2}{r}\frac{\partial}{\partial r}(r^2)$$

$$= 6.$$

Now, as on p. 34, we have that

$$\text{grad}\,(r^2) = 2\mathbf{r} = (2x, 2y, 2z)$$

and so, using cartesian coordinates, we see that

$$\nabla^2(r^2) = \text{div}\,(\text{grad}\,r^2) = \text{div}\,(2x, 2y, 2z) = 6.$$

Example   If $\nabla^2 \varphi = 0$ and φ is a function of the distance r from a fixed point only, find the general form of φ.

Use spherical polar coordinates and put $\varphi = f(r)$. Therefore

$$\nabla^2 \varphi = 0 \quad \text{becomes} \quad \frac{1}{r^2}\frac{\mathrm{d}}{\mathrm{d}r}\left(r^2 \frac{\mathrm{d}f}{\mathrm{d}r}\right) = 0$$

or

$$\frac{\mathrm{d}}{\mathrm{d}r}\left(r^2 \frac{\mathrm{d}f}{\mathrm{d}r}\right) = 0$$

Therefore

$$r^2 \frac{\mathrm{d}f}{\mathrm{d}r} = A$$

where A is a constant and so

$$\frac{\mathrm{d}f}{\mathrm{d}r} = \frac{A}{r^2} \quad \text{and} \quad f = B - \frac{A}{r}$$

where B is also a constant.

Example   If $\nabla^2 \varphi = 0$ and φ is a function of the distance ρ from a fixed line only, find the general form of φ.

Use cylindrical polar coordinates and put $\varphi = f(\rho)$. Therefore

$$\nabla^2 \varphi = 0 \quad \text{becomes} \quad \frac{1}{\rho} \frac{\mathrm{d}}{\mathrm{d}\rho} \left(\rho \frac{\mathrm{d}f}{\mathrm{d}\rho} \right) = 0$$

Hence

$$\rho \frac{\mathrm{d}f}{\mathrm{d}\rho} = A$$

where A is a constant and

$$f = A \ln \rho + B$$

where B is also a constant.

1.94 Curl a

The component of curl **a** in the direction normal to the plane area s is defined as

$$\lim_{s \to 0} \frac{\displaystyle\int_L \mathbf{a} \cdot \mathrm{d}\mathbf{l}}{s}$$

where L is the bounding curve of the area s. We shall obtain the component of curl **a** in the direction of $\mathbf{k_1}$ by taking L to be the boundary of the curvilinear figure PBDC (figure 35) lying entirely in the surface

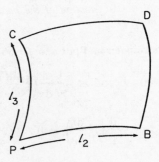

FIGURE 35

$u_1 = \text{constant} = u_1(P)$ where the curves PB and CD are given by

$$u_1 = u_1(P), \qquad u_3 = u_3(P)$$

and

$$u_1 = u_1(P), \qquad u_3 = u_3(P) + \gamma$$

respectively, and the curves PC and BD are given by

$$u_1 = u_1(P), \qquad u_2 = u_2(P)$$

and

$$u_1 = u_1(P), \qquad u_2 = u_2(P) + \beta$$

respectively. If β and γ are small, then the area s enclosed is approximately $l_2 l_3 = h_2 \beta h_3 \gamma$ where h_2 and h_3 are to be evaluated at P. Then if L_1 consists of the two curves PB and DC,

$$\int_{L_1} \mathbf{a} \cdot d\mathbf{l} = (a_2 h_2 \beta)_P - (a_2 h_2 \beta)_C + \text{smaller terms}$$

$$= (a_2 h_2 \beta)_P - \left\{ a_2 h_2 \beta + \frac{\partial}{\partial u_3} (a_2 h_2 \beta) \gamma \right\}_P + \text{smaller terms}.$$

Therefore

$$\int_{L_1} \mathbf{a} \cdot d\mathbf{l} = -\frac{\partial}{\partial u_3} (a_2 h_2) \beta \gamma + \text{smaller terms}.$$

Similarly, we obtain

$$\int_{L_2} \mathbf{a} \cdot d\mathbf{l} = \left\{ a_3 h_3 \gamma + \frac{\partial}{\partial u_2} (a_3 h_3 \gamma) \beta \right\} - a_3 h_3 \gamma + \text{smaller terms}$$

$$= \frac{\partial}{\partial u_2} (a_3 h_3) \beta \gamma + \text{smaller terms}$$

where L_2 consists of the curves BD and CP.

Then, since

$$\int_L \mathbf{a} \cdot d\mathbf{l} = \int_{L_1} \mathbf{a} \cdot d\mathbf{l} + \int_{L_2} \mathbf{a} \cdot d\mathbf{l},$$

we obtain on dividing by s and taking the limit as $s \to 0$ (that is as β, $\gamma \to 0$) that the $\mathbf{k}_1$ component of curl $\mathbf{a}$ is given by

$$\frac{1}{h_2 h_3} \left\{ \frac{\partial}{\partial u_2} (a_3 h_3) - \frac{\partial}{\partial u_3} (a_2 h_2) \right\}.$$

Similarly the expressions

$$\frac{1}{h_3 h_1} \left\{ \frac{\partial}{\partial u_3} (a_1 h_1) - \frac{\partial}{\partial u_1} (a_3 h_3) \right\} \quad \text{and} \quad \frac{1}{h_1 h_2} \left\{ \frac{\partial}{\partial u_1} (a_2 h_2) - \frac{\partial}{\partial u_2} (a_1 h_1) \right\}$$

are obtained for the k_2 and k_3 components respectively. Curl a may then be written symbolically, using the notation of determinants, in the form

$$\text{curl } a = \begin{vmatrix} \dfrac{k_1}{h_2 h_3} & \dfrac{k_2}{h_3 h_1} & \dfrac{k_3}{h_1 h_2} \\[2mm] \dfrac{\partial}{\partial u_1} & \dfrac{\partial}{\partial u_2} & \dfrac{\partial}{\partial u_3} \\[2mm] a_1 h_1 & a_2 h_2 & a_3 h_3 \end{vmatrix}.$$

For example, in spherical polar coordinates, if $a = (a_r, a_\theta, a_\psi)$

$$\text{curl } a = \begin{vmatrix} \dfrac{k_1}{r^2 \sin\theta} & \dfrac{k_2}{r \sin\theta} & \dfrac{k_3}{r} \\[2mm] \dfrac{\partial}{\partial r} & \dfrac{\partial}{\partial \theta} & \dfrac{\partial}{\partial \psi} \\[2mm] a_r & a_\theta r & a_\psi r \sin\theta \end{vmatrix}$$

$$= \frac{k_1}{r^2 \sin\theta} \left\{ \frac{\partial}{\partial \theta} (a_\psi r \sin\theta) - \frac{\partial}{\partial \psi} (a_\theta r) \right\}$$

$$+ \frac{k_2}{r \sin\theta} \left\{ \frac{\partial}{\partial \psi} (a_r) - \frac{\partial}{\partial r} (a_\psi r \sin\theta) \right\} + \frac{k_3}{r} \left\{ \frac{\partial}{\partial r} (a_\theta r) - \frac{\partial}{\partial \theta} (a_r) \right\}.$$

In cylindrical polar coordinates, if $a = (a_\rho, a_\psi, a_z)$

$$\text{curl } a = \begin{vmatrix} \dfrac{k_1}{\rho} & \dfrac{k_2}{1} & \dfrac{k_3}{\rho} \\[2mm] \dfrac{\partial}{\partial \rho} & \dfrac{\partial}{\partial \psi} & \dfrac{\partial}{\partial z} \\[2mm] a_\rho & a_\psi \rho & a_z \end{vmatrix}$$

$$= \frac{k_1}{\rho} \left\{ \frac{\partial a_z}{\partial \psi} - \frac{\partial}{\partial z} (a_\psi \rho) \right\} + k_2 \left\{ \frac{\partial a_\rho}{\partial z} - \frac{\partial a_z}{\partial \rho} \right\} + \frac{k_3}{\rho} \left\{ \frac{\partial}{\partial \rho} (a_\psi \rho) - \frac{\partial a_\rho}{\partial \psi} \right\}.$$

From the elementary properties of partial derivatives it is easily proved that

$$\text{curl } (\alpha a + \beta b) = \alpha \text{ curl } a + \beta \text{ curl } b$$

if α and β are scalar constants.

Example Show that the vector $r^n r$ where r is the position vector of the general point relative to a fixed origin and $r = |r|$ is irrotational for all values of n.

In spherical polar coordinates $r^n\mathbf{r} = (r^{n+1},0,0)$.
Therefore

$$\operatorname{curl} r^n\mathbf{r} = \frac{\mathbf{k}_2}{r\sin\theta}\frac{\partial}{\partial\psi}(r^{n+1}) - \frac{\mathbf{k}_3}{r}\frac{\partial}{\partial\theta}(r^{n+1})$$

on putting $a_r = r^{n+1}$, $a_\theta = a_\psi = 0$ in the above formula. Hence $\operatorname{curl} r^n\mathbf{r} \equiv \mathbf{0}$.
Thus $r^n\mathbf{r}$ is irrotational for all values of n.

Example   Show that if vector fields $\mathbf{F}_1$ and $\mathbf{F}_2$ are defined by

$$\mathbf{F}_1 = f(r)\mathbf{r},$$

and

$$\mathbf{F}_2 = r\mathbf{c} + \frac{\mathbf{c}.\mathbf{r}}{r}\mathbf{r}$$

where $\mathbf{c}$ is a constant vector, $\mathbf{r}$ is the position vector of the general point and f is an arbitrary differentiable function, then both $\mathbf{F}_1$ and $\mathbf{F}_2$ are irrotational.

$$\operatorname{curl}\mathbf{F}_1 = \operatorname{curl}\left(f(r)\mathbf{r}\right) = \operatorname{grad} f(r) \times \mathbf{r} + f(r)\operatorname{curl}\mathbf{r}$$

$$= \left(\frac{df}{dr},0,0\right) \times \mathbf{r} \quad\text{since}\quad \operatorname{curl}\mathbf{r} \equiv \mathbf{0}$$

$$= \frac{df}{dr}\frac{1}{r}(\mathbf{r}\times\mathbf{r}) = 0.$$

Therefore $\mathbf{F}_1$ is irrotational.

$$\operatorname{curl}\mathbf{F}_2 = \operatorname{curl}\left\{r\mathbf{c} + \frac{\mathbf{c}.\mathbf{r}}{r}\mathbf{r}\right\}$$

$$= \operatorname{curl} r\mathbf{c} + \operatorname{curl}\left(\frac{\mathbf{c}.\mathbf{r}}{r}\mathbf{r}\right)$$

$$= (\operatorname{grad} r) \times \mathbf{c} + r\operatorname{curl}\mathbf{c} + \operatorname{grad}\left(\frac{\mathbf{c}.\mathbf{r}}{r}\right) \times \mathbf{r} + \frac{\mathbf{c}.\mathbf{r}}{r}\operatorname{curl}\mathbf{r}$$

$$= (1,0,0) \times \mathbf{c} + \left\{\frac{1}{r}\operatorname{grad}(\mathbf{c}.\mathbf{r}) + (\mathbf{c}.\mathbf{r})\operatorname{grad}\left(\frac{1}{r}\right)\right\} \times \mathbf{r}$$

$$= \frac{1}{r}\mathbf{r}\times\mathbf{c} + \left\{\frac{1}{r}\mathbf{c} + (\mathbf{c}.\mathbf{r})\left(-\frac{1}{r^2},0,0\right)\right\} \times \mathbf{r}$$

$$= \frac{1}{r}\mathbf{r}\times\mathbf{c} + \frac{1}{r}\mathbf{c}\times\mathbf{r} - \frac{1}{r^3}(\mathbf{c}.\mathbf{r})\mathbf{r}\times\mathbf{r}$$

$$= 0$$

since $\mathbf{r}\times\mathbf{c} = -\mathbf{c}\times\mathbf{r}$ and $\mathbf{r}\times\mathbf{r} = 0$. Therefore $\mathbf{F}_2$ is irrotational.

Example   Prove that, in terms of cylindrical polar coordinates ρ, ψ, z

(i) $\text{curl}\,(z\,\text{grad}\,\psi) = -\text{grad}\,(\ln \rho)$,

(ii) $\text{curl}\,(\psi\,\text{grad}\,\rho) = -\dfrac{1}{\rho}\,\text{grad}\,(z)$.

(i) $\text{grad}\,\psi = \left(\dfrac{\partial \psi}{\partial \rho}, \dfrac{1}{\rho}\dfrac{\partial \psi}{\partial \psi}, \dfrac{\partial \psi}{\partial z}\right) = \left(0, \dfrac{1}{\rho}, 0\right)$

Therefore

$$\text{curl}\,(z\,\text{grad}\,\psi) = \left(-\dfrac{1}{\rho}\dfrac{\partial}{\partial z}\left(\dfrac{z}{\rho}\rho\right), 0, \dfrac{1}{\rho}\dfrac{\partial}{\partial \rho}(z)\right) = \left(-\dfrac{1}{\rho}, 0, 0\right),$$

$$\text{grad}\,(\ln \rho) = \left(\dfrac{\partial}{\partial \rho}\ln \rho, \dfrac{1}{\rho}\dfrac{\partial}{\partial \psi}\ln \rho, \dfrac{\partial}{\partial z}\ln \rho\right) = \left(\dfrac{1}{\rho}, 0, 0\right).$$

Hence

$$\text{curl}\,(z\,\text{grad}\,\psi) = -\text{grad}\,(\ln \rho).$$

(ii) $\text{curl}\,(\psi\,\text{grad}\,\rho) = \text{curl}\left\{\psi\left(\dfrac{\partial \rho}{\partial \rho}, \dfrac{1}{\rho}\dfrac{\partial \rho}{\partial \psi}, \dfrac{\partial \rho}{\partial z}\right)\right\} = \text{curl}\,(\psi, 0, 0)$

$$= \left(0, \dfrac{\partial \psi}{\partial z}, -\dfrac{1}{\rho}\dfrac{\partial \psi}{\partial \psi}\right)$$

$$= \left(0, 0, -\dfrac{1}{\rho}\right).$$

$$\text{grad}\,z = \left(\dfrac{\partial z}{\partial \rho}, \dfrac{1}{\rho}\dfrac{\partial z}{\partial \psi}, \dfrac{\partial z}{\partial z}\right)$$

$$= (0, 0, 1).$$

Hence

$$\text{curl}\,(\psi\,\text{grad}\,\rho) = -\dfrac{1}{\rho}\,\text{grad}\,z.$$

Exercises

1. Evaluate $\displaystyle\int_{S} z^3\,ds$ where S is the surface of the hemisphere represented by $x^2 + y^2 + z^2 = a^2$, $z \geqslant 0$.

2. Evaluate $\displaystyle\int_{S} \dfrac{z}{\sqrt{x^2 + y^2 + z^2}}\,ds$ where S is the surface of the hemisphere represented by $x^2 + y^2 + z^2 = a^2$, $z \geqslant 0$.

3. Determine the area of the curved surface of the right circular cone with semivertical angle $\frac{1}{3}\pi$ and vertical height 2. [The required surface area is $\int_S ds$ where S is the surface of the cone $3z^2 = x^2 + y^2$ between the planes $z = 0$ and $z = 2$.]

4. Evaluate $\int_V \sqrt{x^2 + y^2}\, dv$ where V is the region bounded by the planes $x = 0$, $y = 0$, $z = 0$, $z = h$ and the cylindrical surface $x^2 + y^2 = a^2$ and in which x and y are both positive.

5. Evaluate $\int_V (x^2 + y^2 + z^2)^{\frac{1}{2}}\, dv$ where V is the region bounded by the plane $z = 0$ and the spherical surfaces $x^2 + y^2 + z^2 = a^2$ and $x^2 + y^2 + z^2 = 2a^2$ and for which z is positive.

6. Obtain the vector grad φ when $\varphi = 1/r$ and r is distance from a fixed point 0.

7. Obtain the vector grad φ when $\varphi = r(\cos\theta - \sin\theta)$ and r is the distance of a variable point P from a fixed origin O and OP makes an angle θ with a fixed direction Oz. Hence obtain unit vector normal to the surface $\varphi = \text{constant}$ at any point on it for which $\theta = \alpha\ (\neq 0)$.

8. Evaluate div **a** when **a**, expressed in terms of spherical polar coordinates, is given by $(r\cos\theta, r\sin\theta, r\psi)$ and show that div **a** $= 0$ when $\theta = \frac{1}{2}\sin^{-1}(-\frac{2}{5})$.

9. If **a** $= (-\rho\cos\psi, \rho\sin\psi, z\cos\psi)$ and ρ, ψ, z are cylindrical polar coordinates, show that **a** is solenoidal.

10. Evaluate div **a** when **a**, expressed in terms of cylindrical polar coordinates, is given by $(e^\rho \sin\psi, \rho e^\rho \cos\psi, \rho^2)$.

11. Evaluate $\nabla^2\varphi$ when $\varphi = r(\cos\theta - \sin\theta)$ and r and θ are spherical polar coordinates.

12. If $\varphi = \rho^2 \cos^2\psi$ in terms of cylindrical polar coordinates ρ and ψ, evaluate $\nabla^2\varphi$.

13. Evaluate $\nabla^2\varphi$ when $\varphi = [(x^2 + y^2)/(x^2 + y^2 + z^2)]^{\frac{1}{2}}$. [Use spherical polar coordinates.]

14. Evaluate curl **a** when **a**, expressed in terms of spherical polar coordinates, is given by $(r\cos\theta, r\sin\theta, r\psi)$.

15. If **a** $= (-\rho\cos\psi, \rho\sin\psi, z\cos\psi)$ and ρ, ψ, z are cylindrical polar coordinates, show that curl **a** $= 0$ when $\psi = 0$.

16. Show that the vector $(2r \cos \theta \sin \psi, -r \sin \theta \sin \psi, r \cot \theta \cos \psi)$ in which r, θ, ψ are spherical polar coordinates is irrotational.

17. Show that if r, θ, ψ are spherical polar coordinates, then
   (i) curl $(\nabla r \times \nabla \theta) = -(\cot \theta)\nabla(1/r)$,
   (ii) curl $(\nabla r \times \nabla \psi) = 0$.

1.10 Gauss' theorem

This theorem is deduced as a consequence of the definition of div **a** as

$$\lim_{v \to 0} \frac{\int_S \mathbf{a} \cdot d\mathbf{s}}{v}$$

where the surface S encloses the volume v. Consider a region V having a volume v and bounded by a surface S. Divide this volume into n smaller volumes $v_1, v_2, \ldots, v_n$ bounded by surfaces $S_1, S_2, \ldots, S_n$ respectively. Now for any point inside S_k we can write

$$\int_{S_k} \mathbf{a} \cdot d\mathbf{s} = (\text{div } \mathbf{a})\, v_k + \varepsilon_k v_k$$

where $\varepsilon_k \to 0$ as $v_k \to 0$. Now if the volumes v_i and v_j have an element of surface S_{ij} in common, then the integral of **a** over the surface S_{ij} when regarded as part of the surface S_i will cancel with the integral of **a** over S_{ij} when regarded as part of S_j since in evaluating these surface integrals the positive sense of the normal is taken outwards from the volume enclosed in each case.

    Hence

$$\sum_{k=1}^{n} \int_{S_k} \mathbf{a} \cdot d\mathbf{s} = \int_S \mathbf{a} \cdot d\mathbf{s}.$$

Therefore

$$\int_S \mathbf{a} \cdot d\mathbf{s} = \sum_{k=1}^{n} (\text{div } \mathbf{a})v_k + \sum_{k=1}^{n} \varepsilon_k v_k.$$

Now let the maximum of $v_1, v_2, \ldots, v_n$ tend to zero (and so $n \to \infty$).
    Therefore

$$\sum_{k=1}^{n} (\text{div } \mathbf{a})v_k \to \int_V \text{div } \mathbf{a}\, dv$$

and

$$\left| \sum_{k=1}^{n} \varepsilon_k v_k \right| \leqslant \max_k |\varepsilon_k| \sum_{k=1}^{n} v_k = v \max_k |\varepsilon_k| \to 0$$

and so we obtain the result that

$$\int_S \mathbf{a} \cdot d\mathbf{s} = \int_V \text{div } \mathbf{a}\, dv = \int_V \nabla \cdot \mathbf{a}\, dv.$$

This is known as *Gauss' theorem* (or simply the *divergence theorem*). It is a relation which converts a surface integral into a volume integral.

In terms of cartesian coordinates it has the form

$$\iint (a_x \, dy \, dz + a_y \, dz \, dx + a_z \, dx \, dy) = \iiint \left(\frac{\partial a_x}{\partial x} + \frac{\partial a_y}{\partial y} + \frac{\partial a_z}{\partial z} \right) dx \, dy \, dz$$

with appropriate limits of integration. In terms of orthogonal curvilinear coordinates u_1, u_2, u_3 it has the form

$$\iint (a_1 h_2 h_3 \, du_2 \, du_3 + a_2 h_3 h_1 \, du_3 \, du_1 + a_3 h_1 h_2 \, du_1 \, du_2)$$

$$= \iiint \left(\frac{1}{h_1 h_2 h_3} \left\{ \frac{\partial}{\partial u_1} (a_1 h_2 h_3) + \frac{\partial}{\partial u_2} (a_2 h_3 h_1) \right. \right.$$

$$\left. \left. + \frac{\partial}{\partial u_3} (a_3 h_1 h_2) \right\} \right) h_1 h_2 h_3 \, du_1 \, du_2 \, du_3$$

with appropriate limits of integration.

Example   Show that if $\mathbf{F} = ax\mathbf{i} + by\mathbf{j} + cz\mathbf{k}$ *where a, b and c are constants and S is any closed surface enclosing a volume v, then*

$$\int_S \mathbf{F} \cdot d\mathbf{s} = (a + b + c)v.$$

$$\int_S \mathbf{F} \cdot d\mathbf{s} = \int_V \text{div } \mathbf{F} \, dv \text{ where V is the region enclosed by S}$$

$$= \int_V (a + b + c) \, dv = (a + b + c) \int_V dv = (a + b + c)v.$$

Example   Evaluate $\int_S \mathbf{a} \cdot d\mathbf{s}$ *where S is the sphere, centre the origin and radius c and* $\mathbf{a} = x^3\mathbf{i} + y^3\mathbf{j} + z^3\mathbf{k}$.

$$\int_S \mathbf{a} \cdot d\mathbf{s} = \int_V \text{div } \mathbf{a} \, dv = \int_V 3(x^2 + y^2 + z^2) \, dv$$

$$= 3 \int_V r^2 \, dv = 3 \int_0^{2\pi} \int_0^{\pi} \int_0^c r^2 r^2 \sin \theta \, dr \, d\theta \, d\psi$$

$$= 3 \int_0^{2\pi} \int_0^{\pi} \left[\frac{r^5}{5} \right]_0^c \sin \theta \, d\theta \, d\psi$$

$$= \frac{3c^5}{5} \int_0^{2\pi} \left[-\cos \theta \right]_0^{\pi} d\psi = \frac{6}{5} c^5 \left[\psi \right]_0^{2\pi} = \frac{12}{5} \pi c^5.$$

Example   Evaluate $\int_S \mathbf{r} \cdot d\mathbf{s}$ *and* $\int_S d\mathbf{s}$ *where* $\mathbf{r}$ *is the position vector and* S *is any closed surface enclosing a region* V *of volume v.*

$$\int_S \mathbf{r} \cdot d\mathbf{s} = \int_V \text{div } \mathbf{r} \, dv = 3 \int_V dv = 3v.$$

Now Gauss' theorem cannot be applied to $\int_S d\mathbf{s}$ directly, but if $\mathbf{c}$ is *any* constant vector,

$$\mathbf{c} \cdot \int_S d\mathbf{s} = \int_S \mathbf{c} \cdot d\mathbf{s} = \int_V \text{div } \mathbf{c} \, dv = 0.$$

That is, if $\int_S d\mathbf{s} = \mathbf{a}$ say then $\mathbf{c} \cdot \mathbf{a} = 0$ for *any* constant vector $\mathbf{c}$ and so $\mathbf{a} = \mathbf{0}$. Hence

$$\int_S d\mathbf{s} = \mathbf{0}.$$

Note the distinction between this latter result and the result $\int_S ds = s$ where s is the area of the surface S.

Example   If φ_1, φ_2, φ_3 *are scalar functions of position,* $\mathbf{a}$ *is a vector function of position and* S *is a closed surface, enclosing a region* V, *show that*

(i) $\qquad \int_S (\text{grad } \varphi_1 \times \mathbf{a}) \cdot d\mathbf{s} = - \int_V \text{grad } \varphi_1 \cdot \text{curl } \mathbf{a} \, dv,$

(ii) $\int_V \varphi_1 (\text{grad } \varphi_2 \cdot \text{grad } \varphi_3) \, dv$

$$= \int_S \varphi_1 \varphi_2 \text{ grad } \varphi_3 \cdot d\mathbf{s} - \int_V \varphi_2 \text{ div } (\varphi_1 \text{ grad } \varphi_3) \, dv$$

$$= \int_S \varphi_1 \varphi_3 \text{ grad } \varphi_2 \cdot d\mathbf{s} - \int_V \varphi_3 \text{ div } (\varphi_1 \text{ grad } \varphi_2) \, dv.$$

Hence deduce that if $\nabla^2 \alpha = 0$ *and* $\nabla^2 \beta = 0$, *then*

$$\int_S (\alpha \text{ grad } \beta - \beta \text{ grad } \alpha) \cdot d\mathbf{s} = 0.$$

(i) $\quad \int_S (\text{grad } \varphi_1 \times \mathbf{a}) \cdot d\mathbf{s} = \int_V \text{div } (\text{grad } \varphi_1 \times \mathbf{a}) \, dv$

$$= \int_V \{ \mathbf{a} \cdot \text{curl } (\text{grad } \varphi_1) - \text{grad } \varphi_1 \cdot \text{curl } \mathbf{a} \} \, dv$$

$$= - \int_V \text{grad } \varphi_1 \cdot \text{curl } \mathbf{a} \, dv$$

since curl grad $\varphi \equiv 0$.

(ii) $\int_S \varphi_1\varphi_2 \operatorname{grad} \varphi_3 . \, \mathrm{d}\mathbf{s}$

$$= \int_V \operatorname{div} (\varphi_1\varphi_2 \operatorname{grad} \varphi_3) \, \mathrm{d}v$$

$$= \int_V \{\varphi_2 \operatorname{div} (\varphi_1 \operatorname{grad} \varphi_3) + (\varphi_1 \operatorname{grad} \varphi_3) . \operatorname{grad} \varphi_2\} \, \mathrm{d}v.$$

Hence

$$\int_V \varphi_1(\operatorname{grad} \varphi_3 . \operatorname{grad} \varphi_2) \, \mathrm{d}v = \int_S \varphi_1\varphi_2 \operatorname{grad} \varphi_3 . \, \mathrm{d}\mathbf{s} - \int_V \varphi_2 \operatorname{div} (\varphi_1 \operatorname{grad} \varphi_3) \, \mathrm{d}v.$$

Similarly we obtain

$$\int_V \varphi_1(\operatorname{grad} \varphi_2 . \operatorname{grad} \varphi_3) \, \mathrm{d}v = \int_S \varphi_1\varphi_3 \operatorname{grad} \varphi_2 . \, \mathrm{d}\mathbf{s} - \int_V \varphi_3 \operatorname{div} (\varphi_1 \operatorname{grad} \varphi_2) \, \mathrm{d}v.$$

Now put $\varphi_1 = c$ where c is a non-zero constant, $\varphi_2 = \alpha$ and $\varphi_3 = \beta$. Then

$$\int_S c\alpha \operatorname{grad} \beta . \, \mathrm{d}\mathbf{s} - \int_V \alpha \operatorname{div} (c \operatorname{grad} \beta) \, \mathrm{d}v$$

$$= \int_S c\beta \operatorname{grad} \alpha . \, \mathrm{d}\mathbf{s} - \int_V \beta \operatorname{div} (c \operatorname{grad} \alpha) \, \mathrm{d}v.$$

Hence

$$c\int_S \alpha \operatorname{grad} \beta . \, \mathrm{d}\mathbf{s} - c\int_V \alpha\nabla^2\beta \, \mathrm{d}v = c\int_S \beta \operatorname{grad} \alpha . \, \mathrm{d}\mathbf{s} - c\int_V \beta\nabla^2\alpha \, \mathrm{d}v.$$

Therefore

$$\int_S (\alpha \operatorname{grad} \beta - \beta \operatorname{grad} \alpha) . \, \mathrm{d}\mathbf{s} = 0$$

since $\nabla^2\alpha = \nabla^2\beta = 0$.

Example *Show that* $\int_S \varphi \, \mathrm{d}\mathbf{s} = \int_V \operatorname{grad} \varphi \, \mathrm{d}v$ *where* S *is a closed surface enclosing the region* V *of volume* v *and* φ *is a scalar function of position.*

Hence show that if $\mathbf{r}$ *is the position vector and* $\mathbf{c}$ *is a constant vector, then*

$$\int_S \mathbf{r} \times (\mathbf{c} \times \mathrm{d}\mathbf{s}) = 2v\mathbf{c}.$$

Let **a** be any constant vector. Then

$$\mathbf{a} \cdot \int_S \varphi \, d\mathbf{s} = \int_S \varphi \mathbf{a} \cdot d\mathbf{s} = \int_V \operatorname{div} (\varphi \mathbf{a}) \, dv$$

$$= \int_V (\varphi \operatorname{div} \mathbf{a} + \mathbf{a} \cdot \operatorname{grad} \varphi) \, dv.$$

$$= \mathbf{a} \cdot \int_V \operatorname{grad} \varphi \, dv$$

since div **a** = 0.

Hence, since **a** was chosen arbitrarily, we have

$$\int_S \varphi \, d\mathbf{s} = \int_V \operatorname{grad} \varphi \, dv.$$

$$\int_S \mathbf{r} \times (\mathbf{c} \times d\mathbf{s}) = \int_S \mathbf{c}(\mathbf{r} \cdot d\mathbf{s}) - \int_S (\mathbf{r} \cdot \mathbf{c}) \, d\mathbf{s}$$

$$= \mathbf{c} \int_V \operatorname{div} dv - \int_V \operatorname{grad} (\mathbf{r} \cdot \mathbf{c}) \, dv$$

$$= 3\mathbf{c} \int_V dv - \int_V \mathbf{c} \, dv$$

$$= 2v\mathbf{c}.$$

Exercises

1. If $\mathbf{F} = -2x\mathbf{i} + 4y\mathbf{j} - z\mathbf{k}$ and S is any closed surface enclosing a volume v, show that

$$\int_S \mathbf{F} \cdot d\mathbf{s} = v.$$

2. Evaluate $\int_S \mathbf{a} \cdot d\mathbf{s}$ where S is the sphere with centre the origin and radius c and $\mathbf{a} = xy^2\mathbf{i} + yz^2\mathbf{j} + zx^2\mathbf{k}$.

3. Use Gauss' theorem to show that $\int_S \mathbf{r} \cdot d\mathbf{s} = 2\pi a^3$ where **r** is the position vector and S is the surface of the hemisphere $x^2 + y^2 + z^2 = a^2$, $z \geqslant 0$.

4. Verify Gauss' theorem for the cube given by $0 \leqslant x \leqslant 1$, $0 \leqslant y \leqslant 1$, $0 \leqslant z \leqslant 1$ and the vector $\mathbf{u} = (2xy - y^2, xy^2 + yz, 3xyz - z^2)$.

1.11 Stokes' theorem

This theorem is deduced as a consequence of the definition of curl **a**.

$$\operatorname{curl}_n \mathbf{a} = \lim_{s \to 0} \frac{\int_L \mathbf{a} \cdot d\mathbf{l}}{s}$$

where L is a closed curve enclosing a plane area s and $\text{curl}_n\, \mathbf{a}$ is the component of curl $\mathbf{a}$ normal to this plane area.

Consider a surface S, not necessarily plane, having total area s, enclosed by a curve L and divide it into m smaller surfaces $S_1, S_2, \ldots, S_m$ having areas $s_1, s_2, \ldots, s_m$ bounded by curves $L_1, L_2, \ldots, L_m$ respectively.

Now we can write

$$\int_{L_k} \mathbf{a} \cdot d\mathbf{l} = (\text{curl}_n\, \mathbf{a})s_k + \varepsilon_k s_k$$

where $\varepsilon_k \to 0$ as $s_k \to 0$ and $\text{curl}_n\, \mathbf{a}$ is evaluated at some point P in the surface S_k.

Therefore

$$\int_{L_k} \mathbf{a} \cdot d\mathbf{l} = [(\text{curl}\, \mathbf{a}) \cdot \hat{\mathbf{n}}]s_k + \varepsilon_k s_k$$

where $\hat{\mathbf{n}}$ is the unit normal to the surface S_k at P.

By a similar argument to that used in deducing Gauss' theorem, we obtain that

$$\sum_{k=1}^{m} \int_{L_k} \mathbf{a} \cdot d\mathbf{l} = \int_{L} \mathbf{a} \cdot d\mathbf{l}.$$

Therefore

$$\int_{L} \mathbf{a} \cdot d\mathbf{l} = \sum_{k=1}^{m} [(\text{curl}\, \mathbf{a}) \cdot \hat{\mathbf{n}}]s_k + \sum_{k=1}^{m} \varepsilon_k s_k.$$

Now let the maximum of $s_1, s_2, \ldots, s_m$ tend to zero (and so $m \to \infty$). Then

$$\sum_{k=1}^{m} [(\text{curl}\, \mathbf{a}) \cdot \hat{\mathbf{n}}]s_k \to \int_{S} (\text{curl}\, \mathbf{a} \cdot \hat{\mathbf{n}})\, ds = \int_{S} \text{curl}\, \mathbf{a} \cdot (\hat{\mathbf{n}}\, ds) = \int_{S} \text{curl}\, \mathbf{a} \cdot d\mathbf{s}$$

and

$$\left| \sum_{k=1}^{m} \varepsilon_k s_k \right| \leqslant \max_{k} |\varepsilon_k| \sum_{k=1}^{m} s_k = s \max_{k} |\varepsilon_k| \to 0.$$

Thus we obtain

$$\int_{L} \mathbf{a} \cdot d\mathbf{l} = \int_{S} \text{curl}\, \mathbf{a} \cdot d\mathbf{s} = \int_{S} \nabla \times \mathbf{a} \cdot d\mathbf{s}.$$

This result is known as *Stokes' theorem*. It is a relation which converts a line integral into a surface integral.

In terms of cartesian coordinates it has the form

$$\int (a_x\,dx + a_y\,dy + a_z\,dz)$$

$$= \iint \left[\left(\frac{\partial a_z}{\partial y} - \frac{\partial a_y}{\partial z} \right) dy\,dz + \left(\frac{\partial a_x}{\partial z} - \frac{\partial a_z}{\partial x} \right) dz\,dx + \left(\frac{\partial a_y}{\partial x} - \frac{\partial a_x}{\partial y} \right) dx\,dy \right].$$

Example   Use Stokes' theorem to evaluate the line integral

$$\int_L e^{-x}\,\mathbf{a}\,.\,d\mathbf{l}$$

where $\mathbf{a} = (\sin y)\mathbf{i} + (\cos y)\mathbf{j}$ *and* L *is the rectangle with vertices* (0,0,0), $(\pi,0,0)$, $(\pi,\frac{1}{2}\pi,0)$, $(0,\frac{1}{2}\pi,0)$.

$$\operatorname{curl} e^{-x}\mathbf{a} = \begin{vmatrix} \mathbf{i} & \mathbf{j} & \mathbf{k} \\ \dfrac{\partial}{\partial x} & \dfrac{\partial}{\partial y} & \dfrac{\partial}{\partial z} \\ e^{-x}\sin y & e^{-x}\cos y & 0 \end{vmatrix} = (0,0,-e^{-x}\cos y - e^{-x}\cos y).$$

Now

$$\int_L e^{-x}\mathbf{a}\,.\,d\mathbf{l} = \int_S \operatorname{curl}(e^{-x}\mathbf{a})\,.\,d\mathbf{s}$$

where S is the surface bounded by L. But S lies entirely in the x,y-plane and so

$$\int_L e^{-x}\mathbf{a}\,.\,d\mathbf{l} = \int_0^{\frac{1}{2}\pi} \int_0^{\pi} - 2e^{-x}\cos y\,dx\,dy$$

$$= -2\int_0^{\frac{1}{2}\pi} \left[-e^{-x} \right]_0^{\pi} \cos y\,dy$$

$$= 2(e^{-\pi} - 1)\int_0^{\frac{1}{2}\pi} \cos y\,dy$$

$$= 2(e^{-\pi} - 1).$$

Exercises

1. Verify Stokes' theorem by evaluating $\int_L y\mathbf{i}\,.\,d\mathbf{l}$ directly and by evaluating the corresponding surface integral where L is the circle $x^2 + y^2 = a^2$, $z = 0$ and $\mathbf{i}$ is unit vector in the direction of the x-axis.

2. Prove that if **c** is a constant vector and **r** is the position vector of the point (x,y,z), then

$$\nabla\left(\frac{\mathbf{c}\cdot\mathbf{r}}{r^3}\right) = -\nabla\times\left(\frac{\mathbf{c}\times\mathbf{r}}{r^3}\right).$$

By considering

$$\mathbf{c}\cdot\int_L\frac{\mathbf{r}\times d\mathbf{l}}{r^3}$$

where **c** is an arbitrary constant vector, and using Stokes' theorem and the above result, show that

$$\int_L\frac{\mathbf{r}\times d\mathbf{l}}{r^3} = -\int_S\frac{\partial}{\partial n}\left(\frac{\mathbf{r}}{r^3}\right)\cdot d\mathbf{s}$$

where the surface S is bounded by the simple closed curve L and $(\partial/\partial n)(\mathbf{r}/r^3)$ is the derivative of $(\mathbf{r}/r^3)$ with respect to distance along the normal to S.

1.12 Scalar potential

Previously it was shown how to obtain a vector field starting from a scalar field ($\mathbf{a} = \text{grad }\varphi$). We now investigate the conditions under which the converse can be carried out, that is the conditions under which a given vector field **a** can be expressed as the gradient of a scalar field.

If $\mathbf{a} = \text{grad }\varphi$ then $\text{curl }\mathbf{a} = \text{curl grad }\varphi \equiv \mathbf{0}$. Thus a *necessary* condition for **a** to be expressed in the required form is that $\text{curl }\mathbf{a} \equiv \mathbf{0}$.

We shall now show that this is also a sufficient condition. Let PAQ and PBQ be any two paths between the points P and Q (figure 36). Then PAQBP is a closed curve which we shall denote by L.

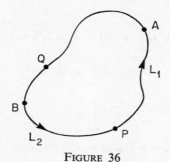

FIGURE 36

From Stokes' theorem, we have

$$\int_L\mathbf{a}\cdot d\mathbf{l} = \int_S\text{curl }\mathbf{a}\cdot d\mathbf{s}$$

where S is a surface bounded by L. Therefore if curl $\mathbf{a} \equiv 0$ on the surface S then

$$\int_S \text{curl } \mathbf{a} \cdot d\mathbf{s} = 0$$

and so

$$\int_L \mathbf{a} \cdot d\mathbf{l} = 0.$$

Therefore

$$\int_{L_1} \mathbf{a} \cdot d\mathbf{l} + \int_{L_2} \mathbf{a} \cdot d\mathbf{l} = 0$$

where L_1 is the curve PAQ and L_2 is the curve QBP (figure 36).

Therefore

$$\int_{L_1} \mathbf{a} \cdot d\mathbf{l} = -\int_{L_2} \mathbf{a} \cdot d\mathbf{l}$$

$$= \int_{L'_2} \mathbf{a} \cdot d\mathbf{l}$$

where L'_2 is L_2 described in the opposite sense, that is the curve PBQ. Thus $\int \mathbf{a} \cdot d\mathbf{l}$ between points P and Q is independent of the path taken and so can be written as $\int_P^Q \mathbf{a} \cdot d\mathbf{l}$. Then if P is regarded as fixed and Q as variable, we can put

$$\int_P^Q \mathbf{a} \cdot d\mathbf{l} = \varphi$$

where φ is a function of the position of Q only. Now let $Q \equiv (x_0, y_0, z_0)$, $Q' \equiv (x_0 + \alpha, y_0 + \beta, z_0 + \gamma)$ and $\varphi(Q') = \varphi(Q) + h$.

Therefore

$$\varphi(Q) + h = \int_P^{Q'} \mathbf{a} \cdot d\mathbf{l} = \int_P^Q \mathbf{a} \cdot d\mathbf{l} + \int_Q^{Q'} \mathbf{a} \cdot d\mathbf{l}$$

$$= \varphi(Q) + \int_Q^{Q'} \mathbf{a} \cdot d\mathbf{l}.$$

Therefore

$$h = \int_Q^{Q'} \mathbf{a} \cdot d\mathbf{l}.$$

In particular, taking $\beta, \gamma = 0$ and $\alpha \neq 0$ and considering the straight line joining Q and Q' we obtain

$$h = \alpha \left(\frac{\partial \varphi}{\partial x} \right)_{x = x_0 + \varepsilon} \quad \text{and} \quad \int_Q^{Q'} \mathbf{a} \cdot d\mathbf{l} = a_x(x_0 + \eta, y_0, z_0)\alpha$$

where $0 \leqslant \varepsilon \leqslant \alpha$ and $0 \leqslant \eta \leqslant \alpha$.

Therefore

$$\left(\frac{\partial \varphi}{\partial x}\right)_{x=x_0+\varepsilon} = a_x(x_0 + \eta, y_0, z_0).$$

Then, letting $\alpha \to 0$, we obtain

$$\frac{\partial \varphi}{\partial x} = a_x$$

both sides being evaluated at the point Q. Similarly, taking $\alpha, \gamma = 0$, $\beta \neq 0$ and $\alpha, \beta = 0$, $\gamma \neq 0$ in turn we obtain

$$\frac{\partial \varphi}{\partial y} = a_y, \qquad \frac{\partial \varphi}{\partial z} = a_z$$

where all expressions are evaluated at the point Q. Then in general, for the point (x,y,z) we have

$$\frac{\partial \varphi}{\partial x} = a_x, \qquad \frac{\partial \varphi}{\partial y} = a_y, \qquad \frac{\partial \varphi}{\partial z} = a_z.$$

These equations enable φ to be calculated uniquely except for an arbitrary constant.

Thus we can find a φ such that $\mathbf{a} = \text{grad } \varphi$ provided curl $\mathbf{a} \equiv \mathbf{0}$. Such a φ is known as a *scalar potential* of the vector field given by $\mathbf{a}$. (Sometimes the scalar potential is defined by $\mathbf{a} = -\text{grad } \varphi$, depending on the particular physical vector field being considered.)

Now the point P was used as the fixed point or origin. If some other point, say O, had been used, then this would simply have introduced the constant $\int_{P}^{O} \mathbf{a} \cdot d\mathbf{l}$.

It is now shown that any two solutions φ_1, φ_2 of $\mathbf{a} = \text{grad } \varphi$ can differ only by a constant.

We have

$$\mathbf{a} = \text{grad } \varphi_1 \quad \text{and} \quad \mathbf{a} = \text{grad } \varphi_2.$$

Therefore

$$\text{grad } (\varphi_1 - \varphi_2) = \mathbf{0}.$$

Therefore

$$\frac{\partial}{\partial x}(\varphi_1 - \varphi_2) = 0, \qquad \frac{\partial}{\partial y}(\varphi_1 - \varphi_2) = 0, \qquad \frac{\partial}{\partial z}(\varphi_1 - \varphi_2) = 0.$$

That is $(\varphi_1 - \varphi_2)$ is independent of x, y and z and so must be a constant.

Thus scalar potentials differing only by a constant can be associated with the same vector field.

If **F** is a force field and curl $\mathbf{F} \equiv \mathbf{0}$ then **F** is said to be *conservative* and it is possible to define a scalar potential energy Ω such that

$$\mathbf{F} = -\text{grad } \Omega.$$

Example *Determine whether a scalar potential exists for the vector field given by*

$$\mathbf{V} = (y^2 + 2xz^2)\mathbf{i} + (2xy - z)\mathbf{j} + (2x^2z - y + 2z)\mathbf{k}$$

and find one if it does.

$$\text{curl } \mathbf{V} = \left(\frac{\partial}{\partial y}(2x^2z - y + 2z) - \frac{\partial}{\partial z}(2xy - z)\right)\mathbf{i}$$

$$+ \left(\frac{\partial}{\partial z}(y^2 + 2xz^2) - \frac{\partial}{\partial x}(2x^2z - y + 2z)\right)\mathbf{j}$$

$$+ \left(\frac{\partial}{\partial x}(2xy - z) - \frac{\partial}{\partial y}(y^2 + 2xz^2)\right)\mathbf{k}.$$

$$= \mathbf{0}.$$

Therefore a scalar potential exists. Let φ be a scalar potential. Then

$$\mathbf{V} = \text{grad } \varphi = \left(\frac{\partial \varphi}{\partial x}, \frac{\partial \varphi}{\partial y}, \frac{\partial \varphi}{\partial z}\right).$$

Therefore

$$\frac{\partial \varphi}{\partial x} = y^2 + 2xz^2$$

and so

$$\varphi = xy^2 + x^2z^2 + f(y,z)$$

for an arbitrary differentiable function f. Also

$$\frac{\partial \varphi}{\partial y} = 2xy - z$$

and so

$$2xy + \frac{\partial f}{\partial y} = 2xy - z.$$

Therefore

$$\frac{\partial f}{\partial y} = -z$$

and so

$$f(y,z) = -yz + g(z)$$

for an arbitrary differentiable function g. Also

$$\frac{\partial \varphi}{\partial z} = 2x^2 z - y + 2z$$

and so

$$2x^2 z - y + \frac{dg}{dz} = 2x^2 z - y + 2z.$$

Therefore

$$\frac{dg}{dz} = 2z$$

and so

$$g(z) = z^2 + c$$

where c is an arbitrary constant. Therefore

$$\varphi = xy^2 + x^2 z^2 - yz + z^2 + c$$

for any constant c is a scalar potential for the given vector field.

Example   Find scalar potentials, if they exist, for the following vector fields

(i) $$\mathbf{F_1} = -\mathbf{r}/r^3,$$

(ii) $$\mathbf{F_2} = \frac{\mathbf{c}}{\mathbf{c} \cdot \mathbf{r}}$$

where $\mathbf{r}$ *is the position vector and* $\mathbf{c}$ *is a constant vector.*

(i) Use spherical polar coordinates.

$$\mathbf{F_1} = -\frac{1}{r^3}(r,0,0) = \left(-\frac{1}{r^2},0,0\right).$$

Then

$$\text{curl } \mathbf{F_1} = \begin{vmatrix} \dfrac{\mathbf{k_1}}{r^2 \sin \theta} & \dfrac{\mathbf{k_2}}{r \sin \theta} & \dfrac{\mathbf{k_3}}{r} \\[2mm] \dfrac{\partial}{\partial r} & \dfrac{\partial}{\partial \theta} & \dfrac{\partial}{\partial \psi} \\[2mm] -1/r^2 & 0 & 0 \end{vmatrix} = \mathbf{0}$$

Therefore a scalar potential exists. Let φ_1 be a scalar potential for $\mathbf{F_1}$. Then

$$\mathbf{F_1} = \text{grad } \varphi_1 = \left(\frac{\partial \varphi_1}{\partial r}, \frac{1}{r}\frac{\partial \varphi_1}{\partial \theta}, \frac{1}{r \sin \theta}\frac{\partial \varphi_1}{\partial \psi}\right).$$

Hence

$$\frac{\partial \varphi_1}{\partial r} = -\frac{1}{r^2}$$

and so

$$\varphi_1 = \frac{1}{r} + f_1(\theta,\psi)$$

for an arbitrary differentiable function f_1. Also

$$\frac{1}{r} \frac{\partial \varphi_1}{\partial \theta} = 0$$

and so

$$\frac{\partial f_1}{\partial \theta} = 0.$$

Therefore

$$f_1(\theta,\psi) = g_1(\psi)$$

for an arbitrary differentiable function g_1. Also

$$\frac{1}{r \sin \theta} \frac{\partial \varphi_1}{\partial \psi} = 0$$

and so

$$\frac{dg_1}{\partial \psi} = 0.$$

Therefore

$$g_1(\psi) = c_1$$

where c_1 is an arbitrary constant. Hence

$$\varphi_1 = \frac{1}{r} + c_1$$

is a scalar potential for all constants c_1.

(ii) At first sight, it would appear that spherical polar coordinates should again be used here. However, since the orientation of the axes of an orthogonal curvilinear system, such as spherical polars, varies with position, it is generally easier when constant vectors are involved to use a conveniently orientated cartesian system as follows.

Take the x-axis to be in the direction of the constant vector $\mathbf{c}$. Then

$$\mathbf{c} = (c,0,0) \quad \text{where} \quad c = |\mathbf{c}|$$

and

$$\mathbf{c} \cdot \mathbf{r} = (c,0,0) \cdot (x,y,z) = cx.$$

Therefore

$$\mathbf{F}_2 = \frac{1}{cx}(c,0,0) = \left(\frac{1}{x},0,0\right).$$

Hence

$$\text{curl } \mathbf{F}_2 = 0$$

and so a scalar potential exists. Let φ_2 be a scalar potential. Therefore

$$\mathbf{F}_2 = \text{grad } \varphi_2 = \left(\frac{\partial \varphi_2}{\partial x}, \frac{\partial \varphi_2}{\partial y}, \frac{\partial \varphi_2}{\partial z} \right).$$

Hence

$$\frac{\partial \varphi_2}{\partial x} = \frac{1}{x}$$

and so

$$\varphi_2 = \ln x + f_2(y,z)$$

for an arbitrary differentiable function f_2. Also

$$\frac{\partial \varphi_2}{\partial y} = 0 \quad \text{and so} \quad \frac{\partial f_2}{\partial y} = 0.$$

Therefore

$$f_2(y,z) = g_2(z)$$

for an arbitrary differentiable function g_2. Also

$$\frac{\partial \varphi_2}{\partial z} = 0 \quad \text{and so} \quad \frac{\partial g_2}{\partial z} = 0.$$

Therefore

$$g_2(z) = c_2$$

where c_2 is an arbitrary constant. Hence $\varphi_2 = \ln x + c_2$ is a scalar potential for the given vector field. Now to complete the problem, φ_2 should be expressed in terms of the given variables namely $\mathbf{c}$ and $\mathbf{r}$.

This is easily done since

$$x = \frac{\mathbf{c} \cdot \mathbf{r}}{c}$$

Therefore

$$\varphi_2 = \ln \frac{\mathbf{c} \cdot \mathbf{r}}{c} + c_2$$

$$= \ln (\mathbf{c} \cdot \mathbf{r}) - \ln c + c_2$$

$$= \ln (\mathbf{c} \cdot \mathbf{r}) + K$$

where $K = c_2 - \ln c$.

Example   Show that if ρ, ψ, z are cylindrical polar coordinates, then the vector field given by

$$\mathbf{F} = \frac{e^{-z}}{\rho^{3}} (2 \cos \psi, \sin \psi, \rho \cos \psi)$$

expressed in terms of a cylindrical polar coordinate system is irrotational, and find a φ such that $\mathbf{F} = \text{grad } \varphi$.

$$\text{curl } \mathbf{F} = \begin{vmatrix} \dfrac{\mathbf{k}_1}{\rho} & \mathbf{k}_2 & \dfrac{\mathbf{k}_3}{\rho} \\[2mm] \dfrac{\partial}{\partial \rho} & \dfrac{\partial}{\partial \psi} & \dfrac{\partial}{\partial z} \\[2mm] \dfrac{2e^{-z} \cos \psi}{\rho^3} & \dfrac{e^{-z} \sin \psi}{\rho^2} & \dfrac{e^{-z} \cos \psi}{\rho^2} \end{vmatrix}$$

$$= \left(-\frac{e^{-z} \sin \psi}{\rho^3} + \frac{e^{-z} \sin \psi}{\rho^3}, \; -\frac{2e^{-z} \cos \psi}{\rho^3} + \frac{2e^{-z} \cos \psi}{\rho^3}, \right.$$

$$\left. -\frac{2e^{-z} \sin \psi}{\rho^4} + \frac{2e^{-z} \sin \psi}{\rho^4} \right)$$

$$= 0.$$

Thus a scalar potential φ exists. Then

$$\frac{\partial \varphi}{\partial \rho} = \frac{2e^{-z} \cos \psi}{\rho^3}$$

and so

$$\varphi = -\frac{e^{-z} \cos \psi}{\rho^2} + f(\psi,z)$$

for an arbitrary differentiable function f. Also

$$\frac{1}{\rho} \cdot \frac{\partial \varphi}{\partial \psi} = \frac{e^{-z} \sin \psi}{\rho^3}$$

and so

$$\frac{e^{-z} \sin \psi}{\rho^3} + \frac{1}{\rho} \frac{\partial f}{\partial \psi} = \frac{e^{-z} \sin \psi}{\rho^3}.$$

Therefore

$$\frac{\partial f}{\partial \psi} = 0 \quad \text{and so} \quad f(\psi,z) = g(z)$$

for an arbitrary differentiable function g. Also

$$\frac{\partial \varphi}{\partial z} = \frac{e^{-z} \cos \psi}{\rho^2}$$

and so

$$\frac{e^{-z} \cos \psi}{\rho^2} + \frac{dg}{dz} = \frac{e^{-z} \cos \psi}{\rho^2}.$$

Therefore

$$\frac{dg}{dz} = 0 \quad \text{and so} \quad g(z) = c$$

where c is an arbitrary constant. Hence $\varphi = (-e^{-z}\cos\psi)/\rho^2 + c$ satisfies $\mathbf{F} = \text{grad } \varphi$.

Exercises

1. Show that the vector $\mathbf{F}_1 = (2x + z^2, z, y + 2xz)$ possesses a scalar potential while the vector $\mathbf{F}_2 = (2y + x^2, x, z + 2xy)$ does not. Find a function φ such that $\mathbf{F}_1 = \text{grad } \varphi$.

2. Given that the function f satisfies $(\partial f/\partial z) = 0$ and $f = 0$ when $x = y = 0$, find f such that the vector

$$\mathbf{F} = (x^3 + 3y^2z)\mathbf{i} + 6xyz\mathbf{j} + f\mathbf{k}$$

has a scalar potential φ and find φ.
  Show also that, with the same function f the vector $\mathbf{F}$ is not solenoidal.

3. Determine which of the following vector fields have scalar potentials (i) $\mathbf{c} \times \mathbf{r}$, (ii) $(\mathbf{c} \cdot \mathbf{r})\mathbf{c}$, (iii) $r^{-1}[r^2\mathbf{c} + (\mathbf{c} \cdot \mathbf{r})\mathbf{r}]$, (iv) $r^{-1}[r^2\mathbf{c} - (\mathbf{c} \cdot \mathbf{r})\mathbf{r}]$ where $\mathbf{r}$ is the position vector and $\mathbf{c}$ is a constant vector.
  Find suitable scalar potentials in appropriate cases.

4. Determine for which of the following vector fields, specified in cartesian, cylindrical and spherical polar coordinates (x,y,z), (ρ,ψ,z) and (r,θ,ψ) respectively, a scalar potential exists and find it when it does. (c is a constant).
  (i) $\mathbf{a} = (2cx(z^3 + y^3), 2cy(z^3 + y^3) + 3cy^2(x^2 + y^2), 3cz^2(x^2 + y^2))$
  (ii) $\mathbf{a} = (-3c\rho^2\cos\psi, c\rho^2\sin\psi, 2cz^2)$
  (iii) $\mathbf{a} = (-2cr\sin\theta\cos\psi, -cr\cos\theta\cos\psi, cr\sin\psi)$.

1.13 Vector potential

We now investigate the conditions under which, given a vector field $\mathbf{a}$ it is possible to find another vector field $\mathbf{b}$ such that $\mathbf{a} = \text{curl } \mathbf{b}$.
  Since div curl $\mathbf{b} \equiv 0$, a *necessary* condition is that div $\mathbf{a} \equiv 0$. To show that this is also a *sufficient* condition, we start with any given vector field $\mathbf{a}$ satisfying the condition div $\mathbf{a} \equiv 0$ and determine a vector field $\mathbf{b}$ such that $\mathbf{a} = \text{curl } \mathbf{b}$.
  Let $\mathbf{a} = (a_x, a_y, a_z)$. Then

$$\text{div } \mathbf{a} = \frac{\partial a_x}{\partial x} + \frac{\partial a_y}{\partial y} + \frac{\partial a_z}{\partial z} = 0.$$

Let $\mathbf{b} = (b_x, b_y, b_z)$. Then

$$\operatorname{curl} \mathbf{b} = \left(\frac{\partial b_z}{\partial y} - \frac{\partial b_y}{\partial z}, \; \frac{\partial b_x}{\partial z} - \frac{\partial b_z}{\partial x}, \; \frac{\partial b_y}{\partial x} - \frac{\partial b_x}{\partial y} \right).$$

The problem is then to determine $b_x(x,y,z)$, $b_y(x,y,z)$ and $b_z(x,y,z)$ such that

$$a_x = \frac{\partial b_z}{\partial y} - \frac{\partial b_y}{\partial z}, \; a_y = \frac{\partial b_x}{\partial z} - \frac{\partial b_z}{\partial x} \quad \text{and} \quad a_z = \frac{\partial b_y}{\partial x} - \frac{\partial b_x}{\partial y}.$$

Try putting one component of $\mathbf{b}$, say b_z, equal to zero. A solution may of course not now be possible, but if it is, then the sufficiency of the condition div $\mathbf{a} = 0$ will have been proved.

Thus we now have

$$a_x = - \frac{\partial b_y}{\partial z}, \; a_y = \frac{\partial b_x}{\partial z} \quad \text{and} \quad a_z = \frac{\partial b_y}{\partial x} - \frac{\partial b_x}{\partial y}.$$

Hence

$$b_x(x,y,z) = \int_{z_0}^{z} a_y \, dz + f(x,y)$$

where z_0 is a constant and f is an arbitrary differentiable function. Also

$$b_y(x,y,z) = - \int_{z_0}^{z} a_x \, dz + g(x,y)$$

where g is another arbitrary differentiable function. Then

$$\frac{\partial b_y}{\partial x} - \frac{\partial b_x}{\partial y} = - \frac{\partial}{\partial x} \int_{z_0}^{z} a_x \, dz + \frac{\partial g}{\partial x} - \frac{\partial}{\partial y} \int_{z_0}^{z} a_y \, dz - \frac{\partial f}{\partial y}$$

$$= - \int_{z_0}^{z} \left(\frac{\partial a_x}{\partial x} + \frac{\partial a_y}{\partial y} \right) dz + \frac{\partial g}{\partial x} - \frac{\partial f}{\partial y}$$

since $(\partial/\partial x)$ and $(\partial/\partial y)$ are independent of the integration with respect to z. Therefore

$$\frac{\partial b_y}{\partial x} - \frac{\partial b_x}{\partial y} = \int_{z_0}^{z} \frac{\partial a_z}{\partial z} \, dz + \frac{\partial g}{\partial x} - \frac{\partial f}{\partial y}$$

since div $\mathbf{a} = 0$. Therefore

$$\frac{\partial b_y}{\partial x} - \frac{\partial b_x}{\partial y} = a_z(x,y,z) - a_z(x,y,z_0) + \frac{\partial g}{\partial x} - \frac{\partial f}{\partial y}$$

$$= a_z(x,y,z)$$

provided

$$-a_z(x,y,z_0) + \frac{\partial g}{\partial x} - \frac{\partial f}{\partial y} = 0.$$

This latter condition is satisfied by taking

$$g(x,y) \equiv 0$$

and

$$f(x,y) = -\int_{y_0}^{y} a_z(x,y,z_0)\, dy$$

where y_0 is a constant.

Hence it has been shown that div $\mathbf{a} \equiv 0$ is a necessary and sufficient condition for the existence of a vector field $\mathbf{b}$ such that $\mathbf{a} = $ curl $\mathbf{b}$.

However, since as above, we can find a $\mathbf{b}$ having any one of its components zero, we see that $\mathbf{b}$ satisfying the above conditions is not unique. Therefore, impose a further restriction on $\mathbf{b}$, namely that div $\mathbf{b} \equiv 0$. A vector field $\mathbf{b}$ satisfying both the equations $\mathbf{a} = $ curl $\mathbf{b}$ and div $\mathbf{b} \equiv 0$ is called a *vector potential* of the vector field $\mathbf{a}$.

Now in general, $\mathbf{b}$ found as above will not have zero divergence and so we continue as follows.

Let $\mathbf{b}'$ be a vector field satisfying the condition $\mathbf{a} = $ curl $\mathbf{b}'$ but not necessarily satisfying div $\mathbf{b}' = 0$.

Let div $\mathbf{b}' = f(x,y,z)$.

Then if we can find another vector field $\mathbf{b}''$ such that div $\mathbf{b}'' = -f(x,y,z)$ and curl $\mathbf{b}'' = \mathbf{0}$, the vector $\mathbf{b} = \mathbf{b}' + \mathbf{b}''$ satisfies

$$\text{curl } \mathbf{b} = \text{curl } \mathbf{b}' + \text{curl } \mathbf{b}'' = \mathbf{a}$$

and

$$\text{div } \mathbf{b} = \text{div } \mathbf{b}' + \text{div } \mathbf{b}'' = f(x,y,z) - f(x,y,z) = 0.$$

But since curl $\mathbf{b}'' = \mathbf{0}$ there exists a scalar function φ such that

$$\mathbf{b}'' = \text{grad } \varphi$$

and so

$$\nabla^2 \varphi = \text{div grad } \varphi = \text{div } \mathbf{b}'' = -f(x,y,z).$$

Thus summing up, a vector $\mathbf{b}$ satisfying the single equation $\mathbf{a} = $ curl $\mathbf{b}$ is unique except to the extent of the addition of the gradient of an arbitrary scalar field. This arbitrary function can be eliminated by insisting also on the additional condition div $\mathbf{b} = 0$. Then if

$$\mathbf{b} = \mathbf{b}' + \text{grad } \varphi,$$
$$\text{div } \mathbf{b}' + \nabla^2 \varphi = 0$$

and this equation determines φ to within an arbitrary additive constant.

Example   Find a vector potential for the vector field given by

$$\mathbf{a} = x\mathbf{i} - y\mathbf{j} + 0\mathbf{k}.$$

div $\mathbf{a} = 1 - 1 = 0$ and so a vector potential exists. Let

$$\mathbf{b} = 0\mathbf{i} + f(x,y,z)\mathbf{j} + g(x,y,z)\mathbf{k}.$$

Since $\mathbf{a}$ has zero $\mathbf{k}$ component, it is easier to consider $\mathbf{b}$ with either a zero $\mathbf{i}$ component or a zero $\mathbf{j}$ component, but not a zero $\mathbf{k}$ component. Then

$$\text{curl } \mathbf{b} = \left(\frac{\partial g}{\partial y} - \frac{\partial f}{\partial z}\right)\mathbf{i} - \frac{\partial g}{\partial x}\mathbf{j} + \frac{\partial f}{\partial x}\mathbf{k} = \mathbf{a}$$

if

$$\frac{\partial f}{\partial x} = 0, \quad \frac{\partial g}{\partial x} = y \quad \text{and} \quad \frac{\partial g}{\partial y} - \frac{\partial f}{\partial z} = x.$$

Therefore

$$f(x,y,z) \equiv f(y,z), \qquad g(x,y,z) = xy + h(y,z)$$

where h is an arbitrary differentiable function and

$$x + \frac{\partial h}{\partial y} - \frac{\partial f}{\partial z} = x.$$

This last condition is satisfied if both $f(y,z)$ and $h(y,z)$ are zero.
    Therefore $\mathbf{b} = (0,0,xy)$ satisfies curl $\mathbf{b} = \mathbf{a}$. Also div $\mathbf{b} = 0$ and so $\mathbf{b}$ is a vector potential for $\mathbf{a}$.

Example   Show that $\mathbf{a} = \nabla(1/r)$, where r is distance from a fixed point, has a vector potential and find it.

Using spherical polar coordinates,

$$\text{div } \mathbf{a} = \text{div grad } \frac{1}{r} = \nabla^2\left(\frac{1}{r}\right) = \frac{\partial^2}{\partial r^2}\left(\frac{1}{r}\right) + \frac{2}{r}\frac{\partial}{\partial r}\left(\frac{1}{r}\right) = \frac{2}{r^3} - \frac{2}{r^3} = 0.$$

Therefore a vector potential exists.
    $\mathbf{a} = (-1/r^2,0,0)$ in spherical polar coordinates. Try $\mathbf{b} = (0, f(r,\theta,\psi), g(r,\theta,\psi))$ as a vector potential for $\mathbf{a}$. Therefore

$$\mathbf{a} = \text{curl } \mathbf{b}$$

$$= \left(\frac{1}{r^2 \sin \theta}\left(\frac{\partial}{\partial \theta}(r \sin \theta g) - \frac{\partial}{\partial \psi}(rf)\right), -\frac{1}{r \sin \theta}\frac{\partial}{\partial r}(r \sin \theta g), \frac{1}{r}\frac{\partial}{\partial r}(rf)\right)$$

Therefore

$$\frac{\partial}{\partial r}(r \sin \theta g) = 0 \quad \text{and so} \quad r \sin \theta g = \alpha(\theta,\psi)$$

where α is an arbitrary differentiable function. Also

$$\frac{\partial}{\partial r}(rf) = 0$$

and so

$$rf = \beta(\theta,\psi)$$

where β is an arbitrary differentiable function. Also

$$-\frac{1}{r^2} = \frac{1}{r^2 \sin\theta}\left(\frac{\partial\alpha}{\partial\theta} - \frac{\partial\beta}{\partial\psi}\right)$$

or

$$\sin\theta = \frac{\partial\beta}{\partial\psi} - \frac{\partial\alpha}{\partial\theta}.$$

This is satisfied by $\beta(\theta,\psi) = 0$ and $\alpha(\theta,\psi) = \cos\theta$.
    Hence

$$\mathbf{b} = (0, 0, \cos\theta).$$

Now

$$\text{div } \mathbf{b} = \frac{1}{r^2 \sin\theta}\frac{\partial}{\partial\psi}(r^2 \sin\theta \cos\theta) = 0$$

and so $\mathbf{b} = (0,0,\cos\theta)$ is a vector potential for $\mathbf{a}$.

*Example   In terms of cylindrical polar coordinates ρ, ψ, z a vector field $\mathbf{F}$ is
given by*

$$\mathbf{F} = (z\rho\cos\psi\sin\psi, z\rho\cos^2\psi, 0).$$

Show that $\text{div }\mathbf{F} = 0$ *and find a function* $f(\rho,\psi,z)$ *such that the vector*
$\mathbf{b} = (0,0,f)$ *satisfies* $\mathbf{F} = \text{curl }\mathbf{b}$. *Hence obtain a vector potential for the vector
field given by* $\mathbf{F}$.

$$\text{div }\mathbf{F} = \frac{1}{\rho}\left\{\frac{\partial}{\partial\rho}(z\rho^2\cos\psi\sin\psi) + \frac{\partial}{\partial\psi}(z\rho\cos^2\psi)\right\}$$

$$= \frac{1}{\rho}\{2z\rho\cos\psi\sin\psi - 2z\rho\cos\psi\sin\psi\}$$

$$= 0.$$

$$\text{curl }\mathbf{b} = \left(\frac{1}{\rho}\frac{\partial f}{\partial\psi}, -\frac{\partial f}{\partial\rho}, 0\right) = \mathbf{F}$$

if $\partial f/\partial\psi = z\rho^2\cos\psi\sin\psi$ and $-\partial f/\partial\rho = z\rho\cos^2\psi$. Therefore

$$f = -\tfrac{1}{2}z\rho^2\cos^2\psi + g(\rho,z)$$

where g is an arbitrary differentiable function, and so

$$z\rho \cos^2 \psi - \frac{\partial g}{\partial \rho} = z\rho \cos^2 \psi.$$

Therefore

$$\frac{\partial g}{\partial \rho} = 0$$

and so

$$g(\rho,z) \equiv g(z).$$

Since g is arbitrary, it can be taken to be identically zero, giving

$$f(\rho,\psi,z) = -\tfrac{1}{2}z\rho^2 \cos^2 \psi$$

such that $\mathbf{b} = (0,0,f)$ satisfies curl $\mathbf{b} = \mathbf{F}$. Now

$$\text{div } \mathbf{b} = \frac{1}{\rho}\left\{\frac{\partial}{\partial z}(\rho f)\right\} = \frac{\partial f}{\partial z} = -\tfrac{1}{2}\rho^2 \cos^2 \psi.$$

Put

$$\text{div }(\mathbf{b} + \mathbf{b}') = \text{div } \mathbf{b} + \text{div } \mathbf{b}' = \text{div } \mathbf{b} + \nabla^2\varphi = 0$$

where $\mathbf{b}' = \text{grad } \varphi$ and so

$$\text{curl } \mathbf{b}' = \mathbf{0}.$$

Then

$$\nabla^2\varphi = -\text{div } \mathbf{b} = \tfrac{1}{2}\rho^2 \cos^2 \psi.$$

This equation is difficult to solve in terms of cylindrical polars, but in terms of cartesian coordinates it becomes

$$\nabla^2\varphi = \tfrac{1}{2}x^2$$

which has a solution $\varphi = \frac{1}{24}x^4$. On changing back to cylindrical coordinates, this gives

$$\varphi = \tfrac{1}{24}\rho^4 \cos^4 \psi.$$

Then

$$\mathbf{b}' = \text{grad } \varphi = \left(\frac{\rho^3}{6}\cos^4 \psi, -\frac{\rho^3}{6}\cos^3 \psi \sin \psi, 0\right).$$

Hence the vector

$$\mathbf{b} + \mathbf{b}' = \left(\frac{\rho^3}{6}\cos^4 \psi, -\frac{\rho^3}{6}\cos^3 \psi \sin \psi, -\tfrac{1}{2}z\rho^2\cos^2\psi\right)$$

provides a vector potential for the vector field $\mathbf{F}$.

Exercises

1. Find a function $f(x,z)$ such that

$$\mathbf{F} = (3x^2 + 7y - 6xy - 3z^2, f, 6z(y - x))$$

is the gradient of a certain scalar function φ and find φ.

Show also that, with the same function $f(x,z)$, the vector **F** is solenoidal and find a vector potential of the form $(g(x,y,z), h(x,y,z), 0)$ for **F**.

2. If ρ, ψ, z are cylindrical polar coordinates, show that $\nabla\psi$ and $\nabla \ln \rho$ are solenoidal and find appropriate vector potentials.

1.14 Green's theorem

If φ_1 and φ_2 are any two finite scalar fields which together with their first and second order derivatives are continuous within a region V enclosed by a surface S, then the theorem states that

$$\int_V (\varphi_1 \nabla^2 \varphi_2 - \varphi_2 \nabla^2 \varphi_1)\, dv = \int_S \left(\varphi_1 \frac{\partial \varphi_2}{\partial n} - \varphi_2 \frac{\partial \varphi_1}{\partial n}\right) ds$$

where $\hat{n}$ is unit vector normal to the surface S and $\partial/\partial n$ is the derivative in the direction of $\hat{n}$.

Proof:

$$\int_S \varphi_1 \operatorname{grad} \varphi_2 \cdot ds = \int_V \operatorname{div} (\varphi_1 \operatorname{grad} \varphi_2)\, dv$$

$$= \int_V (\varphi_1 \nabla^2 \varphi_2 + \operatorname{grad} \varphi_1 \cdot \operatorname{grad} \varphi_2)\, dv.$$

Similarly,

$$\int_S \varphi_2 \operatorname{grad} \varphi_1\, ds = \int_V (\varphi_2 \nabla^2 \varphi_1 + \operatorname{grad} \varphi_2 \cdot \operatorname{grad} \varphi_1)\, dv.$$

Hence, subtracting these results gives

$$\int_S (\varphi_1 \operatorname{grad} \varphi_2 - \varphi_2 \operatorname{grad} \varphi_1) \cdot ds = \int_V (\varphi_1 \nabla^2 \varphi_2 - \varphi_2 \nabla^2 \varphi_1)\, dv$$

or

$$\int_V (\varphi_1 \nabla^2 \varphi_2 - \varphi_2 \nabla^2 \varphi_1)\, dv = \int_S \left(\varphi_1 \frac{\partial \varphi_2}{\partial n} - \varphi_2 \frac{\partial \varphi_1}{\partial n}\right) ds.$$

Special Cases

1. If

$$\varphi_1 = \varphi_2 = \varphi$$

then we have

$$\int_S \varphi \operatorname{grad} \varphi \cdot ds = \int_V \varphi \nabla^2 \varphi\, dv + \int_V (\operatorname{grad} \varphi)^2\, dv.$$

Hence, if φ is a known function at all points inside a region V, the right hand side of the above equation can be evaluated and so the value of φ grad φ integrated over the bounding surface of the region may be determined.

2. If $\varphi_1 = \varphi$ and φ_2 is constant, then we have

$$\int_V \nabla^2 \varphi \, dv = \int_S \frac{\partial \varphi}{\partial n} \, ds.$$

The integral on the right hand side is known as the flux across the surface S. Thus if $\nabla^2 \varphi$ is known everywhere within a given region then the flux across its bounding surface may be calculated.

3. If both φ_1 and φ_2 satisfy Laplace's equation then $\nabla^2 \varphi_1 = 0$ and $\nabla^2 \varphi_2 = 0$. Therefore

$$\int_S \varphi_1 \frac{\partial \varphi_2}{\partial n} \, ds = \int_S \varphi_2 \frac{\partial \varphi_1}{\partial n} \, ds.$$

This result is known as Green's Reciprocal theorem.

Example   Given that $\nabla^2 \varphi = 0$, determine an expression for φ at any point P inside a region V bounded by a surface S in terms of the values of φ and $\partial \varphi / \partial n$ on the surface S. Hence show that

$$4\pi = -\int_S \frac{\partial}{\partial n} \left(\frac{1}{r} \right) ds$$

where r is the distance of a variable point of V from P.

The first part of this example is essentially to determine a solution of Laplace's equation subject to given boundary conditions.

Let S$'$ be the surface of a sphere V$'$ with centre at P and radius ε lying wholly within the region V.

Now considering the region enclosed between the surfaces S and S$'$ and taking $\varphi_1 = \varphi$ and $\varphi_2 = 1/r$, Green's Reciprocal theorem gives

$$\int_S \varphi \frac{\partial}{\partial n} \left(\frac{1}{r} \right) ds + \int_{S'} \varphi \frac{\partial}{\partial n} \left(\frac{1}{r} \right) ds = \int_S \frac{1}{r} \frac{\partial \varphi}{\partial n} \, ds + \int_{S'} \frac{1}{r} \frac{\partial \varphi}{\partial n} \, ds$$

since $\nabla^2 \varphi_2 = 0$ in the region being considered as the origin is excluded.

On S$'$ we have $\partial / \partial n \equiv -(\partial / \partial r)$ ($\hat{n}$ being outwards from the region enclosed is towards P) and $r = \varepsilon$. Therefore

$$\int_{S'} \varphi \frac{\partial}{\partial n} \left(\frac{1}{r} \right) ds = \int_{S'} \varphi \frac{1}{\varepsilon^2} \, ds \simeq \frac{1}{\varepsilon^2} \, \varphi_P 4\pi \varepsilon^2 \text{ for small } \varepsilon,$$

$\rightarrow 4\pi \varphi_P$ as $\varepsilon \rightarrow 0$ (φ_P is the value of φ at the point P). Also

$$\int_{S'} \frac{1}{r} \frac{\partial \varphi}{\partial n} \, ds = \frac{1}{\varepsilon} \int_{S'} -\frac{\partial \varphi}{\partial r} \, ds = -\frac{1}{\varepsilon} \int_{V'} \text{div (grad } \varphi) \, dv$$

$$= 0$$

since div grad $\varphi = \nabla^2 \varphi = 0$. Hence in the limit as $\varepsilon \to 0$ we obtain

$$\int_S \varphi \frac{\partial}{\partial n}\left(\frac{1}{r}\right) \, ds + 4\pi \varphi_P = \int_S \frac{1}{r} \frac{\partial \varphi}{\partial n} \, ds$$

and so

$$\varphi_P = \frac{1}{4\pi} \int_S \left(\frac{1}{r} \frac{\partial \varphi}{\partial n} - \varphi \frac{\partial}{\partial n}\left(\frac{1}{r}\right)\right) \, ds.$$

Now putting $\varphi = 1$ everywhere, we obtain

$$4\pi = -\int_S \frac{\partial}{\partial n}\left(\frac{1}{r}\right) \, ds.$$

Exercises

1. If $\mathbf{F} = \nabla \varphi$ and $\mathbf{F}$ is solenoidal, show that

$$\int_V (\mathbf{F} \cdot \mathbf{F}) \, dv = \int_S \varphi \mathbf{F} \cdot d\mathbf{s}$$

where S is a closed surface enclosing the region V.

2. If $\mathbf{F} = \nabla \varphi$ and div $\mathbf{F} = -4\pi\rho$, show that

$$\int_S \mathbf{F} \cdot d\mathbf{s} = -4\pi \int_V \rho \, dv$$

where S is a closed surface enclosing the region V.

2

Physical fields

2.1 Fluid dynamics

Fluid dynamics is the study of the behaviour of fluids in motion. A fluid is an aggregate of molecules which when at rest yields to any shearing stress however small, and which when studied macroscopically appears to have a continuous structure. Fluids fall into two broad categories—gases and liquids.

Gases tend to fill any space to which they have access and are highly compressible. Liquids tend to retain their original density and are almost incompressible. (A more precise statement of what is meant by 'incompressible' follows in section 2.15.)

All real fluids are *viscous*, that is they can exert a tangential reaction on a surface with which they make contact. There are however many situations in which the viscous effect of the fluid can be neglected and so it is useful to introduce the theoretical concept of a non-viscous fluid, that is a fluid which cannot exert a tangential reaction on any surface with which it makes contact.

In what follows we shall consider fluids with the idealized properties of being completely incompressible and non-viscous. Such fluids are usually referred to as *ideal* or *perfect* fluids.

2.11 Pressure

Consider a plane surface S of area s immersed in an ideal fluid at rest. Let the total thrust on one side of this surface be F. Since the fluid is inviscid, this total thrust will be normal to S. (The thrust on the other side of the surface will of course be F in the opposite direction if S is to remain in equilibrium.) Then, if as $s \to 0$ the surface S shrinks to the point P in the fluid, $\lim_{s \to 0} \dfrac{F}{s}$ is called the pressure p at the point P (in the direction of the normal to S).

94

The pressure can be shown to be the same in all directions in an ideal fluid.

2.12 Streamlines and trajectories

Consider a particle in a fluid and let its position vector relative to some fixed origin be $\mathbf{r}(t)$ at time t. After a further small time τ the position vector of the same particle is $\mathbf{r}(t + \tau)$. Then

$$\lim_{\tau \to 0} \frac{\mathbf{r}(t + \tau) - \mathbf{r}(t)}{\tau} = \frac{d\mathbf{r}}{dt}$$

is called the velocity of the particle of fluid or simply the velocity of the fluid at the point $\mathbf{r}$ and is denoted by $\mathbf{q}$. In cartesian coordinates

$$\mathbf{q} = \left(\frac{dx}{dt}, \frac{dy}{dt}, \frac{dz}{dt} \right).$$

The vector $\mathbf{q}$ will in general depend on both the position in the fluid (given by $\mathbf{r}$) and the time t. Thus $\mathbf{q} = \mathbf{q}(\mathbf{r}, t)$.

At a given instant we can start from a point P in the fluid and construct a line to represent the velocity at P in magnitude and direction. If this line is PQ (figure 37), we can construct at the same instant in time a similar

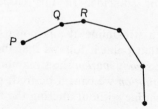

FIGURE 37

line QR at Q and then another similar line at R and so on. As the lengths of the segments tend to zero the figure approaches a smooth curve having the property that the tangent at any point is parallel to the instantaneous fluid velocity at that point. This curve is called a *streamline*. In rectangular cartesian coordinates, such a curve is given in differential form by

$$\frac{dx}{q_x} = \frac{dy}{q_y} = \frac{dz}{q_z}$$

where $\mathbf{q} = (q_x, q_y, q_z)$. The solutions of these differential equations when t has the constant value t_0 give the streamlines at the time t_0.

In terms of orthogonal curvilinear coordinates u_1, u_2, u_3 the streamlines are given by the solutions of the differential equations

$$\frac{h_1\, du_1}{q_1} = \frac{h_2\, du_2}{q_2} = \frac{h_3\, du_3}{q_3}$$

where $\mathbf{q} = (q_1, q_2, q_3)$ in terms of the coordinate system (u_1, u_2, u_3). The streamlines give an instantaneous picture of the fluid motion.

The paths followed by individual fluid elements are called *trajectories*. These are given by the solutions of the differential equations

$$\frac{dx}{dt} = q_x, \qquad \frac{dy}{dt} = q_y, \qquad \frac{dz}{dt} = q_z$$

in cartesian coordinates, or the differential equations

$$h_1 \frac{du_1}{dt} = q_1, \qquad h_2 \frac{du_2}{dt} = q_2, \qquad h_3 \frac{du_3}{dt} = q_3$$

in orthogonal curvilinear coordinates.

When the motion is *steady*, that is not varying with time and so $\partial \mathbf{q}/\partial t = 0$ and $\mathbf{q} \equiv \mathbf{q}(r)$, the trajectories and streamlines coincide.

When the motion is not steady, but varying with time, the trajectories and streamlines do not coincide. In this case, the streamlines are continually changing.

Since the tangent to a streamline at any point on it is parallel to the direction of motion at that point it follows that there is no flow across a streamline. Further, for *steady motion* the streamlines do not change with time. Hence for *steady motion* we can, if desired, place a solid boundary to coincide with a streamline without altering the flow.

2.13 Sources and sinks

A uniform point *source* in a fluid is a point from which the fluid flows outwards uniformly in all directions. The velocity will then be a function of the distance r from the source. If the volume of fluid flowing outwards across any closed surface surrounding the source is $4\pi m$ ($m > 0$) per unit time then the source is said to have strength m. If m is negative the velocity will be everywhere inwards towards the point which is said to be a *sink*.

Now the velocity $\mathbf{q}$ at any point in the fluid having position vector $\mathbf{r}$ relative to a source or sink must be of the form

$$\mathbf{q} = f(r)\mathbf{r}$$

for some function f.

Then

$$\text{curl } \mathbf{q} = \text{grad} f(r) \times \mathbf{r} + f(r) \text{ curl } \mathbf{r}$$
$$= \mathbf{0}$$

since curl $\mathbf{r} \equiv \mathbf{0}$ and grad $f(r)$ is parallel to $\mathbf{r}$.

Thus $\mathbf{q}$ is irrotational and so possesses a scalar potential φ, called the velocity potential, such that $\mathbf{q} = \text{grad } \varphi$.

Consider

$$\varphi = -\frac{m}{r}.$$

Then

$$\mathbf{q} = \left(\frac{m}{r^2}, 0, 0\right)$$

in spherical polars.

Thus, if m is positive, the velocity is everywhere radially outwards from the origin. Also the volume of fluid flowing outwards across the surface S of a sphere with centre the origin and radius a is

$$\int_S \mathbf{q} \cdot d\mathbf{s}^* = \int_S \frac{m}{r^2} ds = \frac{m}{a^2} \int_S ds$$

$$= \frac{m}{a^2} 4\pi a^2 = 4\pi m.$$

$\varphi = -m/r$ is taken as the velocity potential for a source of strength m at the origin. (If m is negative, then $\varphi = -m/r$ represents a source of negative strength, that is a sink.)

2.14 Differentiation following the fluid

Frequently, instead of considering the behaviour of some property of the fluid such as density or temperature at a *particular point* in the fluid we shall be more interested in the behaviour of the property as we move

---

* Consider a plane area s. Let unit normal to this area be $\hat{\mathbf{n}}$ and let the velocity vector at all points of s be $\mathbf{q}$ (a constant vector). Then the component of $\mathbf{q}$ normal to the given plane area is $\mathbf{q} \cdot \hat{\mathbf{n}}$ and so the volume of fluid which will flow across the area s in unit time is $(\mathbf{q} \cdot \hat{\mathbf{n}})s = \mathbf{q} \cdot \mathbf{s}$.

Now, for the surface S, if the element δs is sufficiently small to be regarded as plane and for $\mathbf{q}$ to be taken as constant over it, the volume outflow across it in unit time is $\mathbf{q} \cdot \delta \mathbf{s}$. Then summing for all such elements of the surface and taking the limit as their magnitudes all tend to zero, we obtain that the mean outflow across the surface S in unit time is $\int_S \mathbf{q} \cdot d\mathbf{s}$.

around with a *particular element of fluid*. The value of the property will then vary with the position of the particular element of fluid being considered and with time. The position of the fluid element will itself also vary with time so that, using cartesian axes, its coordinates x, y, z will be functions of time $x(t)$, $y(t)$, $z(t)$.

Then, if the function $f(x(t),y(t),z(t),t)$ gives the value of some property of the fluid associated with the fluid element at the point $(x(t),y(t),z(t))$ at time t, we have

$$\frac{df}{dt} = \frac{\partial f}{\partial t} + \frac{\partial f}{\partial x}\frac{dx}{dt} + \frac{\partial f}{\partial y}\frac{dy}{dt} + \frac{\partial f}{\partial z}\frac{dz}{dt}$$

$$= \frac{\partial f}{\partial t} + u\frac{\partial f}{\partial x} + v\frac{\partial f}{\partial y} + w\frac{\partial f}{\partial z}$$

where $u = dx/dt$, $v = dy/dt$, $w = dz/dt$ are the components of the velocity vector $\mathbf{q}$. Therefore

$$\frac{df}{dt} = \frac{\partial f}{\partial t} + \mathbf{q} \cdot \operatorname{grad} f.$$

Thus we have the equivalence of the operators

$$\frac{d}{dt} \quad \text{and} \quad \frac{\partial}{\partial t} + \mathbf{q} \cdot \nabla$$

For a vector function $\mathbf{a}$, in terms of cartesian coordinates,

$$\frac{d\mathbf{a}}{dt} = \frac{da_x}{dt}\mathbf{i} + \frac{da_y}{dt}\mathbf{j} + \frac{da_z}{dt}\mathbf{k}$$

$$= \left(\frac{\partial a_x}{\partial t} + \mathbf{q} \cdot \nabla a_x\right)\mathbf{i} + \left(\frac{\partial a_y}{\partial t} + \mathbf{q} \cdot \nabla a_y\right)\mathbf{j} + \left(\frac{\partial a_z}{\partial t} + \mathbf{q} \cdot \nabla a_z\right)\mathbf{k}$$

which is also written as

$$\left(\frac{\partial}{\partial t} + \mathbf{q} \cdot \nabla\right)\mathbf{a}.$$

We shall use the notation

$$\frac{D}{Dt} \equiv \frac{\partial}{\partial t} + \mathbf{q} \cdot \nabla$$

The operation here implied is called differentiation following the fluid, implying that we are calculating the rate of change of some quantity associated with the same fluid element as it moves about.

Example *Verify that* $(\mathbf{q} \cdot \mathrm{grad})\mathbf{q} = \mathrm{grad}\,(\tfrac{1}{2}q^2) - \mathbf{q} \times \mathrm{curl}\,\mathbf{q}$.

Use cartesian coordinates.

$$(\mathbf{q} \cdot \mathrm{grad})\mathbf{q} = (\mathbf{q} \cdot \nabla u, \mathbf{q} \cdot \nabla v, \mathbf{q} \cdot \nabla w) \quad \text{where} \quad \mathbf{q} = (u,v,w).$$

Therefore the x-component is

$$\mathbf{q} \cdot \nabla u = (u,v,w) \cdot \left(\frac{\partial u}{\partial x}, \frac{\partial u}{\partial y}, \frac{\partial u}{\partial z}\right)$$

$$= u\frac{\partial u}{\partial x} + v\frac{\partial u}{\partial y} + w\frac{\partial u}{\partial z}.$$

The x-component of the right hand side of the given equation is

$$\frac{\partial}{\partial x}\{\tfrac{1}{2}(u^2 + v^2 + w^2)\} - \left\{v\left(\frac{\partial v}{\partial x} - \frac{\partial u}{\partial y}\right) - w\left(\frac{\partial u}{\partial z} - \frac{\partial w}{\partial x}\right)\right\}$$

$$= u\frac{\partial u}{\partial x} + v\frac{\partial v}{\partial x} + w\frac{\partial w}{\partial x} - v\frac{\partial v}{\partial x} + v\frac{\partial u}{\partial y} + w\frac{\partial u}{\partial z} - w\frac{\partial w}{\partial x}$$

$$= u\frac{\partial u}{\partial x} + v\frac{\partial u}{\partial y} + w\frac{\partial u}{\partial z} = x\text{-component of left-hand side.}$$

Similarly the y- and z-components on each side of the equation can be shown to be equal so that the given result has been verified.

2.15 The equation of continuity

This is an expression in differential form of the physical condition that mass is conserved.

Consider *any fixed* closed surface S enclosing a region V of the fluid in which there are no sources or sinks. If $\rho(x,y,z,t) \equiv \rho(\mathbf{r},t)$ gives the density at any point in the fluid at any instant and $\mathbf{q}(\mathbf{r},t)$ the velocity, then the mass of fluid *leaving* the region V in unit time is

$$\int_S \rho\mathbf{q} \cdot \mathrm{d}\mathbf{s}.$$

But the mass contained in the region V is $\int_V \rho\,\mathrm{d}v$ and so the *increase* in mass in the region V in unit time is

$$\frac{\partial}{\partial t}\int_V \rho\,\mathrm{d}v = \int_V \frac{\partial \rho}{\partial t}\,\mathrm{d}v.$$

Note that since we are considering a *fixed region* in the fluid it is $\dfrac{\partial}{\partial t}\displaystyle\int_V \rho\ dv$ which is required and not $\dfrac{D}{Dt}\displaystyle\int_V \rho\ dv$.

Hence

$$\int_S \rho\mathbf{q}\cdot d\mathbf{s} = -\int_V \frac{\partial \rho}{\partial t}\ dv.$$

Then using Gauss' theorem we obtain

$$\int_V \operatorname{div} \rho\mathbf{q}\ dv + \int_V \frac{\partial \rho}{\partial t}\ dv = 0.$$

Now since this is satisfied for *any* closed surface in the fluid, we must have

$$\operatorname{div} \rho\mathbf{q} + \frac{\partial \rho}{\partial t} = 0$$

at all points in the fluid.

This is one form of Euler's equation of continuity. Another form is obtained by expanding div $(\rho\mathbf{q})$ to give

$$\rho \operatorname{div} \mathbf{q} + \mathbf{q}\cdot\operatorname{grad} \rho + \frac{\partial \rho}{\partial t} = 0$$

or

$$\rho \operatorname{div} \mathbf{q} + \frac{D\rho}{Dt} = 0$$

or

$$\operatorname{div} \mathbf{q} = -\frac{1}{\rho}\frac{D\rho}{Dt}.$$

Now a fluid is said to be *incompressible* if its density is unaffected by pressure changes. The density of a mass of fluid can however also change due to molecular heat conduction but as these effects are very often negligible, an incompressible fluid is generally taken to imply a fluid for which the density of each mass element remains constant. Thus for an incompressible fluid $D\rho/Dt = 0$ and the equation of continuity becomes

$$\operatorname{div} \mathbf{q} = 0.$$

For irrotational motion $\mathbf{q} = \operatorname{grad} \varphi$. In such flows, the particles of fluid which lie along a straight line in the fluid at any instant will always lie along a straight line having constant orientation. The successive positions of such a straight line in the fluid, as the fluid moves irrotationally round a bend, are shown in figure 38.

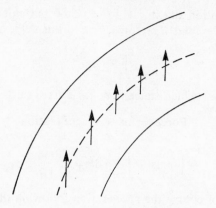

FIGURE 38

Hence the equation of continuity for an incompressible fluid in irrotational motion reduces to

$$\nabla^2 \varphi = 0$$

which is Laplace's equation. In order to solve this equation for a particular problem we will of course require the conditions to be satisfied by φ on the boundary of the region considered. For example, since there can be no flow across a rigid boundary in the fluid, the component of grad φ normal to the boundary must be zero at all points on a rigid boundary. Methods of solving Laplace's equation will be considered in Chapters 3 and 6.

2.16 The equation of fluid motion

We apply Newton's second law of motion to an element V of fluid for which any deformation is negligible compared with its motion as a whole. Let the volume of the element be v and the fluid density be ρ where ρ may be a function of position in the fluid. Suppose also that S is the surface surrounding the element of volume v. The force due to the pressure, on any section of S sufficiently small to be regarded as approximately plane, is pA along the *inward* normal to the section where p is the average pressure on the section and A is its area. That is, the force is $-pA\hat{n}$ where $\hat{n}$ is the unit *outward* normal.

Hence, summing for all such areas and taking the limit as each tends to zero, we obtain that the total force, due to pressure, on the surface S of V is

$$-\int_S p\hat{n}\, ds = -\int_S p\, ds.$$

Suppose in addition an external force field **F** per unit mass (for example gravity) acts on the fluid. The force, due to this field, acting on V is then

$$\int_V \mathbf{F}\rho \, dv.^*$$

Then Newton's law which states that, for a fixed mass,

$$\text{Force} = \text{mass} \times \text{acceleration}$$

gives

$$-\int_S p \, d\mathbf{s} + \int_V \mathbf{F}\rho \, dv = \int_V \frac{D\mathbf{q}}{Dt} \rho \, dv$$

where $D\mathbf{q}/Dt$ is of course the acceleration following the element of fluid. But from an example in section 1.10 (p. 73),

$$\int_S p \, d\mathbf{s} = \int_V \text{grad } p \, dv,$$

and so

$$-\int_V \text{grad } p \, dv + \int_V \mathbf{F}\rho \, dv = \int_V \frac{D\mathbf{q}}{Dt} \rho \, dv.$$

Now since the above equation holds for *any* region within the fluid, it follows that at every point in the fluid we must have

$$-\text{grad } p + \rho\mathbf{F} = \rho \frac{D\mathbf{q}}{Dt}$$

or

$$-\frac{1}{\rho} \text{grad } p + \mathbf{F} = \frac{D\mathbf{q}}{Dt}.$$

Therefore

$$-\frac{1}{\rho} \text{grad } p + \mathbf{F} = \frac{\partial \mathbf{q}}{\partial t} + (\mathbf{q} \cdot \nabla)\mathbf{q}$$

$$= \frac{\partial \mathbf{q}}{\partial t} + \text{grad } (\tfrac{1}{2}q^2) - \mathbf{q} \times \text{curl } \mathbf{q}$$

(see example on p. 99.)

---

* Intuitively, the force acting on the small element δv is $\mathbf{F}\rho \, \delta v$ and the product (mass × acceleration) for the same element is $\rho\delta v(D\mathbf{q}/Dt)$. Then summing over all such elements and taking the limit as each tends to zero, we obtain $\int_V \mathbf{F}\rho \, dv$ and $\int_V \frac{D\mathbf{q}}{Dt} \rho \, dv$ respectively.

Hence

$$\frac{\partial \mathbf{q}}{\partial t} - \mathbf{q} \times \text{curl } \mathbf{q} = \mathbf{F} - \frac{1}{\rho} \text{grad } p - \text{grad } (\tfrac{1}{2}\mathbf{q}^2).$$

Example *Show that $\varphi = (x - t)(y - t)$ represents the velocity potential of a possible incompressible two-dimensional fluid flow. Show also that the streamlines at time t are the curves*

$$(x - t)^2 - (y - t)^2 = \text{constant}$$

and that the trajectories have the equations

$$\ln (x - y)^2 = - (x + y) + \frac{A}{x - y} + B$$

where A and B are constants.

If φ is a velocity potential then $\mathbf{q} = \text{grad } \varphi$. If the fluid is incompressible then the continuity equation gives div $\mathbf{q} = 0$ and so

$$\nabla^2 \varphi = 0.$$

For the given function φ it is easily seen that $\nabla^2 \varphi = 0$ since both $\partial^2 \varphi / \partial x^2$ and $\partial^2 \varphi / \partial y^2$ are zero.

Hence the given φ is a possible velocity potential. Now

$$\mathbf{q} = \text{grad } \varphi = \frac{\partial \varphi}{\partial x} \mathbf{i} + \frac{\partial \varphi}{\partial y} \mathbf{j} \text{ in two dimensions} = (y - t)\mathbf{i} + (x - t)\mathbf{j}.$$

Therefore the streamlines are given by the solutions of

$$\frac{\mathrm{d}x}{y - t} = \frac{\mathrm{d}y}{x - t}.$$

That is the solutions of

$$(x - t) \,\mathrm{d}x = (y - t) \,\mathrm{d}y.$$

Therefore $(x - t)^2 - (y - t)^2 = \text{constant}$.

For the trajectories we have

$$\frac{\mathrm{d}x}{\mathrm{d}t} = (y - t), \qquad \frac{\mathrm{d}y}{\mathrm{d}t} = (x - t).$$

Therefore

$$\frac{\mathrm{d}}{\mathrm{d}t} (x + y) = (x + y) - 2t.$$

The complementary function is the solution of

$$\frac{\mathrm{d}}{\mathrm{d}t} (x + y) - (x + y) = 0$$

which is

$$x + y = A'e^t$$

where A' is a constant. A particular integral is $2t + 2$.
Therefore the general solution is

$$x + y = A'e^t + 2t + 2. \tag{1}$$

Also, we have

$$\frac{d}{dt}(x - y) = -(x - y)$$

which has the solution

$$x - y = B'e^{-t} \tag{2}$$

where B' is a constant.

Equations (1) and (2) are the parametric equations of the trajectories, t being the parameter.

Then eliminating t gives

$$(x + y) = A'\left(\frac{B'}{x - y}\right) + 2\ln\left(\frac{B'}{x - y}\right) + 2$$

$$= \frac{A'B'}{x - y} - \ln(x - y)^2 + \ln B'^2 + 2.$$

Therefore

$$\ln(x - y)^2 = -(x + y) + \frac{A}{x - y} + B$$

where $A = A'B'$ and $B = \ln B'^2 + 2$.

Example   A body of liquid rotates with constant and uniform angular velocity ω about a vertical axis under the action of gravity only. Show that the free surface is a paraboloid of revolution with axis vertical. (The free surface is the interface of the liquid which is assumed to be a surface of constant pressure)

The inherent symmetry of the problem about the vertical suggests that cylindrical polar coordinates should be used.

The external force $\mathbf{F}$ per unit mass acting on the fluid is that due to gravity namely $-g\mathbf{k}$ where $\mathbf{k}$ is unit vector in the vertical. Since the angular velocity is constant, the acceleration per unit mass is $-\omega^2\boldsymbol{\rho}$ where $\boldsymbol{\rho} = (\rho,0,0)$ in terms of cylindrical polars and $\omega = |\boldsymbol{\omega}|$.

Hence the equation of motion gives

$$-\frac{1}{m}\operatorname{grad} p - g\mathbf{k} = -\omega^2\boldsymbol{\rho}$$

where m is the density and p the pressure.

Therefore

$$\frac{1}{m} \operatorname{grad} p = \omega^2 \boldsymbol{\rho} - g\mathbf{k}$$

$$= (\omega^2 \rho, 0, -g).$$

Now curl $(\omega^2 \rho, 0, -g) \equiv \mathbf{0}$ and so a scalar potential Ω exists for the vector $(\omega^2 \rho, 0, -g)$. That is $(\omega^2 \rho, 0, -g) = \operatorname{grad} \Omega$.

Therefore

$$\operatorname{grad} \Omega = \frac{1}{m} \operatorname{grad} p$$

and so we see that the equipotential surfaces for the vector $(\omega^2 \rho, 0, -g)$, that is the surfaces of constant Ω, are also the surfaces of constant p.

Thus the problem has now been reduced to that of finding the form of the surfaces $\Omega = $ constant, which in turn is the problem of finding a scalar potential for the vector $(\omega^2 \rho, 0, -g)$.

$$\frac{\partial \Omega}{\partial \rho} = \omega^2 \rho \quad \text{and so} \quad \Omega = \tfrac{1}{2}\omega^2 \rho^2 + f(z)$$

where f is an arbitrary function of z only since by symmetry Ω is independent of the cylindrical polar coordinate ψ.

Also

$$\frac{\partial \Omega}{\partial z} = -g \quad \text{and so} \quad \frac{df}{dz} = -g.$$

Therefore

$$f = -gz + \text{constant.}$$

Therefore

$$\Omega = \tfrac{1}{2}\omega^2 \rho^2 - gz + \text{constant}$$

and so for the free surface we have

$$\tfrac{1}{2}\omega^2 \rho^2 - gz = \text{constant}$$

or

$$\rho^2 = \left(\frac{2g}{\omega^2}\right) z + \text{constant}$$

which is a paraboloid of revolution with axis vertical.

2.17 *Bernoulli's Equation*

Suppose the external force field $\mathbf{F}$ is conservative, then curl $\mathbf{F} \equiv \mathbf{0}$ and we can define a potential function Ω such that

$$\mathbf{F} = -\operatorname{grad} \Omega.$$

Therefore we have

$$-\frac{1}{\rho}\,\text{grad}\,p - \text{grad}\,\Omega - \text{grad}\,(\tfrac{1}{2}\mathbf{q}^2) = \frac{\partial \mathbf{q}}{\partial t} - \mathbf{q} \times \text{curl}\,\mathbf{q}.$$

Now if also the motion is incompressible, the density $\rho = $ constant and so

$$\frac{1}{\rho}\,\text{grad}\,p = \text{grad}\,\frac{p}{\rho}.$$

Hence

$$-\text{grad}\,\left(\frac{p}{\rho} + \Omega + \tfrac{1}{2}\mathbf{q}^2\right) = \frac{\partial \mathbf{q}}{\partial t} - \mathbf{q} \times \text{curl}\,\mathbf{q}.$$

2.171 Irrotational motion. In this case $\text{curl}\,\mathbf{q} \equiv 0$ and there exists a scalar φ such that $\mathbf{q} = \text{grad}\,\varphi$. Therefore

$$\frac{\partial}{\partial t}(\text{grad}\,\varphi) = -\text{grad}\,\left(\frac{p}{\rho} + \Omega + \tfrac{1}{2}\mathbf{q}^2\right)$$

Therefore

$$\text{grad}\,\left(\frac{\partial \varphi}{\partial t}\right) = -\text{grad}\,\left(\frac{p}{\rho} + \Omega + \tfrac{1}{2}\mathbf{q}^2\right)$$

or

$$\text{grad}\,\left(\frac{p}{\rho} + \Omega + \tfrac{1}{2}\mathbf{q}^2 + \frac{\partial \varphi}{\partial t}\right) = 0$$

and so

$$\frac{p}{\rho} + \Omega + \tfrac{1}{2}\mathbf{q}^2 + \frac{\partial \varphi}{\partial t} = f(t)$$

or

$$\frac{p}{\rho} + \Omega + \tfrac{1}{2}\,|\text{grad}\,\varphi|^2 + \frac{\partial \varphi}{\partial t} = f(t)$$

where $f(t)$ is an arbitrary function of time t.

2.172 Steady motion. In this case $\partial \mathbf{q}/\partial t \equiv 0$ and so

$$-\text{grad}\,\left(\frac{p}{\rho} + \Omega + \tfrac{1}{2}\mathbf{q}^2\right) = -\mathbf{q} \times \text{curl}\,\mathbf{q}(\neq 0).$$

Hence

$$\mathbf{q} \cdot \text{grad}\,\left(\frac{p}{\rho} + \Omega + \tfrac{1}{2}\mathbf{q}^2\right) = 0.$$

Therefore the velocity vector $\mathbf{q}$ is everywhere normal to

$$\text{grad}\,\left(\frac{p}{\rho} + \Omega + \tfrac{1}{2}\mathbf{q}^2\right)$$

and so **q** is everywhere tangential to the surface

$$\frac{p}{\rho} + \Omega + \tfrac{1}{2}\mathbf{q}^2 = \text{constant.}$$

Hence

$$\frac{p}{\rho} + \Omega + \tfrac{1}{2}\mathbf{q}^2 = \text{constant along a streamline.}$$

2.173 Steady, irrotational motion. Here $\partial \mathbf{q}/\partial t \equiv 0$ and curl $\mathbf{q} \equiv \mathbf{0}$. Therefore

$$\text{grad}\left(\frac{p}{\rho} + \Omega + \tfrac{1}{2}\mathbf{q}^2\right) \equiv \mathbf{0}$$

and so

$$\frac{p}{\rho} + \Omega + \tfrac{1}{2}\mathbf{q}^2 = \text{constant everywhere.}$$

, If the motion is not incompressible, that is if ρ is not constant, then the term p/ρ is replaced by $\int \mathrm{d}p/\rho$ in each of the above equations. For example consider the steady, incompressible, irrotational flow along a horizontal tube with a constriction (figure 39).

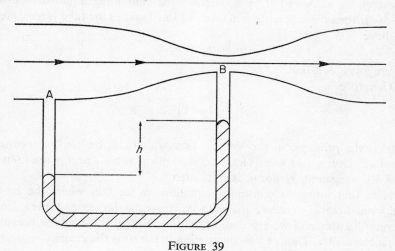

FIGURE 39

If the flow is considered to be approximately horizontal, as will be the case if the constriction in the tube takes place gradually, then the potential Ω due to the force of gravity will be constant and so we have

$$\tfrac{1}{2}\mathbf{q}^2 + \frac{p}{\rho} = \text{constant.}$$

Thus if the pressure and velocity at an unconstricted part A of the tube are p_1 and q_1 respectively while these variables have the respective values p_2, q_2 at B in the constriction,

$$\frac{p_1}{\rho} + \tfrac{1}{2}q_1^2 = \frac{p_2}{\rho} + \tfrac{1}{2}q_2^2.$$

Now, since the flow is incompressible, the volume rate of flow of fluid across the section of the tube at A must be equal to the volume rate of flow across the section at B.

Therefore

$$s_1 q_1 = s_2 q_2$$

where s_1 and s_2 are the cross-sectional areas at A and B respectively.

Therefore

$$p_1 - p_2 = \tfrac{1}{2}\rho q_1^2\left(\left(\frac{s_1}{s_2}\right)^2 - 1\right).$$

Now this pressure difference can be measured by connecting together the two sections at A and B by a narrow tube containing liquid of density ρ' say. If then the difference h in levels of the liquid in the tube is measured we have

$$p_1 - p_2 = g\rho'h$$

where g is a constant.

Therefore

$$\tfrac{1}{2}\rho q_1^2\left(\left(\frac{s_1}{s_2}\right)^2 - 1\right) = g\rho'h.$$

This is the principle of the Venturi tube which can be used to measure wind velocity. s_1 and s_2 are characteristics of the tube, ρ and ρ' are known and h is measured. Hence q_1 is calculated.

Note that using the continuity equation we see that where the cross-sectional area is smaller, the velocity must be larger and then from Bernoulli's theorem we see that where the velocity is larger the pressure must be smaller. Thus at the constriction in the tube the velocity is greater and the pressure smaller. This principle can be used to construct a rather inefficient and wasteful pump described in the next example.

Example   Liquid of density ρ flows along a horizontal pipe past a contraction at which the cross-sectional area is s_1 and is delivered at atmospheric pressure from the end of the pipe which has cross-sectional area s_2. Show that if the flow is assumed to be steady, irrotational and incompressible, then liquid of the same *density ρ will be sucked up into the pipe from a reservoir at a depth*

$$\frac{M^2}{2g\rho^2}\left(\frac{1}{s_1^2} - \frac{1}{s_2^2}\right)$$

below the pipe through a side tube connected with the pipe at the contraction. M is the mass rate of flow of liquid in the pipe.

Let p_1 and q_1 be the pressure and velocity at the contraction in the pipe and p_2 and q_2 be pressure and velocity at the end of the pipe.
Therefore Bernoulli's equation gives

$$\frac{p_1}{\rho} + \tfrac{1}{2}q_1^2 = \frac{p_2}{\rho} + \tfrac{1}{2}q_2^2$$

and so

$$p_2 - p_1 = \tfrac{1}{2}\rho(q_1^2 - q_2^2).$$

From continuity we have

$$\frac{M}{\rho} = s_1 q_1 = s_2 q_2.$$

Therefore

$$p_2 - p_1 = \frac{1}{2}\frac{M^2}{\rho}\left(\frac{1}{s_1^2} - \frac{1}{s_2^2}\right).$$

But $p_2 - p_1$ = weight of liquid in a column of unit cross-sectional area which can be sucked up = $h\rho g$ if h is the height of such a column.
Therefore

$$h\rho g = \frac{1}{2}\frac{M^2}{\rho}\left(\frac{1}{s_1^2} - \frac{1}{s_2^2}\right)$$

and so

$$h = \frac{M^2}{2g\rho^2}\left(\frac{1}{s_1^2} - \frac{1}{s_2^2}\right).$$

Exercises

1. Show that $\varphi = x^2 - 2y^2 + z^2$ is the velocity potential of a possible incompressible fluid motion and find the streamlines.

2. The velocity potential at time t of an incompressible fluid flowing two-dimensionally is given by

$$\varphi = (x + t)(y + t).$$

Show that the streamlines are rectangular hyperbolae and that the paths of the particles have parametric equations of the form

$$x + y = Ae^t - 2t - 2$$
$$x - y = Be^{-t}$$

where A and B are constants.

Show also that the path of the particle which is at (2,1) when $t = 0$ has the equation,

$$\frac{x^2 - y^2 - 5}{x - y} = \ln(x - y)^2 - 2.$$

3. Show that $\varphi = (x - t)^2 - (y + t)^2$ is the velocity potential of a possible two-dimensional incompressible, inviscid fluid motion.

Obtain the equations of the streamlines and find the parametric equations of the trajectories for this motion.

Show that the particle of fluid which is at the point $(1,\frac{1}{2})$ at time $t = 0$ moves along the curve

$$x + y = \tfrac{1}{2}e^{1-2y} + 1.$$

4. A body of liquid rotates under the action of gravity only with constant and uniform angular velocity $\boldsymbol{\omega} = \alpha \mathbf{k}/r^2$ where α is a constant, $\mathbf{k}$ is unit vector in the vertical and r is distance from the axis of $\mathbf{k}$. Find the form of the free surface of the liquid.

5. An incompressible, non-viscous fluid of density ρ flows steadily and irrotationally through a tube of varying cross-section. The pressures at points where the cross-sectional area of the tube is s_1 and s_2 are respectively p_1 and p_2. A differential pressure guage connected between these points reads the pressure $p_1 - p_2$. Show that, ignoring gravity, the mass of fluid M flowing through the tube in unit time can be calculated from the formula

$$M = s_1 s_2 \sqrt{\frac{2\rho(p_1 - p_2)}{s_1^2 - s_2^2}}.$$

2.2 Electrostatics

There are two fundamental kinds of electric charges called positive and negative. From experimental observations it is found that like charges repel one another whereas unlike charges attract each other. Electrostatics is the study of the properties of such charges at rest.

2.21 *Electrostatic field and potential*

Coulomb's law states that the magnitude of the mechanical force exerted between two point charges is directly proportional to the product of the magnitudes of the charges and inversely proportional to the square of their distance apart and has a direction along the line joining the charges.

Thus the magnitude of the force exerted between two point charges of magnitudes e_1 and e_2 a distance r apart in free space* or a vacuum has the form

$$\frac{|e_1 e_2|}{4\pi\varepsilon_0 r^2}$$

in which, for historical reasons, the constant of proportionality has been written as $1/4\pi\varepsilon_0$. The constant ε_0 is called the *permittivity of free space* and its value is such that

$$\frac{1}{4\pi\varepsilon_0} = c^2 \times 10^{-7}$$

where c is the speed of light measured in appropriate units. Using the international system (S.I.) units, charge is measured in coulombs, distance in metres and force in newtons. Then $c \simeq 3 \times 10^8$ metres/second and so

$$\frac{1}{4\pi\varepsilon_0} \simeq 9 \times 10^9.$$

The permittivity ε_0 has dimension

$$(\text{coulombs})^2(\text{newtons})^{-1}(\text{metres})^{-2}.$$

Taking e_2 as unity, we say that there is associated with the charge e_1 a field of force

$$\mathbf{E} = \frac{e_1}{4\pi\varepsilon_0 r^2}\,\hat{\mathbf{r}}$$

where $\hat{\mathbf{r}}$ is unit vector in the direction of $\mathbf{r}$ the vector to any point from e_1 as origin.

Note that if e_1 is positive, then at any point, this gives a force directed away from the origin, that is a repulsion, and that if e_1 is negative, then the force is directed towards the origin, that is an attraction.

$\mathbf{E}$ is called the *electric intensity vector* or *field strength vector* and represents, at any point, the force which would be exerted on unit positive charge placed at that point.

$$\operatorname{curl} \mathbf{E} \equiv 0 \text{ and so } \mathbf{E} \text{ is a conservative field of force.}$$

Suppose now that a unit positive charge moves along any curve from P with position vector $\mathbf{r}_1$ to Q with position vector $\mathbf{r}_2$ in the field $\mathbf{E}$. (figure 40).

* The effect of placing the above charges in different media will be considered later.

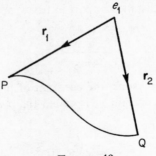

The work done by the field is

$$\int_{r_1}^{r_2} \mathbf{E} \cdot d\mathbf{l}\dagger$$

and is independent of the particular curve PQ since $\mathbf{E}$ is conservative. In terms of spherical polars $d\mathbf{l} = (dr, r\, d\theta, r \sin\theta\, d\psi)$ and so

$$\int_{r_1}^{r_2} \mathbf{E} \cdot d\mathbf{l} = \int_{r_1}^{r_2} \frac{e_1}{4\pi\varepsilon_0 r^2}\, dr$$

$$= -\frac{e_1}{4\pi\varepsilon_0}\left(\frac{1}{r_2} - \frac{1}{r_1}\right).$$

Now letting $r_1 \to \infty$ in the above we see that the work done by the field in bringing unit positive charge from infinity to the point Q is $-e_1/4\pi\varepsilon_0 r_2$. In other words, the work done against the field in bringing unit positive charge from infinity to Q is $e_1/4\pi\varepsilon_0 r_2$. This quantity is called the electrostatic potential at Q due to the point charge e_1. In general, the electrostatic potential φ due to the point charge e, at any point distant r from the charge, is $e/4\pi\varepsilon_0 r$.

† Intuitively, this can be seen as follows:

If $\delta\mathbf{l}$ is an element of L sufficiently small to be regarded as a straight line on which $\mathbf{E}$ is constant, then the component of the force $\mathbf{E}$ in the direction of the element $\delta\mathbf{l}$ is $\mathbf{E} \cdot \delta\mathbf{l}/|\delta\mathbf{l}|$. Hence the work done by this force as its point of application moves in the positive sense from one end of $\delta\mathbf{l}$ to the other is

$$\frac{\mathbf{E} \cdot \delta\mathbf{l}}{|\delta\mathbf{l}|}|\delta\mathbf{l}| = \mathbf{E} \cdot \delta\mathbf{l}.$$

Then, summing over all such elements $\delta\mathbf{l}$ of L and taking the limit as their magnitudes all tend to zero, we obtain for the work done along the path L, $\int_{L} \mathbf{E} \cdot d\mathbf{l}$.

Therefore

$$\varphi = \frac{e}{4\pi\varepsilon_0 r}$$

and so

$$\text{grad } \varphi = -\frac{e}{4\pi\varepsilon_0 r^2}\mathbf{r} = -\mathbf{E}.$$

Hence the electrostatic potential, introduced physically as above, is simply a scalar potential for the electric intensity vector. From the above we see that a positive point charge in electrostatics corresponds to a source in hydrodynamics and a negative point charge corresponds to a sink.

For a number of discrete charges e_1, e_2, e_3, ... the potential is given by

$$\varphi = \frac{1}{4\pi\varepsilon_0}\sum_i \frac{e_i}{r_i}$$

where r_i is the distance of the general point from the charge e_i. The potential becomes infinite at a point charge.

Now starting from any point P in the field we can construct a line PQ to represent the field strength at P in magnitude and direction and then construct a similar line QR at Q and so on. As the lengths of the segments tend to zero the figure approaches a smooth curve having the property that the tangent at any point is parallel to the field-strength vector at that point. This curve is called a *field line* or *line of force*. Thus we see that the field lines or lines of force in electrostatics correspond to the streamlines in hydrodynamics and are found in exactly the same way. Since we are only concerned with charges at rest, all the fields considered are steady and so there is no need to consider anything comparable with trajectories in hydrodynamics. When the electric intensity **E** has constant magnitude, direction and sense, the field is said to be *uniform*. In this case, the field lines will all be parallel straight lines.

The field lines for a positive charge are radially outwards from the charge and the field lines for a negative charge are radially inwards towards the charge.

Example   *Determine the form of the surface of zero potential corresponding to two point charges of opposite sign fixed in space.*

Choose coordinate axes such that the charges are e_1 and $e_2(e_1, e_2 > 0)$ situated at the points A$(-a,0,0)$ and B$(a,0,0)$ respectively where a is a constant. In other words, choose the x-axis along the line joining the charges and let their distance apart be $2a$ (figure 41).

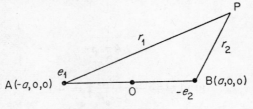

If the point $P(x,y,z)$ lies on the zero equipotential, then

$$\frac{e_1}{r_1} - \frac{e_2}{r_2} = 0$$

where r_1 and r_2 are the lengths of AP and BP respectively.

Therefore

$$e_1 r_2 = e_2 r_1$$

and so

$$e_1^2 r_2^2 = e_2^2 r_1^2.$$

Hence

$$e_1^2((x - a)^2 + y^2 + z^2) = e_2^2((x + a)^2 + y^2 + z^2)$$

and therefore

$$(e_1^2 - e_2^2)(x^2 + y^2 + z^2) - 2ax(e_1^2 + e_2^2) + a^2(e_1^2 - e_2^2) = 0.$$

Now if $e_1 \neq e_2$ this is the equation of a sphere with centre at the point

$$(a(e_1^2 + e_2^2)/(e_1^2 - e_2^2),0,0)$$

on the x-axis, that is a sphere with centre on the line joining the charges.

If $e_1 = e_2$ then the above equation reduces to $x = 0$. That is the zero equipotential is the plane perpendicular to the line joining the charges and through its midpoint.

Example Charges $-5e$, $-\frac{9}{5}e$, $\frac{4}{5}e$ *are fixed at the points* $A(-2a,0,0)$, $B(2a,0,0)$ *and* $C(0,3a,0)$ *respectively in free space. Determine the force field at the origin* O *and at the point* $P(2a,3a,0)$ *(figure 42)*.

The force field at O is composed of

$$\frac{1}{4\pi\varepsilon_0} \frac{9e/5}{(2a)^2} \text{ along OB,}$$

$$\frac{1}{4\pi\varepsilon_0} \frac{5e}{(2a)^2} \text{ along OA}$$

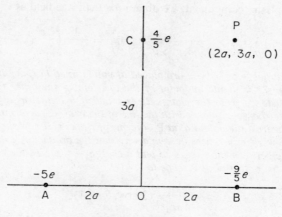

FIGURE 42

and

$$\frac{1}{4\pi\varepsilon_0} \frac{4e/5}{(3a)^2} \text{ along CO.}$$

This is

$$\frac{1}{4\pi\varepsilon_0}\left(\frac{9e}{20a^2} - \frac{5e}{4a^2}\right)$$

in the positive x-direction and

$$\frac{-e}{45\pi a^2 \varepsilon_0}$$

in the positive y-direction. Thus the force at O is

$$\frac{1}{4\pi\varepsilon_0}\left(-\frac{4e}{5a^2}, -\frac{4e}{45a^2}, 0\right).$$

The force field at P is composed of

$$\frac{1}{4\pi\varepsilon_0} \frac{9e/5}{(3a)^2} \text{ along PB,}$$

$$\frac{1}{4\pi\varepsilon_0} \frac{4e/5}{(2a)^2} \text{ along CP}$$

and

$$\frac{1}{4\pi\varepsilon_0} \frac{5e}{(16a^2 + 9a^2)} \text{ along PA.}$$

Resolving this latter component, we obtain for the force field at P

$$\frac{1}{4\pi\varepsilon_0}\left(\frac{e}{5a^2} - \frac{e}{5a^2}\frac{4}{5}, \ -\frac{e}{5a^2} - \frac{e}{5a^2}\frac{3}{5}, \ 0\right) = \frac{1}{4\pi\varepsilon_0}\left(\frac{e}{25a^2}, \ -\frac{8e}{25a^2}, 0\right).$$

Example   Charges $-e$, $2e$, $-e$ are placed at points fixed in space in such a way that the charge $2e$ is at the midpoint of the line joining the other two. If P is a point whose distance from the charge $2e$ is r and if θ is the angle that the line joining P to the charge $2e$ makes with the line of the charges, determine an expression for the electrostatic potential at P in terms of r and θ. If now the charges tend to coincidence while e increases in such a way that the product of e and the square of the distance between the charge $2e$ and a charge $-e$ retains the constant value m, show that the potential at P tends to

$$-\frac{m}{4\pi\varepsilon_0 r^3}(3\cos^2\theta - 1)$$

and that the equations of the lines of force in this limit are

$$\frac{\cos\theta \sin^2\theta}{r^2} = \text{constant.}$$

Choose coordinate axes such that the line of charges is the z-axis and the charge $2e$ is at the origin O. Then r_1 and θ are the first two spherical polar coordinates of the point P.

Let the two charges $-e$ be at the points $A(0,0,-a)$ and $B(0,0,a)$ in terms of rectangular cartesian coordinates (figure 43). The potential φ at P is then

$$\frac{1}{4\pi\varepsilon_0}\left(-\frac{e}{\text{BP}} + \frac{2e}{\text{OP}} - \frac{e}{\text{AP}}\right)$$

$$= \frac{1}{4\pi\varepsilon_0}\left\{\frac{-e}{(r^2 + a^2 - 2ar\cos\theta)^{\frac{1}{2}}} + \frac{2e}{r} - \frac{e}{(r^2 + a^2 - 2ar\cos(\pi - \theta))^{\frac{1}{2}}}\right\}$$

$$= \frac{1}{4\pi\varepsilon_0}\left[-\frac{e}{r}\left\{1 + \left(\frac{a^2}{r^2} - \frac{2a}{r}\cos\theta\right)\right\}^{-\frac{1}{2}} + \frac{2e}{r} - \frac{e}{r}\left\{1 + \left(\frac{a^2}{r^2} + \frac{2a}{r}\cos\theta\right)\right\}^{-\frac{1}{2}}\right]$$

$$= \frac{1}{4\pi\varepsilon_0}\left[-\frac{e}{r}\left\{1 - \frac{1}{2}\left(\frac{a^2}{r^2} - \frac{2a}{r}\cos\theta\right) + \frac{3}{8}\left(\frac{a^2}{r^2} - \frac{2a}{r}\cos\theta\right)^2 + \ldots\right\} + \frac{2e}{r}\right.$$

$$\left.-\frac{e}{r}\left\{1 - \frac{1}{2}\left(\frac{a^2}{r^2} + \frac{2a}{r}\cos\theta\right) + \frac{3}{8}\left(\frac{a^2}{r^2} + \frac{2a}{r}\cos\theta\right)^2 + \ldots\right\}\right] \quad \text{for} \quad r > a.$$

$$= \frac{-e}{4\pi r\varepsilon_0}\left(-\frac{a^2}{r^2} + \frac{3a^2}{r^2}\cos^2\theta + \text{terms with higher powers of } a\right)$$

$$= \frac{-ea^2}{4\pi\varepsilon_0 r^3}(3\cos^2\theta - 1 + \text{terms with positive powers of } a).$$

Now let $a \to 0$ and $e \to \infty$ in such a way that $ea^2 = m$.

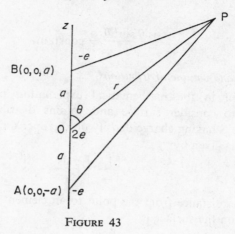

FIGURE 43

Then

$$\varphi \to \frac{-m}{4\pi\varepsilon_0 r^3}(3\cos^2\theta - 1).$$

Now the force field **E** is given by $\mathbf{E} = -\text{grad }\varphi$ and so in this limit, using spherical polar coordinates, we see that

$$\mathbf{E} = -\frac{1}{4\pi\varepsilon_0}\left(\frac{3m}{r^4}(3\cos^2\theta - 1), \frac{6m}{r^4}\cos\theta\sin\theta, 0\right).$$

Thus the lines of force are given by

$$\frac{dr}{-\dfrac{3m}{r^4}(3\cos^2\theta - 1)} = \frac{r\,d\theta}{-\dfrac{6m}{r^4}\cos\theta\sin\theta}.$$

Therefore

$$\frac{2\,dr}{r} = \frac{(3\cos^2\theta - 1)\,d\theta}{\cos\theta\sin\theta} = \frac{\sin\theta(2\cos^2\theta - \sin^2\theta)\,d\theta}{\cos\theta\sin^2\theta}$$

and so

$$2\ln r = \ln(\cos\theta\sin^2\theta) + \text{constant}$$

since

$$\frac{d}{d\theta}(\cos\theta\sin^2\theta) = \sin\theta(2\cos^2\theta - \sin^2\theta).$$

Therefore

$$\ln\left(\frac{\cos\theta\sin^2\theta}{r^2}\right) = \text{constant}$$

and so

$$\frac{\cos \theta \sin^2 \theta}{r^2} = \text{constant}.$$

2.22 Surface and volume distributions

It is useful in the mathematical explanation of certain physical phenomena to consider surface and volume distributions of charges. For a surface S having charge density $\sigma(x,y,z)$ per unit area the potential φ at a point is given by

$$\varphi = \frac{1}{4\pi\varepsilon_0} \int_S \frac{\sigma \, ds}{r}$$

where r is the distance from the point to an element of the surface. The total charge on the surface is

$$\int_S \sigma \, ds.$$

For a region V having charge density $\rho(x,y,z)$ per unit volume the potential φ at a point is given by

$$\varphi = \frac{1}{4\pi\varepsilon_0} \int_V \frac{\rho \, dv}{r}$$

where r is the distance from the point to an element of the region. The total charge within the region is

$$\int_V \rho \, dv.$$

Surface distributions of charge are often associated with *conductors*, that is with substances which permit the free passage of electric charges and within which the field vector $\mathbf{E} = \mathbf{0}$ everywhere.

Volume distributions of charge are often associated with *insulators* or *dielectrics*, that is with substances which do not allow the free passage of electric charges.

Example   Find the electrostatic potential of the surface distribution of charge of constant density σ per unit area on the surface of a sphere of radius a.

Consider P distant R from the centre O of the sphere and take O as origin of a system of spherical polar coordinates and OP as the z-axis. Then the surface S is the surface $r = \text{constant} = a$ and the potential φ at P is given by

$$\varphi = \frac{1}{4\pi\varepsilon_0} \int_S \frac{\sigma \, ds}{d}$$

where d is the distance of the element of the surface S from the point P. Therefore

$$\varphi = \frac{1}{4\pi\varepsilon_0} \int_0^{2\pi} \int_0^\pi \frac{\sigma a^2 \sin\theta \, d\theta \, d\psi}{\sqrt{a^2 + R^2 - 2aR\cos\theta}}$$

$$\left(= \frac{1}{4\pi\varepsilon_0} \iint \frac{\sigma h_2 h_3 \, du_2 \, du_3}{d} \text{ in the general notation}\right).$$

Therefore

$$\varphi = \frac{a^2\sigma}{2\varepsilon_0} \cdot \frac{1}{2aR} \int_0 \frac{2aR\sin\theta \, d\theta}{\sqrt{a^2 + R^2 - 2aR\cos\theta}}$$

$$= \frac{a\sigma}{4R\varepsilon_0} \left[2\sqrt{a^2 + R^2 - 2aR\cos\theta} \right]_0^\pi$$

$$= \begin{cases} \dfrac{a\sigma}{2R\varepsilon_0} \{(R+a) - (R-a)\}, & R > a \\[3mm] \dfrac{a\sigma}{2R\varepsilon_0} \{(R+a) - (a-R)\}, & R < a. \end{cases}$$

The different forms of the solution are obtained by taking the positive value of the square root in both cases.

Hence

$$\varphi = \begin{cases} \dfrac{a^2\sigma}{R\varepsilon_0} \\[3mm] \dfrac{a\sigma}{\varepsilon_0} \end{cases} = \begin{cases} \dfrac{1}{4\pi\varepsilon_0}\left(\dfrac{Q}{R}\right) & \text{for P outside the sphere} \\[3mm] \dfrac{1}{4\pi\varepsilon_0}\left(\dfrac{Q}{a}\right) & \text{for P inside the sphere} \end{cases}$$

where $Q = 4\pi a^2 \sigma$ is the total charge on the surface of the sphere.

Alternatively, divide the surface of the sphere into thin rings as shown in figure 44. Then, in the notation illustrated, if α is small all points of the ring are approximately equidistant from P, this distance being $(a^2 + R^2 - 2aR\cos\theta)^{\frac{1}{2}}$. The total charge on the ring is σ times the area of the ring, that is approximately $\sigma(2\pi a \sin\theta \, a\alpha)$ for small α.

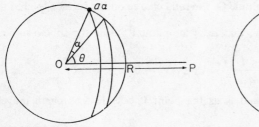

P Outside the sphere (ie. R > a)

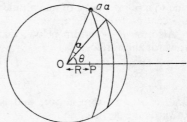

P Inside the sphere (ie. R < a)

FIGURE 44

Hence the potential at P, due to the ring shown is approximately

$$\frac{1}{4\pi\varepsilon_0} \frac{\sigma 2\pi a^2 \sin\theta \, . \, \alpha}{\sqrt{a^2 + R^2 - 2aR\cos\theta}}$$

for small α.

Then the potential at P due to the whole surface distribution is the sum of the potentials due to all such rings which make up the surface of the sphere. In the limit as $\alpha \to 0$ this sum becomes

$$\frac{1}{4\pi\varepsilon_0} \int_0^\pi \frac{\sigma 2\pi a^2 \sin\theta \, d\theta}{\sqrt{a^2 + R^2 - 2aR\cos\theta}} = \frac{a^2\sigma}{2\varepsilon_0} \int_0^\pi \frac{\sin\theta \, d\theta}{\sqrt{a^2 + R^2 - 2aR\cos\theta}}$$

which is the same as obtained previously. Hence, as before,

$$\varphi = \begin{cases} \dfrac{1}{4\pi\varepsilon_0}\left(\dfrac{Q}{R}\right) & \text{for P outside the sphere,} \\[3mm] \dfrac{1}{4\pi\varepsilon_0}\left(\dfrac{Q}{a}\right) & \text{for P inside the sphere.} \end{cases}$$

Thus we see that the potential is constant at all points inside the surface of the sphere.

Hence $\mathbf{E} = -\text{grad } \varphi = \mathbf{0}$ everywhere inside the sphere. This is the condition which applies inside a conductor and so the above expressions may be considered as giving the potential due to a charged solid spherical conductor.

Outside the sphere

$$\mathbf{E} = -\text{grad } \varphi = -\frac{1}{4\pi\varepsilon_0} \frac{\partial}{\partial R}\left(\frac{Q}{R}\right)\hat{\mathbf{R}}$$

(where O is taken as the origin of a system of spherical polar coordinates).

Therefore

$$\mathbf{E} = \frac{1}{4\pi\varepsilon_0} \frac{Q}{R^2}\hat{\mathbf{R}}.$$

Thus outside the sphere the potential and force fields are identical with those due to a point charge of magnitude Q, the total charge on the surface, placed at the centre of the sphere.

Also, as $R \to a$ from below, that is as the point P inside the sphere moves towards the surface,

$$\varphi \to \frac{1}{4\pi\varepsilon_0} \frac{Q}{a} \quad \text{and} \quad \mathbf{E} \to \mathbf{0}$$

and as $R \to a$ from above, that is as the point P outside the sphere moves towards the surface,

$$\varphi \to \frac{1}{4\pi\varepsilon_0} \frac{Q}{a} \quad \text{and} \quad \mathbf{E} \to \frac{1}{4\pi\varepsilon_0}\left(\frac{Q}{a^2}, 0, 0\right).$$

Thus on the surface of the sphere the potential φ is continuous and the component of force normal to the surface (that is $-\partial\varphi/\partial r$) is discontinuous by amount $\{1/(4\pi\varepsilon_0)\}Q/a^2 = \sigma/\varepsilon_0$. That is

$$\lim_{r \to a} \left(\frac{\partial\varphi_1}{\partial r} - \frac{\partial\varphi_2}{\partial r} \right) = \frac{\sigma}{\varepsilon_0}$$

if $\varphi = \varphi_1$ for $r < a$ and $\varphi = \varphi_2$ for $r > a$. We shall see presently that this is the case in general for a surface distribution. That is the derivative $\partial\varphi/\partial n$ in the direction $\mathbf{n}$ normal to the surface decreases by σ/ε_0 on passing through a surface distribution of density σ in the positive direction of the normal, (that is in the direction of increasing n).

2.23 Solid angle

The solid angle ω subtended at a point P by a surface S is defined by

$$\omega = \int_S \frac{\mathbf{r} \cdot \mathbf{ds}}{r^3}$$

where $\mathbf{r}$ is the vector from P to the general element of the surface S, and the positive side of the surface S is chosen so that $\omega \geqslant 0$.

For surfaces which are not closed, the solid angle depends only on the bounding curve of the surface and is otherwise independent of the shape of the surface itself. This result is easily appreciated by considering the following argument. Let S be a given surface (not closed). Form a closed surface not including the point P by constructing any other surface S_1 having the same bounding curve as S. Now apply Gauss' theorem to the region V enclosed between the surfaces S and S_1.

Therefore

$$\int_{S+S_1} \frac{\mathbf{r} \cdot \mathbf{ds}}{r^3} = \int_V \text{div} \left(\frac{\mathbf{r}}{r^3} \right) dv = 0$$

since $\text{div } \mathbf{r}/r^3 \equiv 0$ at all points except P and P is not in V.

Therefore

$$\int_S \frac{\mathbf{r} \cdot \mathbf{ds}}{r^3} + \int_{S_1} \frac{\mathbf{r} \cdot \mathbf{ds}}{r^3} = 0$$

Now one of the above terms must be negative and so equal to minus the appropriate solid angle.

Hence the solid angle subtended by S = the solid angle subtended by S_1.

Now consider the solid angle subtended by a *closed* surface S. If region V enclosed by S does not include the point P, then the solid angle subtended at P by S is

$$\int_S \frac{\mathbf{r} \cdot \mathbf{ds}}{r^3} = \int_V \text{div} \left(\frac{\mathbf{r}}{r^3} \right) dv = 0.$$

Thus the solid angle subtended by any closed surface at a point outside it is zero.

If now the region V enclosed by the surface S includes the point P, surround P also by any other closed surface S_1. Then, as above, since div $(\mathbf{r}/r^3)$ is zero at all points of the region between the surfaces S and S_1 we obtain that the solid angle subtended by S = the solid angle subtended by S_1. But since S_1 is *any* closed surface enclosing the point P, we can in particular choose it to be the sphere with centre P and radius a. Then the solid angle subtended by S is equal to

$$\int_{S_1} \frac{\mathbf{r} \cdot d\mathbf{s}}{r^3} = \int_{S_1} \frac{a\, ds}{a^3} = \frac{1}{a^2} \int_{S_1} ds = 4\pi,$$

since the surface area of the sphere S_1 is $4\pi a^2$.

Thus the solid angle subtended by any closed surface at a point inside it is 4π.

It can also be easily deduced that the solid angle subtended at P by a surface S is equal to the solid angle subtended at P by the radial projection of S (regarding P as origin) onto the unit sphere with centre P.

Example   Determine the solid angle subtended by a circular disc at a point P *on its axis.*

Let the disc have centre O and radius a. Use cylindrical polar coordinates with origin at P and with PO as z-axis. In this system $\mathbf{r}$, the position vector of a general point on the disc, is

$$\mathbf{r} = (\rho, \psi, c)$$

where c is the length of PO, $0 \leqslant \rho \leqslant a$ and $0 \leqslant \psi < 2\pi$. Also

$$r = |\mathbf{r}| = (\rho^2 + c^2)^{\frac{1}{2}}$$

and the vector element of area on the surface S of the disc is $d\mathbf{s} = (0, 0, \rho d\rho\, d\psi)$ since on S the coordinate z is constant. Therefore

$$\omega = \int_S \frac{\mathbf{r} \cdot d\mathbf{s}}{r^3} = \int_0^{2\pi} \int_0^a \frac{c\rho\, d\rho\, d\psi}{(\rho^2 + c^2)^{\frac{3}{2}}}$$

$$= 2\pi c \int_0^a \frac{\rho\, d\rho}{(\rho^2 + c^2)^{\frac{3}{2}}}$$

$$= \pi c \left(\frac{2}{c} - \frac{2}{(a^2 + c^2)^{\frac{1}{2}}} \right)$$

$$= 2\pi (1 - \cos \alpha).$$

where 2α is the angle of the cone formed by joining P to the points of the circumference of the disc (figure 45).

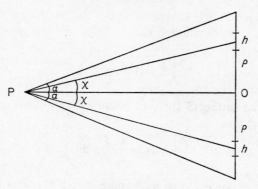

FIGURE 45

Alternatively, divide the disc into thin rings concentric with the disc. Consider the ring which has mean radius ρ and thickness h. If h is small, the area of this ring is approximately $2\pi\rho h$ and the radius vector from P to any point of the ring has approximate magnitude $\sqrt{\rho^2 + c^2}$ and makes a constant angle χ with the normal to the surface of the ring. Hence, if h is small, the solid angle subtended by this ring is approximately

$$\frac{\sqrt{\rho^2 + c^2} \cos\chi\, 2\pi\rho h}{(\rho^2 + c^2)^{\frac{3}{2}}} = \frac{2\pi\rho h}{(\rho^2 + c^2)} \frac{c}{\sqrt{\rho^2 + c^2}}.$$

Then the solid angle subtended at P by the whole disc is the sum of the solid angles due to all such rings which make up the disc. In the limit as $h \to 0$ this sum becomes

$$\int_0^a \frac{2\pi\rho c\, d\rho}{(\rho^2 + c^2)^{\frac{3}{2}}}$$

which is the same result as obtained previously.

2.24 Gauss' law

Consider any closed surface S in space. If this surface encloses a charge e, then the field vector at any point on S having position vector $\mathbf{r}$ relative to the charge e is

$$\mathbf{E} = \frac{e\mathbf{r}}{4\pi\varepsilon_0 r^3}.$$

Therefore the flux (c.f. volume outflow in hydrodynamics) across the surface S is

$$\int_S \mathbf{E} \cdot d\mathbf{s} = \int_S \frac{e\mathbf{r} \cdot d\mathbf{s}}{4\pi\varepsilon_0 r^3}$$

$$= \frac{e\omega}{4\pi\varepsilon_0} = \frac{e}{\varepsilon_0}$$

where ω is the solid angle subtended at the charge. If a number of discrete charges e_1, e_2, ... are enclosed, then

$$\int_S \mathbf{E} \cdot d\mathbf{s} = \frac{1}{\varepsilon_0} \sum_i e_i.$$

Similarly, if the surface S encloses volume distributions of charge, with density $\rho(x,y,z)$ then

$$\int_S \mathbf{E} \cdot d\mathbf{s} = \frac{1}{\varepsilon_0} \int_V \rho \, dv$$

where V is the region enclosed by S and ρ is zero at points within V where there are no volume distributions. Then, using Gauss' theorem in the last equation, we obtain

$$\int_V (\operatorname{div} \mathbf{E}) \, dv = \int_V \frac{\rho}{\varepsilon_0} \, dv.$$

Since this result holds for any region V in space, we must have

$$\operatorname{div} \mathbf{E} = \frac{\rho}{\varepsilon_0}$$

everywhere in space. This result is known as *Gauss' law*. Now since $\mathbf{E} = -\operatorname{grad} \varphi$, we have

$$\nabla^2 \varphi = -\frac{\rho}{\varepsilon_0}$$

which is *Poisson's equation*. If $\rho = 0$ (that is there are no charge distributions present), then we have

$$\nabla^2 \varphi = 0$$

which is Laplace's equation. Thus Laplace's equation holds for all regions in which there are no charges present.

Example   Use Gauss' law to determine the electrostatic field due to a uniformly charged hollow sphere.

Consider the point P distant r from the centre of the sphere. Construct the sphere through P concentric with the given sphere. By symmetry the field strength will have the same magnitude at all points on the surface S of this sphere. Also the field strength vector will be normal to the surface S at all points on it.

Using Gauss' law we have

$$\int_S \mathbf{E} \cdot d\mathbf{s} = \begin{cases} \dfrac{Q}{\varepsilon_0} & \text{if P is outside the sphere, and } Q \text{ is the total} \\ & \quad \text{charge on the sphere.} \\ 0 & \text{if P is inside the sphere.} \end{cases}$$

But

$$\int_S \mathbf{E} \cdot d\mathbf{s} = \int_S E \, ds = E \int_S ds = 4\pi r^2 E.$$

Hence

$$E = \begin{cases} \dfrac{Q}{4\pi\varepsilon_0 r^2} & \text{for P outside the sphere} \\ 0 & \text{for P inside the sphere.} \end{cases}$$

Therefore

$$\mathbf{E} = \begin{cases} \dfrac{Q}{4\pi\varepsilon_0 r^3}\,\mathbf{r} & \text{for P outside the sphere} \\ \mathbf{0} & \text{for P inside the sphere.} \end{cases}$$

(These results are the same as were obtained in a previous example on p. 118, section 2.22).

2.25 Collinear point charges

Suppose charges $e_1, e_2, \ldots, e_n$ are fixed at points $P_1, P_2, \ldots, P_n$ respectively lying on a straight line.

Let XY (figure 46) be a segment of a line of force.

Now the field must be symmetric about the line of charges, so that the surface S obtained by rotating XY about this line will be such that no lines of force will cross it.

Thus

$$\int_S \mathbf{E} \cdot d\mathbf{s} = 0.$$

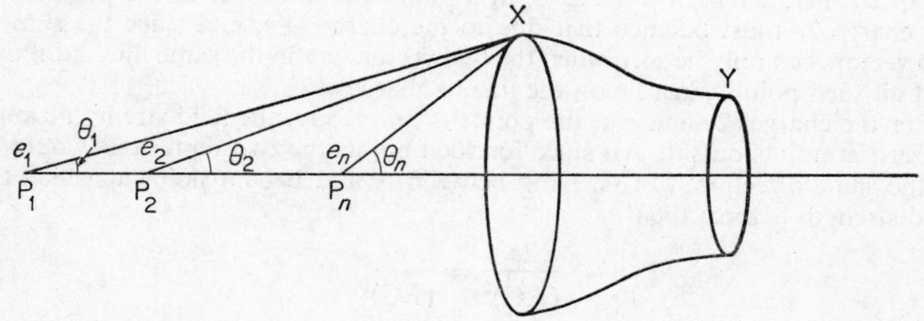

FIGURE 46

Then if there are no charges within the region enclosed by S and the plane discs at X and Y, Gauss' law gives that the integral of **E** over this whole surface must be zero.

Hence the flux across the plane disc at X must be equal to the flux across the plane disc at Y. But the flux across the disc at X is

$$\frac{1}{4\pi\varepsilon_0}(e_1\omega_1 + e_2\omega_2 + \ldots + e_n\omega_n) = \frac{1}{4\pi\varepsilon_0}\sum_{i=1}^{n}e_i\omega_i$$

where ω_i is the solid angle subtended by the disc at the charge e_i (for $i = 1, 2, \ldots, n$). Also $\omega_i = 2\pi(1 - \cos\theta_i)$ for $i = 1, 2, \ldots, n$ where θ_i is the angle between the line of charges and any line joining the point P_i to a point of the circumference of the disc.

Therefore the quantity $\sum_i \dfrac{e_i}{2\varepsilon_0}(1 - \cos\theta_i)$ must have the same value for the disc at X as for the disc at Y.

Thus $\sum_i \dfrac{e_i}{2\varepsilon_0}(1 - \cos\theta_i)$ must be constant along a line of force. But $\sum_i \dfrac{e_i}{2\varepsilon_0}$ is obviously a constant and so we must have

$$\sum_i e_i\cos\theta_i = \text{constant}$$

on a line of force. Thus for different values of the constant this equation represents a line of force.

Example   Show that for systems consisting of a charge 2e at the origin and either ±e at some other point there is only one point at which the field strength is zero. Show also that with the charges 2e and −e, the extreme lines of force which pass from the charge 2e to the charge −e leave the charge 2e at right angles to the line joining the charges. Sketch the lines of force for both systems.

For the field strength to be zero, at a point, the force due to the presence of the charge 2e must balance that due to the charge ±e, and since the sum of two vectors can only be zero when the two vectors are in the same line, it follows that all such points must lie on the line of charges.

For the charges 2e and e at the points A and B say, the field strength cannot be zero at points outside AB since for such points the two repulsions would be in the same direction. Let C, lying between A and B be a point at which the field strength is zero, then

$$\frac{2e}{(AC)^2} = \frac{e}{(BC)^2}$$

and so

$$AC = \sqrt{2}\,BC.$$

This then gives the only point at which the field strength is zero.

The lines of force have the equations

$$2e \cos \theta_1 + e \cos \theta_2 = k$$

for different constants k, where θ_1 and θ_2 are as previously defined. If AB is taken as x-axis and A as origin, then the equations of the lines of force in any plane Oxy through Ox are

$$2e \frac{x}{\sqrt{x^2 + y^2}} + e \frac{(x - a)}{\sqrt{(x - a)^2 + y^2}} = k$$

for different constants k, where a is the length of AB. These curves are sketched in figure 47.

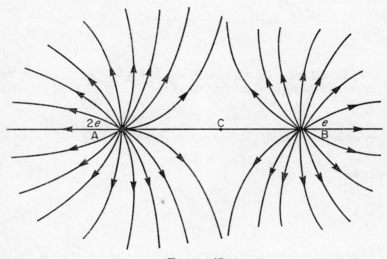

FIGURE 47

For the charges $2e$ and $-e$ the field strength cannot be zero at points between A and B. Thus if the field strength is zero at D then D lies outside AB and once again

$$\frac{2e}{(AD)^2} = \frac{e}{(BD)^2}.$$

Thus AD must be greater than BD. That is D lies on AB produced and is such that

$$AD = \sqrt{2}\, BD.$$

The equations of the lines of force are

$$2e \cos \theta_1 - e \cos \theta_2 = k$$

or

$$2 \cos \theta_1 - \cos \theta_2 = c \quad \left(c = \frac{k}{e}\right).$$

If a line of force starts at A and finishes at B, then as it approaches B, $\theta_1 \to 0$ (figure 48) and so we obtain in the limit

$$2 - \cos \theta_2 = c.$$

Therefore

$$\cos \theta_2 = 2 - c$$

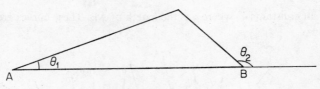

FIGURE 48

and so $1 \leqslant c \leqslant 3$. Now on AB itself $\theta_1 = 0$ and $\theta_2 = \pi$ and so $c = 3$. Thus $c = 1$ must give the required extreme line of force.

Therefore

$$2 \cos \theta_1 - \cos \theta_2 = 1.$$

Now as we approach A, $\theta_2 \to \pi$ and so we obtain in the limit

$$2 \cos \theta_1 + 1 = 1.$$

Therefore

and so

$$\cos \theta_1 = 0$$

$$\theta_1 = \frac{\pi}{2}.$$

That is the line of force leaves A at right angles to AB. As above, the lines of force in a plane Oxy through Ox, the line of charges, are given by

$$\frac{2ex}{\sqrt{x^2 + y^2}} - \frac{e(x - a)}{\sqrt{(x - a)^2 + y^2}} = k.$$

These curves are sketched in figure 49.

Gauss' law can be simply applied to show that at any surface distribution of charge the component of the field strength vector **E** normal to the surface is discontinuous by an amount σ/ε_0 where σ is the charge density.

Consider the small right circular cylinder with plane ends of area s enclosing an element of area s of the charged surface. (This is only possible if the area s is small enough for the element of charged surface

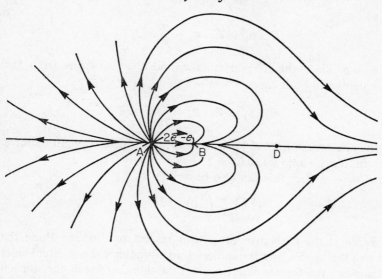

FIGURE 49

enclosed by the cylinder to be regarded as plane.) The ends of the cylinder are both parallel to the tangent plane to the surface at the element, one just on each side of the charged surface (figure 50), so that the area of the curved surface of the cylinder is negligible compared with s.

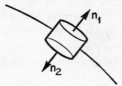

FIGURE 50

Applying Gauss' law to the cylinder, we obtain

$$\frac{\sigma s}{\varepsilon_0} = \mathbf{E}_1 \cdot \mathbf{n}_1 s + \mathbf{E}_2 \cdot \mathbf{n}_2 s + \text{smaller terms},$$

where $\mathbf{E}_1$ and $\mathbf{E}_2$ are the values of $\mathbf{E}$ on the ends of the cylinder and $\mathbf{n}_1$ and $\mathbf{n}_2$ are the unit normals to the ends.

It can be shown that these 'smaller terms' all have the common factor s^2. Thus dividing the last equation by s and then taking the limit as $s \to 0$

we obtain

$$\mathbf{E_1} \cdot \mathbf{n_1} + \mathbf{E_2} \cdot \mathbf{n_2} = \frac{\sigma}{\varepsilon_0}.$$

But $\mathbf{n_2} = -\mathbf{n_1}$ since the normal is taken positive outwards from the volume enclosed in both cases. Therefore

$$\mathbf{E_1} \cdot \mathbf{n_1} - \mathbf{E_2} \cdot \mathbf{n_1} = \frac{\sigma}{\varepsilon_0}.$$

That is the component normal to the surface, of the field strength vector $\mathbf{E}$ is discontinuous by amount σ/ε_0.

Since $\mathbf{E} = -\text{grad } \varphi$, this can also be written as

$$\left(\frac{\partial \varphi}{\partial n}\right)_2 - \left(\frac{\partial \varphi}{\partial n}\right)_1 = \frac{\sigma}{\varepsilon_0}$$

where $\partial\varphi/\partial n$ is the derivative of φ with respect to distance along the normal $\mathbf{n}$ to the surface and the suffices 1 and 2 indicate the different sides of the surface, the former referring to that side for which the normal vector is directed away from the surface. That is $\partial\varphi/\partial n$ decreases by amount σ/ε_0 on passing through the surface in the direction of increasing n.

2.26 Dipoles

Consider charges $+e$ and $-e$ situated a small distance $2a$ apart. Now let $a \to 0$ and $e \to \infty$ such that $2ae \to m$. In the limit we have a *dipole of vector moment* $\mathbf{m}$ where $\mathbf{m}$ has magnitude m, direction along the line joining the charges, and sense from $-e$ to $+e$. In hydrodynamics if the above process is carried out for a source and a sink (both having the same strength) then we obtain a *hydrodynamic doublet*.

The dipole and the doublet are essentially theoretical concepts which enable the potential and force (or velocity) fields to be determined in certain practical situations.

Let O be the midpoint of the line joining the charges $\pm e$ and consider a point P, distant r from O, and such that OP makes an angle θ with this line. Suppose that P is distant r_1 from the charge $+e$ and distant r_2 from the charge $-e$ (figure 51). Then the potential at P due to these charges is

$$\frac{1}{4\pi\varepsilon_0}\left(\frac{e}{r_1} - \frac{e}{r_2}\right)$$

$$= \frac{1}{4\pi\varepsilon_0}\left(\frac{e}{\sqrt{r^2 + a^2 - 2ar\cos\theta}} - \frac{e}{\sqrt{r^2 + a^2 - 2ar\cos(\pi - \theta)}}\right)$$

$$= \frac{1}{4\pi\varepsilon_0}\left[\frac{e}{r}\left\{1 - \left(\frac{2a}{r}\cos\theta - \frac{a^2}{r^2}\right)\right\}^{-\frac{1}{2}} - \frac{e}{r}\left\{1 + \left(\frac{2a}{r}\cos\theta + \frac{a^2}{r^2}\right)\right\}^{-\frac{1}{2}}\right].$$

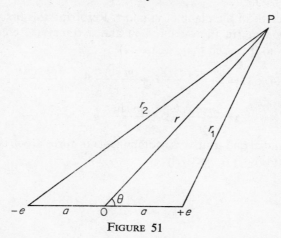

FIGURE 51

Now applying the binomial theorem to this expression we see that the potential is

$$\frac{1}{4\pi\varepsilon_0}\left[\frac{e}{r}\left\{1+\frac{1}{2}\left(\frac{2a}{r}\cos\theta-\frac{a^2}{r^2}\right)+\ldots\right\}-\frac{e}{r}\left\{1-\frac{1}{2}\left(\frac{2a}{r}\cos\theta+\frac{a^2}{r^2}\right)+\ldots\right\}\right]$$

$$=\frac{e}{4\pi\varepsilon_0 r}\left(\frac{2a}{r}\cos\theta+\text{terms with }a^2\text{ as a factor}\right)$$

$$=\frac{ae}{2\pi\varepsilon_0 r}\left(\frac{\cos\theta}{r}+\text{terms with }a\text{ as a factor}\right)$$

Now let $a\to 0$, $e\to\infty$ such that $2ae\to m$. Thus we obtain that the potential φ at P due to the dipole is

$$\frac{m\cos\theta}{4\pi\varepsilon_0 r^2}=\frac{\mathbf{m}\cdot\mathbf{r}}{4\pi\varepsilon_0 r^3}\quad(\mathbf{r}\text{ having sense }\textit{from}\text{ the dipole }\textit{to}\text{ the point P})$$

$$=-\frac{1}{4\pi\varepsilon_0}\mathbf{m}\cdot\nabla\left(\frac{1}{r}\right).$$

Then the field intensity $\mathbf{E}$ at P is given by

$$\mathbf{E}=-\text{grad }\varphi=-\frac{1}{4\pi\varepsilon_0}\text{grad}\left(\frac{\mathbf{m}\cdot\mathbf{r}}{r^3}\right)$$

$$=-\frac{1}{4\pi\varepsilon_0}\left\{\frac{1}{r^3}\text{grad}\,(\mathbf{m}\cdot\mathbf{r})+(\mathbf{m}\cdot\mathbf{r})\text{grad}\left(\frac{1}{r^3}\right)\right\}$$

$$=-\frac{1}{4\pi\varepsilon_0}\left\{\frac{\mathbf{m}}{r^3}+(\mathbf{m}\cdot\mathbf{r})\left(-\frac{3}{r^4}\hat{\mathbf{r}}\right)\right\}$$

$$=\frac{1}{4\pi\varepsilon_0}\left\{-\frac{\mathbf{m}}{r^3}+\frac{3(\mathbf{m}\cdot\mathbf{r})}{r^5}\mathbf{r}\right\}$$

which is a vector in the plane of **m** and **r**. Resolving this first term into a radial component (in the direction $\hat{\mathbf{r}}$) and a transverse component (in the direction $\hat{\mathbf{n}}$) it is seen that (figure 52)

$$4\pi\varepsilon_0 \mathbf{E} = -\frac{m\cos\theta}{r^3}\hat{\mathbf{r}} + \frac{m\sin\theta}{r^3}\hat{\mathbf{n}} + \frac{3m\cos\theta}{r^3}\hat{\mathbf{r}}$$

$$= \frac{2m\cos\theta}{r^3}\hat{\mathbf{r}} + \frac{m\sin\theta}{r^3}\hat{\mathbf{n}}.$$

That is the radial and transverse components of force are $m\cos\theta/(2\pi\varepsilon_0 r^3)$ and $m\sin\theta/(4\pi\varepsilon_0 r^3)$ respectively.

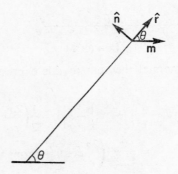

FIGURE 52

This can also be obtained by considering

$$\mathbf{E} = -\operatorname{grad}\varphi = -\frac{1}{4\pi\varepsilon_0}\operatorname{grad}\left(\frac{m\cos\theta}{r^2}\right)$$

and using spherical polar coordinates in the manipulations.

Thus the radial component of **E** is

$$-\frac{\partial\varphi}{\partial r} = \frac{m\cos\theta}{2\pi\varepsilon_0 r^3}$$

and the transverse component is

$$-\frac{1}{r}\frac{\partial\varphi}{\partial\theta} = \frac{m\sin\theta}{4\pi\varepsilon_0 r^3}$$

In particular, the force at a point on the axis of the dipole is

$$\frac{\mathbf{m}}{2\pi\varepsilon_0 r^3}$$

since $\sin\theta = 0$ and $\cos\theta = \pm1$ depending upon whether θ is 0 or π.

Example A square of side a has two equal charges e at opposite corners and a dipole of moment $2\sqrt{2}ae$ at a third corner pointing towards one of the charges. Find the electrostatic potential and field strength at the fourth corner of the square.

Let the square be OABC and choose cartesian coordinates with origin at O, x-axis in the direction OA and y-axis in the direction OC (figure 53).

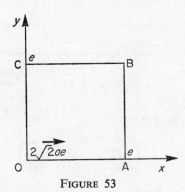

FIGURE 53

Take the charges to be at the points A and C and the dipole at the point O pointing along the x-axis. Then the potential at B is

$$\frac{1}{4\pi\varepsilon_0}\left(\frac{e}{CB}+\frac{e}{BA}+\frac{2\sqrt{2}\,ae\cos\pi/4}{(OB)^2}\right)\left(\text{since } A\hat{O}B=\frac{\pi}{4}\right)$$

$$=\frac{1}{4\pi\varepsilon_0}\left(\frac{e}{a}+\frac{e}{a}+\frac{2ae}{2a^2}\right)=\frac{3e}{4\pi\varepsilon_0 a}.$$

The force at B due to the charge at A is $e/4\pi\varepsilon_0 a^2$ in the direction AB, that is $[e/(4\pi\varepsilon_0 a^2)]\mathbf{j}$.

The force at B due to the charge at C is $e/4\pi\varepsilon_0 a^2$ in the direction CB, that is $[e/(4\pi\varepsilon_0 a^2)]\mathbf{i}$.

The force at B due to the dipole at O is $(-\sqrt{2}ae)/[2\pi\varepsilon_0(2a^2)^{\frac{3}{2}}]$ in the direction OA together with $[3(2\sqrt{2}ae)\cos\pi/4]/[4\pi\varepsilon_0(2a^2)^{\frac{3}{2}}]$ in the direction OB, that is

$$\frac{1}{4\pi\varepsilon_0}\left\{-\frac{e}{a^2}\mathbf{i}+\frac{3e}{\sqrt{2}\,a^2}\left(\frac{1}{\sqrt{2}}\mathbf{i}+\frac{1}{\sqrt{2}}\mathbf{j}\right)\right\}.$$

Hence the resultant force at B is

$$\frac{1}{4\pi\varepsilon_0}\left(\frac{3e}{2a^2}\mathbf{i}+\frac{5e}{2a^2}\mathbf{j}\right).$$

Example A dipole of moment ea and a charge 2e are held at points A and B a distance 2a apart with the axis of the dipole pointing towards the charge. Show

that the field strength is zero at only one point between A *and* B *and find this point.*
Show also that the field strength is zero at a point on BA *produced at a distance*
ξa *from* A *where*

$$\xi^3 - \xi^2 - 4\xi - 4 = 0.$$

Let P lie between A and B and be such that AP $= \eta a$ for $0 < \eta < 2$ (figure 54).

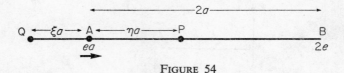

FIGURE 54

Then the field strength at P is

$$\frac{1}{4\pi\varepsilon_0}\left(\frac{2ea}{(\eta a)^3} - \frac{2e}{(2a - \eta a)^2}\right)$$

in the direction $\overline{AB}$. This is zero if

$$\frac{2ea}{\eta^3 a^3} - \frac{2e}{a^2(2 - \eta)^2} = 0.$$

That is

$$(2 - \eta)^2 - \eta^3 = 0.$$

That is

$$\eta^3 - \eta^2 + 4\eta - 4 = 0.$$

That is

$$(\eta - 1)(\eta^2 + 4) = 0.$$

The only real root of this equation is $\eta = 1$. Hence there is only one point
between A and B at which the field strength is zero and that is the point P
where AP $= a$ which is of course the midpoint of AB.

Now consider the point Q on BA produced and such that AQ $= \xi a$ (figure 54).
Then the field strength at Q is zero if

$$\frac{2ea}{(\xi a)^3} - \frac{2e}{(2a + \xi a)^2} = 0.$$

That is

$$(2 + \xi)^2 - \xi^3 = 0.$$

That is

$$\xi^3 - \xi^2 - 4\xi - 4 = 0$$

which is the desired result.

2.27 *Surface distribution of dipoles* (*or double layer*)

Let the density of the distribution be $\boldsymbol{\tau}$ per unit area where $\tau = |\boldsymbol{\tau}|$ is constant and $\boldsymbol{\tau}$ is everywhere normal to the surface S. That is $\boldsymbol{\tau}$ is parallel to the normal vector $\mathbf{n}$ to the surface at all points on it.

Therefore the potential φ due to the distribution on the surface S is given by

$$-\frac{1}{4\pi\varepsilon_0} \int_S \frac{(\boldsymbol{\tau}\,ds)\cdot\mathbf{r}}{r^3}.$$

(The minus sign appears because the vector $\mathbf{r}$ is now taken *from* P *to* a point on the surface) (figure 55).

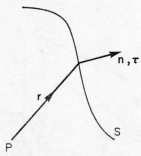

FIGURE 55

Therefore

$$\varphi = -\frac{1}{4\pi\varepsilon_0} \int_S \frac{(\tau\hat{\mathbf{n}}\,ds)\cdot\mathbf{r}}{r^3}$$

$$= -\frac{\tau}{4\pi\varepsilon_0} \int_S \frac{\mathbf{r}\cdot d\mathbf{s}}{r^3}$$

$$= -\frac{\tau\Omega}{4\pi\varepsilon_0}$$

where Ω is the solid angle subtended at P by the surface S. If P is on the other side of the surface S, then the outward normal vector will have the opposite sense to that used above and so the potential φ will be $+\tau\Omega/4\pi\varepsilon_0$.

Now since $\Omega \to 2\pi$ as P approaches S from either side, it is seen that $\varphi \to \pm\tau/2\varepsilon_0$ depending upon which side S is approached from.

Thus φ will be discontinuous by amount τ/ε_0 at the surface.

Example   Determine the potential due to a dipole distribution of density τ per unit area over a disc of radius a at a point on its axis. The direction of τ is everywhere normal to the disc.

The potential at P is

$$\pm \frac{\tau}{4\pi\varepsilon_0} \text{ (solid angle subtended at P by the disc)}$$

$$= \pm \frac{\tau}{2\varepsilon_0} (1 - \cos\theta) \text{ where } \theta \text{ is as shown in figure 56}$$

$$= \pm \frac{\tau}{2\varepsilon_0} \left(1 - \frac{z}{\sqrt{z^2 + a^2}} \right)$$

where z is the distance of P along the axis of the disc from its centre.

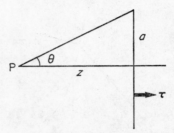

FIGURE 56

If τ points towards P then the positive sign is used and if τ points away from P then the minus sign is used.

2.28 Volume distribution of dipoles

Suppose the dipole strength per unit volume throughout the region V is **p**. Then the potential φ at a point P outside V is

$$-\frac{1}{4\pi\varepsilon_0} \int_V \frac{(\mathbf{p}\, dv) \cdot \mathbf{r}}{r^3}$$

where again **r** is taken *from* P *to* a point in the region V.

Therefore

$$\varphi = \frac{-1}{4\pi\varepsilon_0} \int_V \mathbf{p} \cdot \left(\frac{\mathbf{r}}{r^3} \right) dv = \frac{1}{4\pi\varepsilon_0} \int_V \mathbf{p} \cdot \text{grad} \left(\frac{1}{r} \right) dv.$$

Now

$$\text{div} \left(\mathbf{p}\, \frac{1}{r} \right) = \frac{1}{r} \text{div}\, \mathbf{p} + \mathbf{p} \cdot \text{grad} \left(\frac{1}{r} \right)$$

and so

$$\varphi = \frac{1}{4\pi\varepsilon_0} \int_V \operatorname{div}\left(\frac{\mathbf{p}}{r}\right) dv - \frac{1}{4\pi\varepsilon_0} \int_V \frac{1}{r} \operatorname{div} \mathbf{p} \, dv$$

$$= \frac{1}{4\pi\varepsilon_0} \int_S \frac{\mathbf{p} \cdot \mathbf{ds}}{r} - \frac{1}{4\pi\varepsilon_0} \int_V \frac{\operatorname{div} \mathbf{p}}{r} \, dv$$

where S is the surface enclosing the region V.

The first term on the right-hand side is the potential of a surface distribution of charge of density equal to the outward normal component of $\mathbf{p}$. The second term is the potential of a volume distribution of charge of density $-\operatorname{div} \mathbf{p}$.

These two distributions are called *Poisson's equivalent distributions*.

2.29 Dielectrics

In contrast to conductors in which the electric field is always zero, dielectrics respond to an applied electric field by changing into what may be regarded, for the purpose of determining potential and force fields, as a volume distribution of dipoles. The dielectric is then said to be *polarized* and the volume dipole density $\mathbf{p}$ may or may not be in the same direction as the field vector $\mathbf{E}$. When $\mathbf{p}$ is in the same direction as $\mathbf{E}$, the polarization is said to be *isotropic*. If $\mathbf{p} = \lambda\varepsilon_0\mathbf{E}$ where λ is a scalar constant, and not a function of position, the polarization is also homogeneous and λ is called the *dielectric susceptibility*.

Now for a volume distribution of dipoles of density $\mathbf{p}$ within the region V surrounded by the surface S, the potential φ is given by

$$\varphi = \frac{1}{4\pi\varepsilon_0} \int_S \frac{1}{r} \mathbf{p} \cdot \mathbf{ds} - \frac{1}{4\pi\varepsilon_0} \int_V \frac{\operatorname{div} \mathbf{p}}{r} \, dv$$

$$= \frac{1}{4\pi\varepsilon_0} \int_S \frac{p_n}{r} \, ds + \frac{1}{4\pi\varepsilon_0} \int_V \frac{-\operatorname{div} \mathbf{p}}{r} \, dv$$

where p_n is the component of $\mathbf{p}$ normal to the surface S at any point.

Therefore

$$\varphi = \frac{1}{4\pi\varepsilon_0} \int_S \frac{\sigma' \, ds}{r} + \frac{1}{4\pi\varepsilon_0} \int_V \frac{\rho' \, dv}{r}$$

where $\sigma' = p_n$ and $\rho' = -\operatorname{div} \mathbf{p}$.

In general however there may also be 'true' surface and volume charges of densities σ and ρ respectively.

Then

$$\operatorname{div} \mathbf{E} = \frac{1}{\varepsilon_0}(\rho + \rho') = \frac{\rho}{\varepsilon_0} - \frac{\operatorname{div} \mathbf{p}}{\varepsilon_0}.$$

Therefore

$$\mathrm{div}\,(\varepsilon_0\mathbf{E} + \mathbf{p}) = \rho.$$

The quantity $(\varepsilon_0\mathbf{E} + \mathbf{p})$ is called the *electric flux density* and is denoted by **D**.

Therefore

$$\mathrm{div}\,\mathbf{D} = \rho.$$

Now consider the behaviour of **D** on crossing the surface S from the dielectric into free space.

Let **E** be given by $\mathbf{E}_1$ outside the dielectric and by $\mathbf{E}_2$ within the dielectric.

Therefore **D** is given by $\mathbf{D}_1 = \varepsilon_0\mathbf{E}_1$ outside the dielectric and by $\mathbf{D}_2 = \varepsilon_0\mathbf{E}_2 + \mathbf{p}$ within the dielectric.

Then if E_{n_1}, E_{n_2}, D_{n_1} and D_{n_2} are the components of $\mathbf{E}_1$, $\mathbf{E}_2$, $\mathbf{D}_1$ and $\mathbf{D}_2$ normal to the surface S (figure 57), we have

$$\begin{aligned}
D_{n_1} - D_{n_2} &= \varepsilon_0 E_{n_1} - (\varepsilon_0 E_{n_2} + p_n) \\
&= \varepsilon_0(E_{n_1} - E_{n_2}) - p_n \\
&= (\sigma + \sigma') - \sigma' \\
&= \sigma.
\end{aligned}$$

That is the normal component of **D** is discontinuous by amount σ. Thus it is seen that **D** takes the place of $\varepsilon_0\mathbf{E}$ within dielectrics. Then if $\mathbf{p} = \chi\varepsilon_0\mathbf{E}$,

$$\mathbf{D} = \varepsilon_0(1 + \chi)\mathbf{E} = \varepsilon_0\kappa\mathbf{E} = \varepsilon\mathbf{E}$$

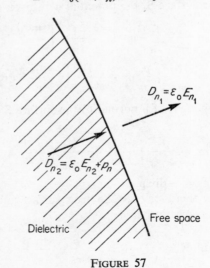

FIGURE 57

where $\kappa = 1 + \chi$ is called the *dielectric constant* or *specific inductive capacity* of the dielectric and $\varepsilon = \varepsilon_0 \kappa$. (In free space $\chi = 0$ and $\kappa = 1$).

Coulomb's law then becomes

$$\mathbf{D} = \frac{e\mathbf{r}}{4\pi r^3} \quad \text{or} \quad \mathbf{E} = \frac{1}{4\pi \varepsilon}\left(\frac{e\mathbf{r}}{r^3}\right).$$

At the boundary of two dielectrics with no free surface charges the normal component of $\mathbf{D}$ is continuous. That is the normal component of $\varepsilon \mathbf{E}$ and so of $\kappa \mathbf{E}$ is continuous, and so $\kappa(\partial \varphi / \partial n)$ is continuous. (If φ is the electrostatic potential).

The potential φ is itself continuous however since there are no double layers present.

The field lines are in fact refracted at the boundary, as illustrated in figure 58.

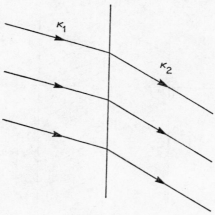

FIGURE 58

Note too that, in this case, since $\operatorname{div} \mathbf{D} = \rho$ and $\mathbf{D} = \varepsilon \mathbf{E}$ we have $\operatorname{div}(\varepsilon \mathbf{E}) = \rho$ or $\operatorname{div}(\kappa \mathbf{E}) = \rho/\varepsilon_0$. Then if κ is constant,

$$\operatorname{div} \mathbf{E} = \frac{\rho}{\kappa \varepsilon_0} = \frac{\rho}{\varepsilon}$$

and

$$\nabla^2 \varphi = -\frac{\rho}{\kappa \varepsilon_0} = -\frac{\rho}{\varepsilon}.$$

When there are no true volume charges present this again reduces to Laplace's equation $\nabla^2 \varphi = 0$.

Example *Two infinite slabs of dielectrics each of unit thickness but with different dielectric constants κ_1 and κ_2 respectively are placed with a pair of faces in contact between two infinite, thin conducting plates maintained at potentials V_1 and V_2 respectively.*

Determine the potential at any point in the dielectrics and show that it is unchanged if the dielectrics are replaced by an insulated conducting plate with surface charge density

$$\frac{\varepsilon_0(V_1 - V_2)(\kappa_1 - \kappa_2)}{(\kappa_1 + \kappa_2)}$$

placed midway between the conducting plates.

Choose x-axis normal to the faces of the slabs with origin O in the common face as shown in figure 59.

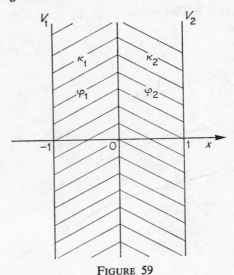

FIGURE 59

Let φ_1 and φ_2 be the potential functions for the dielectrics with dielectric constants κ_1 and κ_2 respectively. By symmetry φ_1 and φ_2 must be functions of x only. Then for the first dielectric, that is for $-1 \leqslant x \leqslant 0$, $\nabla^2 \varphi_1 = 0$ reduces to $\mathrm{d}^2\varphi_1/\mathrm{d}x^2 = 0$ and so we obtain

$$\varphi_1 = a_1 x + b_1,$$

where a_1 and b_1 are constants.

Similarly for the second dielectric, that is for $0 \leqslant x \leqslant 1$, we obtain

$$\varphi_2 = a_2 x + b_2$$

where a_2 and b_2 are constants. Now when $x = -1$, $\varphi_1 = V_1$ and so

$$V_1 = b_1 - a_1$$

and when $x = 1$, $\varphi_2 = V_2$ and so

$$V_2 = b_2 + a_2.$$

Also, φ and $\kappa \partial \varphi / \partial x$ are continuous at $x = 0$ and so

$$b_1 = b_2$$

and

$$\kappa_1 a_1 = \kappa_2 a_2.$$

Hence

$$V_2 - V_1 = a_1 + a_2 = a_1 \left(1 + \frac{\kappa_1}{\kappa_2} \right)$$

and so

$$a_1 = \frac{\kappa_2 (V_2 - V_1)}{\kappa_1 + \kappa_2}.$$

Then

$$a_2 = \frac{\kappa_1 (V_2 - V_1)}{\kappa_1 + \kappa_2},$$

$$b_1 = b_2 = a_1 + V_1 = \frac{\kappa_1 V_1 + \kappa_2 V_2}{\kappa_1 + \kappa_2}.$$

Hence

$$\varphi_1 = \frac{\kappa_2 (V_2 - V_1)}{\kappa_1 + \kappa_2} x + \frac{\kappa_1 V_1 + \kappa_2 V_2}{\kappa_1 + \kappa_2} \qquad -1 \leqslant x \leqslant 0.$$

and

$$\varphi_2 = \frac{\kappa_1 (V_2 - V_1)}{\kappa_1 + \kappa_2} x + \frac{\kappa_1 V_1 + \kappa_2 V_2}{\kappa_1 + \kappa_2} \qquad 0 \leqslant x \leqslant 1.$$

Now replace the dielectrics by the conducting plate. As before

$$\varphi_1 = A_1 x + B_1, \qquad -1 \leqslant x \leqslant 0$$

$$\varphi_2 = A_2 x + B_2, \qquad 0 \leqslant x \leqslant 1.$$

But $\varphi_1 = V_1$ when $x = -1$ and $\varphi_2 = V_2$ when $x = 1$ and so once again we have

$$V_1 = B_1 - A_1 \quad \text{and} \quad V_2 = B_2 + A_2.$$

Also φ is continuous on $x = 0$ and so $B_1 = B_2$. Then if σ is the charge density on the plate at $x = 0$,

$$\left(\frac{\partial \varphi_1}{\partial x} \right)_{x=0} - \left(\frac{\partial \varphi_2}{\partial x} \right)_{x=0} = \frac{\sigma}{\varepsilon_0}.$$

That is

$$A_1 - A_2 = \frac{\sigma}{\varepsilon_0}.$$

Thus it is seen that φ_1 and φ_2 will have the same expressions as for the previous case if

$$\frac{\sigma}{\varepsilon_0} = a_1 - a_2 = \frac{\kappa_2(V_2 - V_1)}{\kappa_1 + \kappa_2} - \frac{\kappa_1(V_2 - V_1)}{\kappa_1 + \kappa_2} = \frac{(V_1 - V_2)(\kappa_1 - \kappa_2)}{\kappa_1 + \kappa_2}.$$

That is

$$\sigma = \frac{\varepsilon_0(V_1 - V_2)(\kappa_1 - \kappa_2)}{\kappa_1 + \kappa_2}.$$

Exercises

1. Determine the electrostatic potential at a corner A of a square of side a due to three equal charges e at the other corners of the square and a charge e' at the centre of the square. What value of e' will make the resultant force field at A zero?

2. Charges e, e and $3e$ are held at the points $A(-a,0,0)$, $B(a,a,0)$ and $C(0,3a,0)$ respectively in free space. Determine the electrostatic force field at the origin and the force required to hold the charge at B in position.

3. Use Gauss' law to determine the electrostatic field due to a uniformly charged, very long (infinite) thin wire when the charge density is λ per unit length of wire. [Hint: Consider a circular cylinder of unit length and axis coincident with the wire].

4. Use Gauss' law to determine the electrostatic field at all points inside a sphere of radius a and uniform charge density ρ per unit volume.

5. If point charges $3e$ and $-e$ are placed at points A and B, determine the greatest angle a line of force leaving A and entering B can make with the line AB.

6. Three charges $-e'$, e, $-e'$ are held at the points A, O, B where AOB is a straight line and O is the midpoint of AB. Prove that if $e > 2e'$ all lines of force which enter A or B must leave O and that if α is the greatest angle a line of force leaving O and entering B can make with OB then $e \sin^2 (\alpha/2) = e'$.

7. A dipole of moment $\sqrt{2}ae\mathbf{j}$ is fixed at the point $a\mathbf{i}$, a dipole of moment $\sqrt{2}ae\mathbf{i}$ is fixed at the point $a\mathbf{j}$ and a point charge $-e$ is fixed at the point $(a\mathbf{i} + a\mathbf{j} + a\mathbf{k})$ where $\mathbf{i}$, $\mathbf{j}$ and $\mathbf{k}$ are unit vectors in the directions of the cartesian axes $Oxyz$. Find the resultant field at the point $a\mathbf{k}$.

8. A dipole of moment $\mathbf{m}$ is held at a fixed point. Find the magnitude of the field strength at points at which the direction of the field is perpendicular to the axis of the dipole.

9. An infinitely long circular, cylindrical shell of internal radius 2 units and external radius 4 units made of material with dielectric constant 2κ is placed within another similar cylindrical shell of internal radius 4 units and external radius 8 units made of material with dielectric constant κ. The inside surface of the smaller cylinder is maintained at the constant potential V_1 while the outside surface of the larger cylinder is maintained at the constant potential $2V_1$. Determine the potential at any point within either shell.

2.3 Magnetostatics

As for electrostatics, we consider mathematically two fundamental kinds of magnetic charges called positive and negative. Like charges repel one another whereas unlike charges attract each other. In contrast to the case of electric charges however this concept only has physical significance when the charges occur in equal and opposite pairs. Hence if Ω is the magnetostatic potential, then $\nabla^2\Omega = 0$ everywhere. With this exception the laws of electrostatics also hold for magnetostatics. A different notation is used however.

The magnetic field vector $\mathbf{H}$ is given by $\mathbf{H} = -\operatorname{grad}\Omega$ and corresponds to $\mathbf{E}$ in electrostatics.

The vector $\mathbf{B}$ given by $\mathbf{B} = \mu\mathbf{H}$ is called the *magnetic induction* and μ is called the *permeability*. $\mathbf{B}$ corresponds to $\mathbf{D}$ and μ corresponds to ε in electrostatics. Also, $\mu = \mu_0\kappa_m$ where κ_m is called the (magnetic) *specific inductive capacity* and μ_0 the *permeability of free space* is given by

$$\mu_0 = 1/\varepsilon_0 c^2$$

where c is the speed of light measured in appropriate units. Using S.I. units μ_0 has the numerical value $4\pi \times 10^{-7}$. Then

$$\operatorname{div} \mathbf{B} = 0 \qquad (\text{c.f. } \operatorname{div} \mathbf{D} = \rho).$$

Corresponding to electrostatic dipoles, we have magnetic dipoles formed in the same way. A surface distribution of magnetic dipoles is generally called a *magnetic shell*.

A substance placed in a magnetic field tends to become magnetized and behaves like a volume distribution of magnetic dipoles.

Example   A magnetic dipole of moment $\mathbf{m}$ *is held in a uniform magnetic field of intensity* $\mathbf{H}_0$ *where* $\mathbf{H}_0$ *is parallel to the axis of the dipole. Determine the points at which the resultant magnetic field vanishes.*

The resultant magnetic field $\mathbf{H}$ is given by

$$\mathbf{H} = \mathbf{H}_0 - \frac{\mathbf{m}}{4\pi\mu_0 r^3} + \frac{3(\mathbf{m}\cdot\mathbf{r})\mathbf{r}}{4\pi\mu_0 r^5}.$$

But $\mathbf{H}_0$ is parallel to $\mathbf{m}$ and so $\mathbf{m} = k\mathbf{H}_0$ for some constant k.

Therefore

$$\mathbf{H} = \mathbf{H_0}\left(1 - \frac{k}{4\pi\mu_0 r^3}\right) + \frac{3k(\mathbf{H_0} \cdot \mathbf{r})\mathbf{r}}{4\pi\mu_0 r^5}.$$

Thus if $\mathbf{r}$ is not parallel to $\mathbf{H_0}$ (that is for points not on the axis of the dipole) $\mathbf{H}$ will only be the zero vector when the coefficient of $\mathbf{H_0}$ and the coefficient of $\mathbf{r}$ are each zero. That is when

$$1 - \frac{k}{4\pi\mu_0 r^3} = 0$$

and

$$\frac{3k}{4\pi\mu_0 r^5}(\mathbf{H_0} \cdot \mathbf{r}) = 0.$$

That is when

$$r^3 = \frac{k}{4\pi\mu_0}$$

and $\mathbf{H_0}$ is perpendicular to $\mathbf{r}$. Hence when $\mathbf{H_0}$ and $\mathbf{m}$ have the same sense, and so k is positive, the magnetic field is zero at all points on the circle of radius

$$\left(\frac{k}{4\pi\mu_0}\right)^{\frac{1}{3}} = \left(\frac{m}{4\pi\mu_0 H_0}\right)^{\frac{1}{3}}$$

in the plane through the dipole and perpendicular to its axis.

When $\mathbf{r}$ is parallel to $\mathbf{H_0}$, put $\mathbf{H_0} = \alpha\mathbf{r}$ where $\alpha \neq 0$.
Therefore

$$\mathbf{H} = \alpha\mathbf{r}\left(1 - \frac{k}{4\pi\mu_0 r^3}\right) + \frac{3k\alpha r^2}{4\pi\mu_0 r^5}\mathbf{r} = \alpha\left(1 + \frac{k}{2\pi\mu_0 r^3}\right)\mathbf{r} = \mathbf{0}$$

when $1 + k/2\pi\mu_0 r^3 = 0$, that is when $r^3 = -k/2\pi\mu_0$. Since $r\ (= |\mathbf{r}|)$ is always positive, this latter equation only has a real solution when k is negative, that is when $\mathbf{H_0}$ and $\mathbf{m}$ have opposite senses. In this case there are two points on the axis of the dipole, each distant $(-k/2\pi\mu_0)^{\frac{1}{3}}$ from the dipole itself, at which the magnetic field is zero.

Example   A solid hemisphere of radius a is uniformly magnetized with intensity
I at right angles to its plane face and directed towards this face. Determine the
potential at any external point Q on the axis of symmetry on the side remote from
the curved face. Show that if Q is at a distance a from the plane face on this side,
the magnetic field intensity at Q has magnitude

$$\frac{I}{12\mu_0}(8 - 5\sqrt{2}).$$

The potential φ is given by

$$4\pi\mu_0\varphi = \int_S \frac{\mathbf{I} \cdot d\mathbf{s}}{r} - \int_V \frac{\text{div } \mathbf{I}}{r}\,dv$$

where S is the surface of the hemisphere (including the plane face) and V is the region occupied by the hemisphere.

Now **I** is constant and so div **I** = 0.

Therefore

$$4\pi\mu_0\varphi = \int_S \frac{\mathbf{I} \cdot \mathbf{ds}}{r} .$$

Consider the plane face S'. If Q is a distance x along the axis from the centre of S'

$$\int_{S'} \frac{\mathbf{I} \cdot \mathbf{ds}}{r} = -\int_0^a \frac{I\, 2\pi z\, \mathrm{d}z}{\sqrt{z^2 + x^2}}$$

on dividing S' into concentric rings of radius z (figure 60).

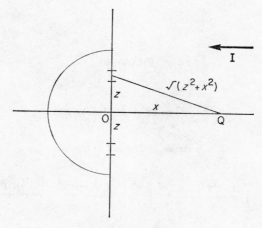

FIGURE 60

Hence

$$\int_S \frac{\mathbf{I} \cdot \mathbf{ds}}{r} = -\pi I \left[2\sqrt{x^2 + z^2} \right]_0^a = -2\pi I (\sqrt{x^2 + a^2} - x).$$

Now consider the curved surface S''. Dividing S'' into rings we see (figure 61) that

$$\int_{S''} \frac{\mathbf{I} \cdot \mathbf{ds}}{r} = \int_{S''} \frac{I\, \mathrm{d}s \cos \psi}{r}$$

where 2ψ is the angle subtended at O by a diameter of a typical ring.

Hence

$$\int_{S''} \frac{\mathbf{I} \cdot \mathbf{ds}}{r} = \int_0^{\frac{\pi}{2}} \frac{I\, 2\pi a \sin \psi a\, \mathrm{d}\psi \cos \psi}{r} .$$

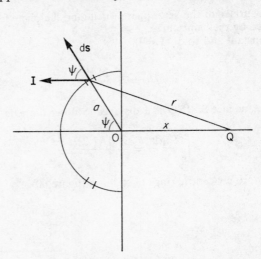

But

$$r^2 = a^2 + x^2 + 2ax \cos \psi$$

and so

$$2r \, dr = -2ax \sin \psi \, d\psi.$$

Therefore

$$\int_{S''} \frac{\mathbf{I} \cdot d\mathbf{s}}{r} = -\frac{\pi I}{x^2} \int_{a+x}^{\sqrt{a^2+x^2}} \frac{(r^2 - (a^2 + x^2))r \, dr}{r}$$

$$= -\frac{\pi I}{x^2} \left[\frac{r^3}{3} - (a^2 + x^2)r \right]_{a+x}^{\sqrt{a^2+x^2}}$$

$$= \frac{2\pi I}{3x^2} \{ (a^2 + x^2)^{\frac{3}{2}} - (a^3 + x^3) \}.$$

Therefore

$$4\pi\mu_0\varphi = \int_S \frac{\mathbf{I} \cdot d\mathbf{s}}{r} = \int_{S'} \frac{\mathbf{I} \cdot d\mathbf{s}}{r} + \int_{S''} \frac{\mathbf{I} \cdot d\mathbf{s}}{r}$$

$$= \frac{2\pi I}{3x^2} \{ -3x^2\sqrt{a^2 + x^2} + 3x^3 + (a^2 + x^2)^{\frac{3}{2}} - (a^3 + x^3) \}$$

$$= \frac{2\pi I}{3} \left\{ \left(\frac{a^2}{x^2} - 2 \right) \sqrt{a^2 + x^2} - \left(\frac{a^3}{x^2} - 2x \right) \right\}.$$

Hence field at Q in the direction OQ is given by

$$
\left(-\frac{\partial \varphi}{\partial x}\right)_{x=a}
$$

$$
= \frac{1}{4\pi\mu_0}\left[-\frac{2\pi I}{3}\left\{-\frac{2a^2}{x^3}\sqrt{a^2+x^2}+\left(\frac{a^2}{x^2}-2\right)\frac{x}{\sqrt{a^2+x^2}}+\frac{2a^3}{x^3}+2\right\}\right]_{x=a}
$$

$$
= \frac{-I}{12\mu_0}(8-5\sqrt{2}).
$$

Exercises

1. A uniform magnetic shell of strength τ is in the form of the hemisphere $x^2 + y^2 + z^2 = a^2$, $x \geqslant 0$. The direction of magnetization is normal to the surface of the hemisphere and its sense is outwards from the origin. Show that the magnetic field intensity at points on the axis is

$$
\frac{\tau a^2}{2\mu_0(x^2+a^2)^{\frac{3}{2}}}.
$$

2. A circular disc of radius c has on its surface a uniform distribution of poles of density σ. Show that the magnetic potential at a point P on the axis of the disc and a distance x from the disc is

$$
\frac{\sigma}{2\mu_0}(\sqrt{x^2+c^2}-x).
$$

By considering the potential due to two parallel circular discs having magnetic pole densities σ and $-\sigma$ and separated by a distance h and then letting $h \to 0$ and $\sigma \to \infty$ in such a way that $h\sigma \to \tau$, find the potential due to a magnetic shell in the form of a circular disc of radius c at a point P on its axis.

3. A solid sphere of radius a is uniformly magnetized with constant intensity **I**. Determine the magnetic potential and field intensity at points outside the sphere (in free space) on a diameter of the sphere parallel to **I**.

4. A solid circular cylinder of radius a and length $2h$ is uniformly magnetized with constant intensity **I** parallel to its length and placed in free space. Find the potential and force fields at points on the axis of the cylinder which are exterior to the cylinder.

2.4 Current electricity

If two conductors at different potentials are connected together by a conducting wire it is found that the wire becomes heated and that a magnetic field is created in the neighbourhood of the wire. An *electric*

6

current is then said to be flowing in the wire, transmitting charge from one conductor to the other.

Let $\mathbf{v}$ be the velocity of the charge along the wire and $\rho(x,y,z)$ the charge density. Then

$$\mathbf{j} = \rho \mathbf{v} \text{ is called the } current\ density.$$

The total charge passing through a given surface S in unit time is called the *electric current* and is denoted by *I*. It is given by

$$I = \int_S \rho \mathbf{v} \cdot \mathbf{ds} = \int_S \mathbf{j} \cdot \mathbf{ds}.$$

If S encloses a region V, then the total charge Q enclosed by S is given by

$$Q = \int_V \rho \, dv$$

and if there are no charge sources or sinks within V then the loss of charge per unit time is

$$-\frac{\partial Q}{\partial t} = -\int_V \frac{\partial \rho}{\partial t} \, dv.$$

Hence

$$\int_S \mathbf{j} \cdot \mathbf{ds} = -\int_V \frac{\partial \rho}{\partial t} \, dv$$

and so

$$\int_V \operatorname{div} \mathbf{j} \, dv = -\int_V \frac{\partial \rho}{\partial t} \, dv.$$

Since this result holds for arbitrary regions V,

$$\operatorname{div} \mathbf{j} = -\frac{\partial \rho}{\partial t}$$

at all points. This is called the *equation of continuity of charge*. It can also be shown experimentally that $\mathbf{j} = \lambda \mathbf{E}$ where $\mathbf{E}$ is the electric field and λ is the electrical conductivity of the conductor. This important result is known as *Ohm's law*. Thus

$$\operatorname{div}(\lambda \mathbf{E}) = -\frac{\partial \rho}{\partial t}$$

and so

$$\operatorname{div}(\lambda \operatorname{grad} \varphi) = \frac{\partial \rho}{\partial t}$$

if φ is the electric potential.

Therefore

$$\lambda \nabla^2 \varphi = \frac{\partial \rho}{\partial t}$$

if the conductor has constant conductivity throughout. Then if we have a steady state in which the charge density ρ remains constant, we again obtain Laplace's equation

$$\nabla^2 \varphi = 0.$$

Now for dielectrics or conductors we have

$$\operatorname{div} \mathbf{D} = \rho$$

and so

$$\operatorname{div} \left(\frac{\partial \mathbf{D}}{\partial t} \right) = \frac{\partial \rho}{\partial t} = -\operatorname{div} \mathbf{j}$$

or

$$\operatorname{div} \left(\mathbf{j} + \frac{\partial \mathbf{D}}{\partial t} \right) = 0.$$

$(\mathbf{j} + \partial \mathbf{D}/\partial t)$ is called the *total current* and $\partial \mathbf{D}/\partial t$ is called the *displacement current*.

Also,

$$\frac{\partial \rho}{\partial t} = -\operatorname{div} \mathbf{j} = -\lambda \operatorname{div} \mathbf{E} = -\frac{\lambda}{\varepsilon} \operatorname{div} \mathbf{D} = -\frac{\lambda}{\kappa \varepsilon_0} \rho$$

when λ and κ are independent of position.

Hence

$$\rho = \rho_0 e^{-\frac{\lambda}{\kappa \varepsilon_0} t}$$

where ρ_0 is the charge density when $t = 0$. Thus, the charge density ρ in a homogeneous dielectric decays exponentially with time. For good conductors, the decay rate is so rapid that the charge density is normally taken to be zero.

For a steady current of magnitude I flowing in a thin wire in the form of a closed circuit L, it is observed that the magnetic potential Ω produced at a point P (in free space) is $\pm kI\omega$ where ω is the solid angle subtended by the circuit L at the point P and k is a constant which in S.I. units is equal to $1/4\pi$. The sign of the potential depends upon the sense of the current flow in the circuit relative to the point P. The above potential is of course equivalent, in S.I. units, to the potential in free space due to a surface distribution of dipoles of strength $\mu_0 I$ where $\mathbf{I}$ is everywhere normal to the surface and the boundary of the surface coincides with the circuit L.

The sense of **I** and the sign of the potential is given by the following consideration. If the current in the circuit appears to an observer to be flowing in a clockwise sense then **I** points away from the observer and the potential at the point of observation will then be $-I\omega/4\pi$. If the current in the circuit appears to an observer to be flowing in an anticlockwise sense then **I** points towards the observer and the potential at the point of observation is $+I\omega/4\pi$. This equivalent surface dipole distribution is usually called a *magnetic shell*.

We can then write, using the usual notation,

$$\Omega = -\frac{I}{4\pi} \int_S \frac{\mathbf{r} \cdot \mathbf{ds}}{r^3}$$

where **r** is the position vector relative to the point P of the general point on the surface S.

Then the magnetic field vector **H** is given by

$$\mathbf{H} = -\text{grad } \Omega = \text{grad}\left(\frac{I\omega}{4\pi}\right) = \frac{I}{4\pi}\text{ grad }\omega$$

and it can be shown that this can be written as

$$\frac{I}{4\pi} \int_L \frac{\mathbf{r} \times \mathbf{dl}}{r^3}$$

where **r** is the vector from P to a point on the circuit L (figure 62). This is known as Biot and Savart's law.

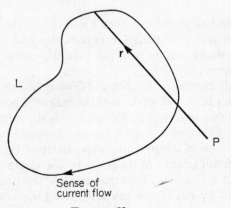

L

Sense of
current flow

FIGURE 62

Example Determine the field and potential due to a steady current I flowing in an infinitely long straight wire.

$$\mathbf{H} = \frac{I}{4\pi} \int_{L} \frac{\mathbf{r} \times \mathbf{dl}}{r^3}$$

$$= \frac{I}{4\pi} \int_{-\infty}^{\infty} \frac{\mathbf{r} \times \mathbf{k} \, dz}{r^3}$$

where $\mathbf{k}$ is unit vector having the same direction and sense as the current flow, and z is distance measured along the wire from an origin O such that OP is

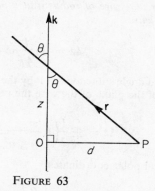

FIGURE 63

perpendicular to the wire (figure 63). Thus, in terms of cylindrical polar co-ordinates with the wire as axis,

$$\mathbf{H} = \left(0, \frac{I}{4\pi} \int_{-\infty}^{\infty} \frac{r \sin \theta \, dz}{r^3}, 0\right)$$

where θ is the angle between $\mathbf{r}$ and $\mathbf{k}$ as shown in the diagram. Now if we put $z = d \tan \varphi$, where d is the distance of the point P from the wire, $r = d \sec \varphi$ and $\sin \theta = \cos \varphi$. Therefore

$$\int_{-\infty}^{\infty} \frac{\sin \theta \, dz}{r^2} = \int_{-\frac{\pi}{2}}^{\frac{\pi}{2}} \frac{\cos \varphi \, d \sec^2 \varphi \, d\varphi}{d^2 \sec^2 \varphi}$$

$$= \frac{1}{d} \int_{-\frac{\pi}{2}}^{\frac{\pi}{2}} \cos \varphi \, d\varphi$$

$$= \frac{2}{d}.$$

Therefore

$$\mathbf{H} = \left(0, \frac{I}{2\pi d}, 0\right).$$

Hence

$$\Omega = \frac{-I}{2\pi}\, \psi \ (+\text{constant})$$

where ψ is the cylindrical polar coordinate.

Example   Determine the potential and field vector due to a steady current I flowing round a circular wire of radius a at a point on the axis of the wire and distant z from its centre.

$$\Omega = \pm \frac{I}{4\pi}\, \omega$$

where ω is the solid angle subtended at P by the circular wire. (The sign depends upon which side of the plane of the wire the point P lies).

Thus

$$\Omega = \pm \frac{I}{2}\left(1 - \frac{z}{\sqrt{a^2 + z^2}}\right).$$

Then, in cylindrical polar coordinates,

$$\mathbf{H} = \left(0, 0, \pm \frac{I}{2}\frac{\partial}{\partial z}\left(\frac{z}{\sqrt{a^2 + z^2}}\right)\right)$$

$$= \left(0, 0, \frac{\pm I a^2}{2(z^2 + a^2)^{\frac{3}{2}}}\right).$$

Exercises

1. Equal currents I amps flow in the same sense along two infinitely long parallel straight wires a distance $2a$ apart. Determine the magnitude, direction and sense of the magnetic field at points P and Q where P lies on a common perpendicular to both wires and is a distance $(d - a)$ from one of the wires and a distance $(d + a)$ from the other, and $d > a$, and the point Q is a distance d from each of the wires.

2. Equal and opposite currents $\pm I$ amps flow in two infinitely long parallel straight wires a distance $2a$ apart. Determine the form of the equipotential surfaces and show that the magnitude of the magnetic field at a point, distant r_1 from one wire and r_2 from the other, is

$$\frac{Ia}{\pi r_1 r_2}.$$

2.5 Gravitation

Newton deduced his Law of Gravitation from a study of the motions of the planets in the solar system. It states that any two point masses m_1 and m_2, distant r apart, attract each other with a force of magnitude

$$\frac{\gamma m_1 m_2}{r}$$

where γ is called the *gravitational constant*.

Measuring mass in kilograms, distance in metres and force in Newtons (that is S.I. units), γ has the approximate numerical value $6 \cdot 66 \times 10^{-5}$. Its dimensions are (Newtons) (metres)² (kilograms)⁻².

Taking m_2 as unity, we say that there is associated with the mass m_1 a field of force

$$\mathbf{F} = \frac{-\gamma m_1}{r^2} \hat{\mathbf{r}}$$

where $\mathbf{r}$ is unit vector in the direction of $\mathbf{r}$ the vector to any point from m_1 as origin.

Thus once again we have an analogy with electrostatics. The essential differences are that there is nothing corresponding to a negative charge and that two point masses *attract* each other whereas two positive charges *repel* each other. This latter point is taken care of by taking the force field $\mathbf{F}$ to be given by $+\text{grad } \varphi$ where φ is the gravitational potential which has the form $\gamma m/r$ at a point distant r from a point mass m. With this exception, the laws of electrostatics then also hold for gravitation with m corresponding to a positive point charge and the constant γ corresponding to $1/4\pi\varepsilon_0$. Gauss' law of course becomes $\text{div } \mathbf{F} = -4\pi\gamma\rho$ where ρ is the volume density of mass. We then obtain, as for electrostatics, that $\partial\varphi/\partial n$ decreases by amount $4\pi\gamma\sigma$ on passing, in the direction of increasing n, through a surface distribution of matter of density σ where n is distance measured along the normal to the surface.

Example   Consider the gravitational potential and force due to a uniform rod of length 2a and mass M at points on the axis and points on the perpendicular bisector of the rod.

For the point P on the axis of the rod and at a distance x ($> a$) from the centre of the rod (figure 64) we have for the potential φ

$$\varphi = \int_{-a}^{a} \frac{\gamma\rho \, \mathrm{d}l}{x - l}$$

where ρ ($= M/2a$) is the line density of the rod.

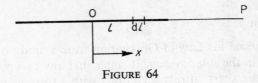

Hence

$$\varphi = -\gamma\rho\Big[\ln|x-l|\Big]_{-a}^{a}$$

$$= \gamma\rho\ln\left|\frac{x+a}{x-a}\right|$$

$$= \gamma\rho\ln\left(\frac{x+a}{x-a}\right).$$

Then

$$\mathbf{F} = \nabla\varphi = \frac{\partial\varphi}{\partial x}\mathbf{i}$$

where **i** is unit vector in the direction of OP.

Hence

$$\mathbf{F} = \gamma\rho\left(\frac{x-a}{x+a}\right)\frac{(x-a)-(x+a)}{(x-a)^2}\mathbf{i}$$

$$= \frac{-2a\rho\gamma}{x^2-a^2}\mathbf{i} = \frac{-\gamma M}{x^2-a^2}\mathbf{i}.$$

The negative sign is consistent with the force being an attraction and so directed from P towards O.

For $x < -a$,

$$\varphi = \int_{-a}^{a}\frac{\gamma\rho\,dl}{l-x} = \gamma\rho\ln\left(\frac{x-a}{x+a}\right)$$

and

$$\mathbf{F} = \frac{\gamma M}{x^2-a^2}\mathbf{i}$$

which once again is an attraction towards O. Combining these results we see that, for $|x| > a$,

$$\varphi = \gamma\rho\ln\left(\frac{|x|+a}{|x|-a}\right)$$

and the attraction towards O is

$$\frac{\gamma M}{x^2-a^2}.$$

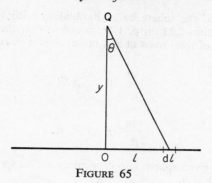

FIGURE 65

Now for the point Q on the perpendicular bisector of the rod and at a distance $y \, (\neq 0)$ from the centre of the rod (figure 65) we have for the potential φ

$$\varphi = \int_{-a}^{a} \frac{\gamma \rho \, dl}{(y^2 + l^2)^{\frac{1}{2}}}$$

$$= \int_{\alpha}^{\beta} \frac{\gamma \rho y \sec^2 \theta \, d\theta}{y \sec \theta}$$

where $l = y \tan \theta$, $\alpha = \tan^{-1}(-a/y)$ and $\beta = \tan^{-1}(a/y)$.
Hence

$$\varphi = \gamma \rho \int_{\alpha}^{\beta} \sec \theta \, \frac{(\sec \theta + \tan \theta)}{(\sec \theta + \tan \theta)} \, d\theta$$

$$= \gamma \rho \left[\ln (\sec \theta + \tan \theta) \right]_{\alpha}^{\beta}$$

$$= \gamma \rho \ln \left\{ \frac{(1 + a^2/y^2)^{\frac{1}{2}} + a/y}{(1 + a^2/y^2)^{\frac{1}{2}} - a/y} \right\}$$

$$= \gamma \rho \ln \left\{ \frac{(y^2 + a^2)^{\frac{1}{2}} + a}{(y^2 + a^2)^{\frac{1}{2}} - a} \right\}.$$

Then

$$\mathbf{F} = \nabla \varphi = \frac{\partial \varphi}{\partial y} \mathbf{j}$$

where $\mathbf{j}$ is unit vector in the direction OQ. Hence

$$\mathbf{F} = \frac{-2a\rho\gamma}{y\sqrt{a^2 + y^2}} \mathbf{j} = \frac{-\gamma M}{y\sqrt{a^2 + y^2}} \mathbf{j}.$$

Again the negative sign signifies an attraction.

Example   *Determine the potential and force fields due to a solid sphere of radius a and mass M at a point* P *distant r from its centre.*

Let the density of the sphere be ρ. By analogy with an example done in electrostatics in section 2.22 on p. 118 we obtain that the potential due to a thin spherical shell of total mass m and radius R at a point distant r from its centre is

$$\frac{\gamma m}{r} \quad \text{if} \quad r > R$$

and

$$\frac{\gamma m}{R} \quad \text{if} \quad r < R.$$

Thus, dividing the given sphere into elementary shells, we obtain for the total potential φ at P if $r > a$,

$$\varphi = \int_0^a \frac{\gamma 4\pi R^2 \, dR \rho}{r}$$

$$= \frac{4\pi \gamma \rho}{r} \left[\tfrac{1}{3} R^3 \right]_0^a$$

$$= \frac{\gamma M}{r} \text{ (since the volume of the sphere is } 4\pi a^3/3).$$

Then the force field **F** is given by

$$\mathbf{F} = \text{grad } \varphi = -\frac{\gamma M}{r^2} \, \hat{\mathbf{r}}.$$

That is an attraction towards the centre of the sphere.

Thus we see that for points outside the sphere the potential and force fields are the same as if the whole mass was concentrated at the centre.

Now if P lies within the sphere, that is $r < a$, the potential at P due to the sphere of radius r is

$$\frac{\gamma \tfrac{4}{3}\pi r^3 \rho}{r} \text{ (as above)}$$

(as above)

$$= \tfrac{4}{3}\pi \gamma \rho r^2.$$

The potential due to the remainder of the sphere is

$$\int_r^a \frac{\gamma 4\pi R^2 \rho \, dR}{R}$$

on dividing into thin shells. Therefore this potential is

$$4\pi \gamma \rho \int_r^a R \, dR = 4\pi \gamma \rho \tfrac{1}{2}(a^2 - r^2).$$

Hence the total potential at P in this case is

$$\varphi = \tfrac{4}{3}\pi\gamma\rho r^2 + 2\pi\gamma\rho(a^2 - r^2)$$

$$= \frac{\gamma M}{2a}\left(3 - \frac{r^2}{a^2}\right).$$

The force field is then given by

$$\mathbf{F} = \text{grad } \varphi = -\frac{\gamma M r}{a^3}\,\hat{\mathbf{r}}.$$

That is an attraction towards the centre directly proportional to distance from the centre.

As expected by analogy with electrostatics, we see, on putting in the various expressions for potential and force, that both the potential and the normal component of force are continuous on the surface of the sphere.

Consider now a particle of mass m a distance h above the surface of the earth. If it is assumed that the earth is a sphere of mass M and radius R, the force of attraction between the earth and the particle will be

$$\frac{\gamma M m}{(R + h)^2}.$$

But this force acting on the particle is what is called its weight and has the value mg where g is the (constant)† acceleration due to gravity.

Therefore

$$mg = \frac{\gamma M m}{(R + h)^2}$$

$$\simeq \frac{\gamma M m}{R^2} \quad \text{if} \quad h \ll R.$$

Therefore

$$g = \frac{\gamma M}{R^2} \quad \text{and so} \quad \gamma = \frac{g R^2}{M}.$$

Example   A sphere of uniform material has a spherical hole in it. If $c\ (> 0)$ is the distance between the centre of the sphere and the centre of the hole, determine the gravitational field within the hole.

† Since the earth is not in fact a perfect sphere, the value of g is not completely constant over the surface of the earth. Its variations however are small enough to be neglected unless the masses and distances involved are comparable with those of the earth itself.

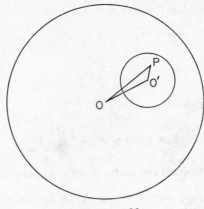

FIGURE 66

Let the sphere have centre O and the cavity have centre O' (figure 66). Then OO' = c. Consider a point P within the cavity. The force field at P is that due to the whole sphere minus that due to the part of the sphere which has been removed to form the cavity.

Hence, using the result of the previous example, the force field at P is

$$-\tfrac{4}{3}\gamma\pi\rho\overline{OP} - (-\tfrac{4}{3}\gamma\pi\rho\overline{O'P})$$

where ρ is the density of the material of the sphere. Therefore the force field at P is

$$\tfrac{4}{3}\pi\gamma\rho(\overline{PO} + \overline{O'P}) = \tfrac{4}{3}\pi\gamma\rho\overline{O'O}.$$

Thus for all points P within the cavity this force has magnitude

$$\tfrac{4}{3}\pi\gamma\rho\,|\overline{O'O}| = \tfrac{4}{3}\pi\gamma\rho c$$

and direction parallel to the line of centres of sphere and cavity. The sense of the force is from the centre of the cavity to the centre of the sphere.

Exercises

1. Determine the gravitational potential and force field at the origin due to masses $\tfrac{1}{2}m$, m, $3m$ at the points $(-a,0,0)$, $(2a,0,0)$ and $(0,3a,0)$ respectively.

2. Two uniform thin rods AB and BC each of mass M and length $2a$ are joined together at B so that $A\hat{B}C$ is a right angle. Determine the gravitational potential and force field at the point which has the same perpendicular distance a from each rod.

3. AOB is a fixed diameter of a thin, uniform spherical shell of centre O and radius a. The shell is cut into two unequal portions by a plane perpendicular to

this fixed diameter, the portion containing A being the larger. The smaller portion is now discarded. If the plane cuts the shell in a circle of radius b, show that the potential at the point P distance x ($> a$) from O on OB produced is

$$\frac{2\pi\gamma\sigma a}{x}\{(a + x) - \sqrt{a^2 + x^2 - 2ax\cos\alpha}\}$$

where σ is the density of the shell, γ the gravitational constant and

$$\alpha = \sin^{-1}(b/a).$$

Hence deduce that the attraction at the point B is

$$2\pi\gamma\sigma(1 - \sin\tfrac{1}{2}\alpha).$$

4. The density of a solid sphere of radius a varies linearly from ρ_0 at the centre to $k\rho_0$ at the surface. Obtain an expression for the gravitational force field at any point within the sphere and show that this has a maximum value within the sphere if $k < \tfrac{1}{3}$.

5. Show that the gravitational potential of a uniform thin hemispherical shell of mass M, radius a, centre O and axis of symmetry OA (where OA $= a$) is given at a point P distant x ($> a$) from O on OA produced by

$$\frac{\gamma M}{ax}\{\sqrt{a^2 + x^2} + (a - x)\}$$

where γ is the gravitational constant.

Hence show that if the shell is replaced by a uniform solid hemisphere of the same mass and radius, the potential at P is

$$\frac{\gamma M}{2a^3 x}\{2(x^2 + a^2)^{\frac{3}{2}} - 2x^3 - 3xa^2 + 2a^3\}$$

and calculate the attraction at the point for which $x = a$.

What is the attraction at the point on the axis distant a from the centre on the opposite side from the hemisphere?

6. A solid sphere of radius a and constant density ρ is surrounded by a concentric shell of internal radius a and external radius $2a$ in which the density is 2ρ. Find expressions for the potential at all points both inside and outside the body and determine the attraction at a distance $5a$ from the centre.

2.6 'Source' distributions

By 'sources' is now meant hydrodynamical sources (or sinks), electrostatic charges, magnetic poles (where appropriate) or particles of matter, and we are now in a position to determine what source distributions will give rise to given potential fields.

Use will be made of the following results which have been obtained earlier:

1. Potential φ is finite everywhere except at isolated sources.
2. Potential φ is continuous everywhere except at a double layer where it is discontinuous by amount τ/ε_0 (or τ/μ_0) where τ is the strength of the double layer. (Since negative masses do not exist and hence gravitational double layers do not exist, φ will be continuous everywhere in a gravitational field.)
3. The components of grad φ are continuous everywhere except at a surface distribution of sources. If the surface source distribution has density σ, then the component $\partial\varphi/\partial n$ of grad φ normal to the surface decreases by amount $k\sigma$, where k is a constant, on passing through the surface in the direction of increasing n. In electrostatic fields $k = 1/\varepsilon$ where ε is the permittivity of the medium and in gravitational fields $k = 4\pi\gamma$ where γ is the gravitational constant.
4. Potential φ satisfies Laplace's equation $\nabla^2\varphi = 0$ everywhere except where there is a volume distribution of sources. If the volume distribution has density ρ, then φ satisfies Poisson's equation $\nabla^2\varphi = -k\rho$ where, as above, $k = 1/\varepsilon$ in electrostatic fields and $k = 4\pi\gamma$ in gravitational fields.

Example   Find the distribution of charge which will give rise to the following potential field.

$$\varphi_1 = \tfrac{1}{2}(b^2 - a^2) \qquad\qquad 0 \leqslant r < a,$$

$$\varphi_2 = \frac{1}{6}\left(3b^2 - r^2 - \frac{2a^3}{r}\right) \qquad a \leqslant r < b,$$

$$\varphi_3 = \frac{1}{3}\left(\frac{b^3 - a^3}{r}\right) \qquad\qquad r \geqslant b.$$

where r is distance measured from a fixed point O.

1. It is easily seen that φ is always finite and so there are no isolated charges.
2. φ_1, φ_2, φ_3 are themselves continuous functions and so double layers can only be present on the surfaces $r = a$ and $r = b$.
   On $r = a$,

$$\varphi_1 \to \tfrac{1}{2}(b^2 - a^2)$$

   and

$$\varphi_2 \to \tfrac{1}{6}(3b^2 - a^2 - 2a^2) = \tfrac{1}{2}(b^2 - a^2).$$

Therefore φ is continuous across the surface $r = a$ and so there is no double layer on this surface. On $r = b$,

$$\varphi_2 \to \frac{1}{6}\left(3b^2 - b^2 - \frac{2a^3}{b}\right)$$

$$= \frac{1}{3b}(b^3 - a^3)$$

and

$$\varphi_3 \to \frac{1}{3b}(b^3 - a^3).$$

Therefore φ is continuous across the surface $r = b$ and so there is no double layer on this surface.

3. In terms of spherical polar coordinates with origin at O we have

$$\text{grad } \varphi_1 = \mathbf{0}$$

$$\text{grad } \varphi_2 = \left(\frac{1}{6}\left(-2r + \frac{2a^3}{r^2}\right), 0, 0\right)$$

$$\text{grad } \varphi_3 = \left(-\frac{1}{3}\frac{(b^3 - a^3)}{r^2}, 0, 0\right).$$

Thus the components of grad φ are continuous in each of the three given regions $0 \leqslant r < a$, $a \leqslant r < b$, $r \geqslant b$. Hence surface distributions of charge can only be present on the surfaces $r = a, r = b$. On these surfaces, the normal component of grad φ is $\partial\varphi/\partial r$. Therefore on the surface $r = a$,

$$\frac{\partial\varphi_1}{\partial r} \to 0 \quad \text{and} \quad \frac{\partial\varphi_2}{\partial r} \to \frac{1}{6}\left(-2a + \frac{2a}{a^2}\right) = 0.$$

Thus $\partial\varphi/\partial r$ is continuous on $r = a$ and so there is no surface distribution of charge on this surface.

On the surface $r = b$,

$$\frac{\partial\varphi_2}{\partial r} \to \frac{1}{6}\left(-2b + \frac{2a^3}{b^2}\right) = -\frac{1}{3}\frac{(b^3 - a^3)}{b^2}$$

and

$$\frac{\partial\varphi_3}{\partial r} \to -\frac{1}{3}\frac{(b^3 - a^3)}{b^2}.$$

Thus $\partial\varphi/\partial r$ is continuous on $r = b$ and so there is no surface distribution of charge on this surface.

4. $\nabla^2\varphi_1 = 0$ and so there is no volume distribution of charge in the sphere $r \leqslant a$.

$$\nabla^2\varphi_2 = \frac{\partial^2\varphi_2}{\partial r^2} + \frac{2}{r}\frac{\partial\varphi_2}{\partial r} = -\frac{1}{3}\left(1 + \frac{2a^3}{r^3}\right) - \frac{2}{3}\left(1 - \frac{a^3}{r^3}\right) = -1.$$

Thus there is a volume distribution of charge of density ε, the permittivity of the medium, in the spherical shell of internal radius a and external radius b.

$$\nabla^2 \varphi_3 = \frac{\partial^2 \varphi_3}{\partial r^2} + \frac{2}{r}\frac{\partial \varphi_3}{\partial r} = \frac{2}{3}\frac{(b^3 - a^3)}{r^3} - \frac{2}{3}\frac{(b^3 - a^3)}{r^3} = 0.$$

Thus there is no volume distribution of charge in the region $r > b$. Hence the given potential field is due entirely to a volume distribution of charge of density ε within a spherical shell of internal radius a and external radius b.

Example   Determine what distribution of matter gives rise to the gravitational potential distribution

$$\varphi_1 = \gamma \rho_0 (13a^2 - 3z^2) \qquad\qquad r < a.$$
$$\varphi_2 = \gamma a^3 \rho_0 (a^2 r^2 - 3z^2 a^2 + 12r^4)/r^5 \qquad r \geqslant a.$$

where

$$r^2 = x^2 + y^2 + z^2.$$

Investigate the behaviour of the potential for large values of r and hence deduce that the total mass present is $12\rho_0 a^3$.

The expressions for φ_1 and φ_2 involve both the cartesian coordinate z and the spherical polar coordinate r. Since the given regions are defined in terms of r, we shall use spherical polar coordinates and put $z = r \cos \theta$.
Therefore

$$\varphi_1 = \gamma \rho_0 (13a^2 - 3r^2 \cos^2 \theta) \qquad\qquad r < a$$

and

$$\varphi_2 = \frac{\gamma \rho_0 a^3}{r^5} (a^2 r^2 - 3a^2 r^2 \cos^2 \theta + 12r^4) \qquad r \geqslant a$$

$$= \frac{\gamma \rho_0 a^3}{r^3} (a^2 - 3a^2 \cos^2 \theta + 12r^2).$$

1. φ is always finite and so there are no isolated particles of matter present.
2. Since double layers do not exist in gravitation it is not necessary to examine the continuity of φ.
3. Once again, the only surface on which there may be a mass distribution is the surface $r = a$, and on this surface the normal component of grad φ is $\partial \varphi / \partial r$.

$$\frac{\partial \varphi_1}{\partial r} = -6\gamma \rho_0 r \cos^2 \theta \quad \text{and} \quad \frac{\partial \varphi_2}{\partial r} = \gamma \rho_0 a^3 \left\{ \frac{-3(1 - 3\cos^2 \theta)a^2}{r^4} - \frac{12}{r^2} \right\}.$$

Therefore on the surface $r = a$,

$$\frac{\partial \varphi_1}{\partial r} \rightarrow -6\gamma \rho_0 a \cos^2 \theta$$

and

$$\frac{\partial \varphi_2}{\partial r} \to \gamma \rho_0 a\{-3(1 - 3\cos^2 \theta) - 12\} = -3\gamma \rho_0 a(5 - 3\cos^2 \theta).$$

Hence, if there is a surface distribution of mass of density σ on the surface $r = a$,

$$4\pi\gamma\sigma = -6\gamma\rho_0 a \cos^2 \theta - \{-3\gamma\rho_0 a(5 - 3\cos^2 \theta)\}$$
$$= 15\gamma\rho_0 a(1 - \cos^2 \theta).$$

Thus there is a surface distribution of mass of density

$$\frac{15a\rho_0 \sin^2 \theta}{4\pi}$$

on the spherical surface $r = a$.

4. In spherical polar coordinates (r, θ, ψ)

$$\nabla^2 \varphi = \frac{\partial^2 \varphi}{\partial r^2} + \frac{2}{r}\frac{\partial \varphi}{\partial r} + \frac{1}{r^2}\frac{\partial^2 \varphi}{\partial \theta^2} + \frac{\cot \theta}{r^2}\frac{\partial \varphi}{\partial \theta}$$

for φ independent of ψ. Therefore

$$\nabla^2 \varphi_1 = -6\gamma\rho_0 \cos^2 \theta - 12\gamma\rho_0 \cos^2 \theta$$
$$+ 6\gamma\rho_0(2\cos^2 \theta - 1) + 6\gamma\rho_0 \cos^2 \theta$$
$$= -6\gamma\rho_0.$$

Therefore there is a volume distribution of charge of density $6\rho_0/4\pi$ within the sphere $r \leqslant a$. Alternatively, since $\varphi_1 = \gamma\rho_0(13a^2 - 3z^2)$, in terms of cartesian coordinates

$$\nabla^2 \varphi_1 = \frac{\partial^2 \varphi_1}{\partial z^2} = -6\gamma\rho_0.$$

Also

$$\nabla^2 \varphi_2 = \gamma\rho_0 a^3 \Bigg\{ \left(\frac{12a^2}{r^5} - \frac{36a^2 \cos^2 \theta}{r^5} + \frac{24}{r^3}\right)$$
$$+ \frac{2}{r}\left(\frac{-3a^2}{r^4} + \frac{9a^2 \cos^2 \theta}{r^4} - \frac{12}{r^2}\right)$$
$$+ \frac{1}{r^2}\left[\frac{6a^2}{r^3}(2\cos^2 \theta - 1)\right] + \frac{\cot \theta}{r^2}\frac{6a^2 \cos \theta \sin \theta}{r^3} \Bigg\}$$
$$= 0.$$

Thus there is no volume distribution of matter in the region $r > a$. Now, for large values of r,

$$\varphi_2 = \gamma\rho_0 a^3 \left(\frac{a^2}{r^3} - \frac{3a^2 \cos^2 \theta}{r^3} + \frac{12}{r}\right)$$

and so it is seen that as $r \to \infty$, $\varphi_2 \to 12\gamma\rho_0 a^3/r$ the potential due to a total mass $12\rho_0 a^3$ at the origin.

It can quite easily be verified, in the following way, that the total mass present is in fact $12\rho_0 a^3$.

If S is the surface of the sphere $r = a$, then the mass of the surface distribution is

$$\int_S \sigma \, ds = \int_S \frac{15a\rho_0 \sin^2 \theta}{4\pi} \, ds$$

$$= \frac{15a\rho_0}{4\pi} \int_0^\pi \int_0^{2\pi} \sin^2 \theta a^2 \sin \theta \, d\psi \, d\theta.$$

(In spherical polars, on the spherical surface $r = a$, $ds = a \, d\theta a \sin \theta \, d\psi$.)
Hence

$$\int_S \sigma \, ds = \frac{15a^3\rho_0}{4\pi} 2\pi \int_0^\pi \sin \theta (1 - \cos^2 \theta) \, d\theta$$

$$= \frac{15a^3\rho_0}{2} \left[-\cos \theta + \frac{\cos^3 \theta}{3} \right]_0^\pi$$

$$= 10a^3\rho_0.$$

If V is the sphere $r \leqslant a$, then the mass of the volume distribution is

$$\int_V \rho \, dv = \int_V \frac{3\rho_0}{2\pi} \, dv = \frac{3\rho_0}{2\pi} \int_V dv$$

$$= \frac{3\rho_0}{2\pi} \frac{4}{3} \pi a^3$$

$$= 2a^3\rho_0.$$

Therefore the total mass of the distribution is $12a^3\rho_0$.

Example　The density of a distribution of charge has value

$$\frac{1}{2} \frac{\rho_0}{a} r(3 \cos^2 \theta - 1)$$

for $r < a$ and is zero for $r > a$ where θ is the angle the radius vector **r** *makes with a fixed direction. Verify that appropriate forms for the potentials φ_1 and φ_2 inside and outside the sphere $r = a$ are respectively*

$$\varphi_1 = \left(Cr^2 - \frac{\rho_0}{12\varepsilon a} r^3 \right) (3 \cos^2 \theta - 1)$$

and

$$\varphi_2 = Dr^{-3}(3 \cos^2 \theta - 1)$$

where ε is the permittivity of the material of the sphere. Determine the constants **C** *and* **D** *in order that the appropriate boundary conditions may be satisfied at* r = a.

$$\nabla^2 \varphi_1 = \left(2C - \frac{\rho_0}{2a\varepsilon} r\right)(3\cos^2\theta - 1) + \frac{2}{r}\left(2Cr - \frac{\rho_0}{4a\varepsilon} r^2\right)(3\cos^2\theta - 1)$$

$$+ \left(C - \frac{\rho_0}{12a\varepsilon} r\right)\{-6(2\cos^2\theta - 1)\}$$

$$+ \frac{\cot\theta}{r^2}\left(Cr^2 - \frac{\rho_0}{12a\varepsilon} r^3\right)(-6\cos\theta\sin\theta)$$

$$= \frac{-\rho_0}{2a\varepsilon} r(3\cos^2\theta - 1).$$

Therefore the volume density of charge for r < a is

$$\frac{1}{2}\frac{\rho_0}{a} r(3\cos^2\theta - 1).$$

$$\nabla^2 \varphi_2 = \frac{12D}{r^5}(3\cos^2\theta - 1) - \frac{6D}{r^5}(3\cos^2\theta - 1)$$

$$+ \frac{D}{r^5}\{-6(2\cos^2\theta - 1)\} + \frac{D\cot\theta}{r^5}(-6\cos\theta\sin\theta) = 0.$$

Therefore volume density of charge is zero for r > a.

Now φ must be continuous across the surface r = a and so

$$\left(Ca^2 - \frac{\rho_0}{12\varepsilon a} a^3\right)(3\cos^2\theta - 1) = \frac{D}{a^3}(3\cos^2\theta - 1) \quad \text{for all } \theta.$$

Therefore

$$Ca^2 - \frac{\rho_0}{12\varepsilon} a^2 = \frac{D}{a^3}.$$

Also, ∂φ/∂r must be continuous across the surface r = a and so

$$\left(2Ca - \frac{\rho_0 a}{4\varepsilon}\right)(3\cos^2\theta - 1) = \frac{-3D}{a^4}(3\cos^2\theta - 1) \quad \text{for all } \theta.$$

Therefore

$$2Ca - \frac{\rho_0 a}{4\varepsilon} = -\frac{3D}{a^4}.$$

Hence

$$\frac{\rho_0 a^2}{12\varepsilon} = \frac{5D}{a^3}$$

and so

$$D = \frac{\rho_0 a^5}{60\varepsilon}.$$

Then

$$C = \frac{D}{a^5} + \frac{\rho_0}{12\varepsilon} = \frac{\rho_0}{10\varepsilon}.$$

2.61 Potential energy (or self-energy) of distributions of charge

Suppose charges $e_1, e_2, \ldots, e_n$ are brought one at a time from infinity (where the potential is zero) to the points $P_1, P_2, \ldots, P_n$ respectively in a medium with permittivity ε. No work is done in bringing in the charge e_1. However, once this charge is in position, there will be set up an electrostatic force field $\mathbf{E}_1 = \dfrac{e_1}{4\pi\varepsilon r^2} \hat{\mathbf{r}}$ where $\hat{\mathbf{r}}$ is unit vector in the direction of $\mathbf{r}$ the position vector with P_1 as origin, and $r = |\mathbf{r}|$. Then the work done against this force field in bringing the charge e_2 from infinity to its position at P_2 in the field is

$$\int_{\infty}^{r_{12}} -e_2 \mathbf{E}_1 \cdot d\mathbf{r}$$

where r_{12} is the length of $P_1 P_2$. Now substituting the above expression for $\mathbf{E}_1$ in this integral we obtain

$$-\frac{e_1 e_2}{4\pi\varepsilon} \int_{\infty}^{r_{12}} \frac{dr}{r^2} = \frac{e_1 e_2}{4\pi\varepsilon} \left[\frac{1}{r} \right]_{\infty}^{r_{12}}$$

$$= \frac{e_1 e_2}{4\pi\varepsilon r_{12}}.$$

It is clearly seen that, if φ_{ij} denotes the electrostatic potential at the point P_i due to the charge e_j at the point P_j, then the above expression for the work done can be written as $e_1 \varphi_{12}$ or $e_2 \varphi_{21}$ or symmetrically as $\frac{1}{2}(e_1 \varphi_{12} + e_2 \varphi_{21})$.

Similarly, the work done in introducing the charge e_3 at the point P_3 in the presence of the charges e_1 and e_2 is

$$\frac{1}{4\pi\varepsilon} \left(\frac{e_1 e_3}{r_{13}} + \frac{e_2 e_3}{r_{23}} \right) \quad \text{or} \quad \tfrac{1}{2}(e_1 \varphi_{13} + e_3 \varphi_{31} + e_2 \varphi_{23} + e_3 \varphi_{32}).$$

Hence the total work done at this stage may be written in either of the alternative forms

$$\frac{1}{4\pi\varepsilon} \left(\frac{e_1 e_2}{r_{12}} + \frac{e_1 e_3}{r_{13}} + \frac{e_2 e_3}{r_{23}} \right) = \frac{1}{8\pi\varepsilon} \sum_{j=1}^{3} \sum_{i=1}^{3} \frac{e_i e_j}{r_{ij}} \left(\text{since } \frac{e_1 e_2}{r_{12}} = \frac{e_2 e_1}{r_{21}} \text{ etc.} \right)$$

or

$$\tfrac{1}{2}(e_1\varphi_{12} + e_2\varphi_{21} + e_1\varphi_{13} + e_3\varphi_{31} + e_2\varphi_{23} + e_3\varphi_{32})$$
$$= \tfrac{1}{2}\{e_1(\varphi_{12} + \varphi_{13}) + e_2(\varphi_{21} + \varphi_{23}) + e_3(\varphi_{31} + \varphi_{32})\}$$
$$= \tfrac{1}{2}\{e_1\varphi_1 + e_2\varphi_2 + e_3\varphi_3\}$$
$$= \tfrac{1}{2}\sum_{i=1}^{3} e_i\varphi_i$$

where φ_1 is the potential at P_1 due to the charges e_2 and e_3 at the points P_2 and P_3 and φ_2 and φ_3 are similarly defined. This work done is known as the *potential energy* or *self-energy* of the system of charges e_1, e_2, e_3.

Continuing this process until all the charges e_1, e_2, ..., e_n have been brought in we see that the potential energy or self-energy W of the whole system is given by

$$W = \frac{1}{8\pi\varepsilon} \sum_{j=1}^{n} \sum_{i=1}^{n} \frac{e_i e_j}{r_{ij}}$$
$$= \tfrac{1}{2}\sum_{i=1}^{n} e_i\varphi_i$$

where now φ_i is the potential at the point P_i due to all the charges except the charge e_i at P_i itself.

If, instead of discrete point charges, the system is composed of surface and volume distributions of charge,

$$W = \tfrac{1}{2}\int \varphi\sigma\,\mathrm{d}s + \tfrac{1}{2}\int \varphi\rho\,\mathrm{d}v$$

where the first integral is taken over the appropriate surfaces and the second is taken throughout the whole region being considered.

In gravitation, since the force between two point masses is an attraction ($\mathbf{F} = +\mathrm{grad}\,\varphi$), the corresponding result is

$$W = - \tfrac{1}{2}\int \varphi\sigma\,\mathrm{d}s - \tfrac{1}{2}\int \varphi\rho\,\mathrm{d}v.$$

Returning our attention to the electrostatic case, since $\mathrm{div}\,\mathbf{D} = \rho$,

$$\tfrac{1}{2}\int \varphi\rho\,\mathrm{d}v = \tfrac{1}{2}\int \varphi\,\mathrm{div}\,\mathbf{D}\,\mathrm{d}v$$

$$= \tfrac{1}{2}\int \{\mathrm{div}\,(\varphi\mathbf{D}) - \mathbf{D}\,.\,\mathrm{grad}\,\varphi\}\,\mathrm{d}v$$

$$= \tfrac{1}{2}\int \varphi\mathbf{D}\,.\,\mathbf{n}\,\mathrm{d}s + \tfrac{1}{2}\int \mathbf{E}\,.\,\mathbf{D}\,\mathrm{d}v.$$

The above surface integral must be taken over the whole boundary of the region. That is over the sphere of radius R where $R \to \infty$ and over the surfaces of the conductors on which the charges lie. Since on the above sphere of radius R, $\varphi\mathbf{D}$ is of order $1/R^3$ and S, the area of the surface of the sphere, is of order R^2, the integral over this sphere $\to 0$ as $R \to \infty$. Now for the second part $\mathbf{n}$ is positive outwards from the volume being considered which in this case means inwards on the surfaces of the conductors.

Thus we now have

$$\tfrac{1}{2}\int \varphi\rho \, dv = - \tfrac{1}{2}\int \varphi\mathbf{D} \cdot \mathbf{n}_1 \, ds + \tfrac{1}{2}\int \mathbf{E} \cdot \mathbf{D} \, dv$$

where $\mathbf{n}_1 = -\mathbf{n}$ is outwards on the surface of the conductors. Also, since $\mathbf{D} = 0$ inside a conductor and σ is equal to the discontinuity in the normal component of $\mathbf{D}$ on moving through the surface,

$$\tfrac{1}{2}\int \varphi\sigma \, ds = \tfrac{1}{2}\int \varphi\mathbf{D} \cdot d\mathbf{s} = \tfrac{1}{2}\int \varphi\mathbf{D} \cdot \mathbf{n}_1 \, ds.$$

Hence

$$W = \tfrac{1}{2}\int \mathbf{E} \cdot \mathbf{D} \, dv.$$

If the medium is isotropic and homogeneous so that $\mathbf{D} = \varepsilon\mathbf{E}$ for constant ε, then

$$W = \frac{\varepsilon}{2}\int E^2 \, dv.$$

The corresponding result in magnetostatics is

$$W = \tfrac{1}{2}\int \mathbf{H} \cdot \mathbf{B} \, dv$$

which reduces, in the same way as above, to $\tfrac{1}{2}\mu \int H^2 \, dv$ for a uniform, homogeneous medium. Now $\int \mathbf{E} \cdot \mathbf{D} \, dv$ is taken throughout the volume outside the conductors. But since $\mathbf{E} = 0$ inside a conductor, it will not make any difference if the integral is in fact evaluated throughout the whole of space. The electrostatic energy may thus be regarded as being distributed throughout the field and as having density $\tfrac{1}{2}\mathbf{E} \cdot \mathbf{D}$ per unit volume. For a homogeneous, isotropic medium the electrostatic energy density is then $\tfrac{1}{2}\varepsilon E^2$ per unit volume.

Example   Find the self-energy of a uniform sphere of mass M and radius a.

The density ρ of the sphere is given by

$$\rho = \frac{M}{4\pi a^3/3}.$$

As we have already seen, the potential at a point distant r ($< a$) from the centre of the sphere is $2\pi\gamma\rho(a^2 - \frac{1}{3}r^2)$.

Therefore the self-energy W is given by

$$W = -\frac{1}{2}\int \varphi\rho \, dv$$

where the integral is to be evaluated throughout the given sphere.

Therefore

$$W = -\frac{1}{2}\int_0^a 2\pi\gamma\rho(a^2 - \frac{1}{3}r^2)\rho 4\pi r^2 \, dr$$

on dividing the sphere into thin shells of radius r and surface area $4\pi r^2$. Hence

$$W = -4\pi^2\gamma\rho^2 \int_0^a (a^2 r^2 - \frac{1}{3}r^4) \, dr$$

$$= -4\pi^2\gamma\rho^2(4/15a^5)$$

$$= \frac{-3\gamma M^2}{5a}.$$

Example Determine the self-energy of a charge Q distributed uniformly over the surface of a conducting sphere of radius a in free space.

The potential φ on the surface of the sphere is $Q/4\pi\varepsilon_0 a$ (see a previous example in section 2.22 on page 120). The surface density σ of charge on the surface of the sphere is $Q/4\pi a^2$. Therefore the self-energy W is given by

$$W = \frac{1}{2}\int_S \varphi\sigma \, ds = \frac{1}{2}\int_S \frac{Q}{4\pi\varepsilon_0 a} \cdot \frac{Q}{4\pi a^2} \, ds$$

where S is the surface of the sphere.

Therefore

$$W = \frac{Q^2}{32\pi^2\varepsilon_0 a^3}\int_S ds = \frac{Q^2}{32\pi^2\varepsilon_0 a^3} 4\pi a^2$$

$$= \frac{Q^2}{8\pi a\varepsilon_0}.$$

Alternatively, the electrostatic field $\mathbf{E}$ is given by

$$\mathbf{E} = \begin{cases} \dfrac{Q}{4\pi\varepsilon_0 r^2} \, \hat{\mathbf{r}} & \text{for} \quad r > a \\ 0 & \text{for} \quad r < a. \end{cases}$$

Then

$$W = \frac{\varepsilon_0}{2} \int_V E^2 \, dv$$

where V is all space excluding the sphere of radius a.

Hence

$$W = \frac{\varepsilon_0}{2} \int_V \left(\frac{Q}{4\pi\varepsilon_0 r^2}\right)^2 dv$$

$$= \lim_{R \to \infty} \frac{Q^2}{32\pi^2\varepsilon_0} \int_a^R \int_0^{2\pi} \int_0^\pi \frac{1}{r^4} r^2 \sin\theta \, d\theta \, d\psi \, dr$$

in terms of spherical polar coordinates r, θ, ψ.

Therefore

$$W = \frac{Q^2}{32\pi^2\varepsilon_0} \int_a^\infty \int_0^{2\pi} \frac{1}{r^2} \left[-\cos\theta\right]_0^\pi d\psi \, dr$$

$$= \frac{Q^2}{16\pi^2\varepsilon_0} \int_a^\infty \frac{1}{r^2} \left[\psi\right]_0^{2\pi} dr$$

$$= \frac{Q^2}{8\pi\varepsilon_0} \left[-\frac{1}{r}\right]_a^\infty$$

$$= \frac{Q^2}{8\pi a\varepsilon_0}.$$

2.62 Potential energy of a dipole in a given field

Let the given field have intensity $\mathbf{E} = -\text{grad } \varphi$.

The work done in bringing charges $-e$, $+e$ to the points A and B respectively in the field is

$$-e\varphi_A + e\varphi_B$$

where φ_A and φ_B are the potentials at the points A and B respectively. Now if $|\overline{AB}|$ is small, this work done is equal to

$$-e\varphi_A + e\{\varphi_A + (\text{grad } \varphi)_A \cdot \overline{AB} + \text{smaller terms}\}.$$

Then, taking the limit of this expression as $e \to \infty$, $|\overline{AB}| \to 0$ such that $e\overline{AB} \to \mathbf{m}$ we obtain $(\text{grad } \varphi)_A \cdot \mathbf{m}$, which is equal to $-\mathbf{E}_A \cdot \mathbf{m}$. That is the self-energy of the dipole $\mathbf{m}$ at the point A in the field $\mathbf{E}$ is $-\mathbf{E}_A \cdot \mathbf{m}$.

2.63 Mutual potential energy of two dipoles

Suppose a dipole of moment $\mathbf{m}$ is at a fixed point O and another dipole of moment $\mathbf{m}'$ is then introduced at the point having position vector $\mathbf{r}$ relative to O.

From the above, the potential energy W of the dipole $\mathbf{m}'$ is given by

$$W = -\mathbf{E}\cdot\mathbf{m}'$$

where $\mathbf{E}$ is the field due to the dipole $\mathbf{m}$. Hence

$$W = \frac{1}{4\pi\varepsilon}\left(\frac{\mathbf{m}\cdot\mathbf{m}'}{r^3} - \frac{3(\mathbf{m}\cdot\mathbf{r})(\mathbf{m}'\cdot\mathbf{r})}{r^5}\right).$$

The potential energy of $\mathbf{m}$ when placed in the field of $\mathbf{m}'$ is obtained by interchanging $\mathbf{m}$ and $\mathbf{m}'$ and putting $-\mathbf{r}$ for $\mathbf{r}$ and is thus seen to be identical with the above. Thus W is called the *mutual potential energy* of the two dipoles.

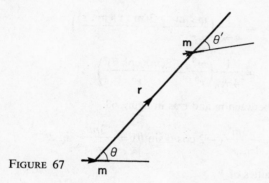

FIGURE 67

Now if α is the angle between $\mathbf{m}$ and $\mathbf{m}'$ and θ, θ' are the angles between the axes of $\mathbf{m}$ and $\mathbf{m}'$ and the line joining them (figure 67), then

$$W = \frac{mm'}{4\pi\varepsilon r^3}(\cos\alpha - 3\cos\theta\cos\theta').$$

If also $\mathbf{m}$ and $\mathbf{m}'$ are coplanar, then $\alpha = \theta - \theta'$ and so

$$W = \frac{mm'}{4\pi\varepsilon r^3}(\sin\theta\sin\theta' - 2\cos\theta\cos\theta').$$

For two dipoles $\mathbf{m}$ and $\mathbf{m}'$ at points A and B respectively in a field $\mathbf{E}$ the potential energy is

$$W - \mathbf{E}_{\mathrm{A}}\cdot\mathbf{m} - \mathbf{E}_{\mathrm{B}}\cdot\mathbf{m}'.$$

Similarly, if two magnetic dipoles (or small magnets) $\mathbf{m}$ and $\mathbf{m}'$ are at points A and B respectively in a magnetic field $\mathbf{H}$, the potential energy is

$$W - \mathbf{H}_{\mathrm{A}}\cdot\mathbf{m} - \mathbf{H}_{\mathrm{B}}\cdot\mathbf{m}'.$$

Example   Two electrostatic dipoles each with moment of magnitude m have their centres a fixed distance r apart in free space. They are free to rotate about their centres in the same plane and in such a way that they always remain parallel and in the same sense. Determine the positions of equilibrium and investigate their stability.

If a uniform field of strength E ($< 3m/4\pi\varepsilon_0 r^3$) is now applied in the direction of the line joining the centres of the dipoles, show that the above positions of stable equilibrium are still positions of stable equilibrium.

In textbooks on mechanics it is shown that for positions of equilibrium of a given system the potential energy has a stationary value and that the equilibrium is then stable or unstable according as the stationary value is a minimum or a maximum. We therefore proceed by determining the potential energy V of the given system.

$$V = \frac{1}{4\pi\varepsilon_0}\left(\frac{\mathbf{m}\cdot\mathbf{m}}{r^3} - \frac{3(\mathbf{m}\cdot\mathbf{r})(\mathbf{m}\cdot\mathbf{r})}{r^5}\right)$$

$$= \frac{1}{4\pi\varepsilon_0}\left(\frac{m^2}{r^3} - \frac{3(mr\cos\theta)^2}{r^5}\right)$$

where θ is the angle between $\mathbf{m}$ and $\mathbf{r}$ as in figure 68.

$$\frac{\partial V}{\partial\theta} = \frac{-3m^2}{4\pi\varepsilon_0 r^3}(-2\cos\theta\sin\theta) = \frac{3m^2}{4\pi\varepsilon_0 r^3}\sin 2\theta$$

and for stationary values of V,

$$\frac{\partial V}{\partial\theta} = 0.$$

Therefore $\sin 2\theta = 0$ and so $\theta = 0, \tfrac{1}{2}\pi, \pi, \tfrac{3}{2}\pi$. (These are the only physically distinct roots.) Now

$$\frac{\partial^2 V}{\partial\theta^2} = \frac{3m^2}{2\pi\varepsilon_0 r^3}\cos 2\theta$$

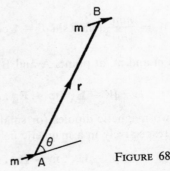

FIGURE 68

which is positive for $\theta = 0, \pi$ and negative for $\theta = \frac{1}{2}\pi, \frac{3}{2}\pi$. Thus V is a minimum, and so the equilibrium stable, when $\theta = 0, \pi$.

Let the potential energy, after the additional field is applied, be V'. Then

$$V' = V - \mathbf{E}_A \cdot \mathbf{m} - \mathbf{E}_B \cdot \mathbf{m}$$

$$= \frac{1}{4\pi\varepsilon_0}\left(\frac{m^2}{r^3} - \frac{3m^2\cos^2\theta}{r^3}\right) - 2Em\cos\theta \qquad (|\mathbf{E}_A| = |\mathbf{E}_B| = E).$$

Now

$$\frac{\partial V'}{\partial\theta} = \frac{3m^2}{4\pi\varepsilon_0 r^3}\sin 2\theta + 2Em\sin\theta = 0$$

when

$$\sin\theta\left(\frac{3m}{4\pi\varepsilon_0 r^3}\cos\theta + E\right) = 0.$$

That is, when $\sin\theta = 0$ (and so $\theta = 0, \pi$) or

$$\cos\theta = \frac{4\pi\varepsilon_0 r^3 E}{3m}.$$

Then

$$\frac{\partial^2 V'}{\partial\theta^2} = \frac{3m^2}{2\pi\varepsilon_0 r^3}\cos 2\theta + 2Em\cos\theta.$$

Thus when

$$\theta = 0, \qquad \frac{\partial^2 V'}{\partial\theta^2} = \frac{3m^2}{2\pi\varepsilon_0 r^3} + 2Em > 0$$

and when

$$\theta = \pi, \qquad \frac{\partial^2 V'}{\partial\theta^2} = \frac{3m^2}{2\pi\varepsilon_0 r^3} - 2Em > 0 \quad \text{if} \quad E < \frac{3m}{4\pi\varepsilon_0 r^3}.$$

Therefore $\theta = 0, \pi$ still represent positions of stable equilibrium.

Exercises

1. Find the distribution of matter which gives rise to the gravitational potential

$$\gamma\{9a^3 - r^3 + ar^2(3\cos^2\theta - 1)\}, \qquad (r < a)$$

$$\gamma\left\{\frac{8a^4}{r} + \frac{a^6}{r^3}(3\cos^2\theta - 1)\right\}, \qquad (r > a)$$

where $r^2 = x^2 + y^2 + z^2$ and γ is the gravitational constant.

2. Find the distribution of charge which gives rise to the potential

$$\frac{ea^2}{r} + \frac{ea^4}{5}\left(\frac{3z^2}{r^5} - \frac{1}{r^3}\right), \qquad r > a$$

and

$$ea + \frac{e}{5a}(3z^2 - r^2), \qquad r < a,$$

where $r^2 = x^2 + y^2 + z^2$. The permittivity of the medium is ε. Calculate also the self-energy of the distribution.

3. The attraction at a point inside a sphere of radius a in which the density is a function of the distance r from the centre is $\frac{4}{3}\pi\gamma k r^3/a^2$ ($r \leqslant a$) where k is a constant. Find the mean density of the sphere and show that its self potential energy is

$$-\frac{5}{9}\frac{\gamma M^2}{a}$$

where M is the total mass of the sphere.

4. Determine the charge distribution which gives rise to the potential field

$$\varphi_1 = Ar \cos \theta \quad \text{for} \quad r < a,$$

$$\varphi_2 = \frac{Aa^3}{r^2} \cos \theta \quad \text{for} \quad r > a$$

where r and θ are spherical polar coordinates and A, a are constants. Calculate the self-energy of the distribution.

5. Show that the self potential energy of a distribution of charge of uniform density ρ within the spherical shell with internal radius b and external radius c is

$$\frac{\rho^2}{30\varepsilon^2}(2c^5 - 5c^2b^3 + 3b^5)$$

where ε is the permittivity of the material of the shell.

6. Show that a dipole of moment $\mathbf{m}$ free to rotate about its centre in a uniform electric field $\mathbf{E}$ has two possible positions of equilibrium, one of which is stable and one of which is unstable.

7. Two small magnets each having moment of magnitude m are held in free space at the points A and B distant a apart in the uniform magnetic field $\mathbf{H}$ where $\mathbf{H}$ has the direction and sense of $\overline{BA}$. The magnets make small angles α and β on the same side of AB and in the same plane. Show that to the second order in the small quantities α and β their potential energy may be taken as

$$\frac{m}{4\pi\mu_0 a^3}\{(m - 2\pi\mu_0 a^3 H)(\alpha^2 + \beta^2) + m\alpha\beta\} + V_0$$

where V_0 is a constant independent of α and β. Hence show that both magnets can rest in stable equilibrium along AB if $m > 4\pi\mu_0 a^3 H$.

8. Two dipoles having moments of magnitude $2m$ and $5m$ are free to rotate, in a uniform field of intensity **E**, about their fixed centres. If **E** is perpendicular to the line of centres, determine the condition that a position of stable equilibrium should exist with the axes of the two dipoles pointing in the direction of **E**.

2.7 Soil mechanics

2.71 *Flow of water through soils* (Seepage equation).

In order to cause the water to flow through the soil it is necessary for a pressure gradient ∇p to exist in the water. Then the quantity of water which percolates per unit time through unit area normal to the direction of flow is called the *discharge velocity*. For fine sands, and for soils finer than sand, it is found that the discharge velocity **v** is given very accurately by

$$\mathbf{v} = -\frac{K}{\eta}\nabla p$$

where K is a constant called the *coefficient of permeability* of the soil and η is the *coefficient of viscosity* of water. Very often civil engineers write this in the form

$$\mathbf{v} = -\left(\frac{K\rho g}{\eta}\right)\left(\frac{1}{\rho g}\nabla p\right)$$
$$= -k\left(\frac{1}{\rho g}\nabla p\right)$$

where $(1/\rho g)\nabla p$ is called the *hydraulic gradient* and k is also often referred to as the coefficient of permeability. Within the range of temperatures encountered under field conditions the values of η and of ρ are almost constant so that k can be regarded as constant, giving the result that discharge velocity is proportional to hydraulic gradient. This result is known as *Darcy's law*.

Now, assuming that water is perfectly incompressible, the continuity equation gives that

$$\text{div } \mathbf{v} = 0.$$

Put $\varphi = (-K/\eta)p$ where φ is called the potential, then

$$\mathbf{v} = \nabla\varphi$$

and

$$\text{div } \mathbf{v} = \nabla^2\varphi = 0.$$

That is, we again obtain Laplace's equation.

When considering foundation problems the civil engineer is however normally interested only in two-dimensional types of flow. For example the flow of water out of a storage reservoir through the soil located beneath the foundation of a dam can be regarded as two dimensional and the solution of this problem will be considered in detail in a later section.

2.72 The consolidation equation

When deriving the seepage equation use was made of the continuity equation div $\mathbf{v} = 0$. This implies that the quantity of water flowing out of any element of the soil in a given time is equal to the quantity of water flowing into the element in the same time. We now consider the case when, due perhaps to increased pressure on top of the soil, water is in fact squeezed out. That is the volume occupied by water in the soil is decreased. Such a process is called a process of consolidation.

Formally, any process involving a decrease of the water content of a saturated soil without replacement of the water by air is called a *process of consolidation.*

The external compressive pressure applied to a mass of soil is called the *total pressure P.* The part of the total pressure which, during the consolidation process of a soil, is taken up by water is called the *neutral pressure, p.* The other part of the total pressure, which is transferred intergranularly on the soil particles, is called the *effective pressure p_e.* Then

$$P = p + p_e.$$

During a consolidation process a gradual decrease in neutral pressure takes place along with a gradual increase in effective pressure in such a way that the sum $(p + p_e)$ remains constant. Now let τ be the volume of water per unit volume of soil. Then $-(\partial \tau / \partial t)$ gives the volume of water flowing out of unit volume of soil in unit time and the continuity equation now takes the form

$$\operatorname{div} \mathbf{v} = -\frac{\partial \tau}{\partial t}.$$

Also, $(\partial \tau / \partial t)$ is put equal to $m(\partial p / \partial t)$ where m is called the *coefficient of volume decrease.*†

---

† It is shown in engineering text books that if e is the void ratio, that is the ratio of the volume of water to the volume of solid in the soil, then $m = \dfrac{-\dfrac{\partial e}{\partial P}}{1 + e}$.

Therefore

$$\text{div } \mathbf{v} = -m\,\frac{\partial p}{\partial t}$$

But

$$\mathbf{v} = \frac{-K}{\eta}\,\nabla p$$

and so

$$\nabla^2 p = \frac{m\eta}{K}\,\frac{\partial p}{\partial t}$$

or

$$\frac{\partial p}{\partial t} = c\nabla^2 p$$

where $c = K/m\eta$ is called the *coefficient of consolidation*.

This is the consolidation equation and in cartesian coordinates x, y and z it has the form

$$\frac{\partial p}{\partial t} = c\left(\frac{\partial^2 p}{\partial x^2} + \frac{\partial^2 p}{\partial y^2} + \frac{\partial^2 p}{\partial z^2}\right).$$

Solutions of this equation (which has the same form as the heat conduction equation derived later) will be considered in two space dimensions in section 7.21.

3

Method of Images

We have seen that in electrostatics for regions in which there are no charges and in hydrodynamics for regions in which there are no sources or sinks, the potential function φ satisfies Laplace's equation $\nabla^2 \varphi = 0$.

In this chapter we consider a particular method of solution of Laplace's equation which is useful in certain special cases. In order to obtain a complete solution however it is necessary to have suitable boundary conditions to be satisfied by the potential φ. This means that we will normally be given (in the physics of a problem) values of φ or of its normal derivative at all points of the boundary of the region for which the solution is required.

For example, in electrostatics, the boundary of the given region may be maintained at a constant potential in which case φ is given on the boundary. Alternatively, in fluid dynamics, since there can be no flow across a rigid boundary in the fluid, the derivative of φ normal to such a boundary must be zero.

In the example on p. 199 Laplace's equation has to be solved in two separate regions having a common boundary. The required boundary conditions are then not given explicitly as above, but are obtained by fitting together, on the common boundary, the two solutions for the separate regions.

3.1 Fundamental theorem

If at each point of the boundary of a given region V in which Laplace's equation is satisfied, φ or its normal derivative $\partial \varphi / \partial n$ is known, then there is only one solution φ which satisfies the given conditions unless φ itself is not specified anywhere, in which case the solution is indeterminate to the extent of an additive constant. This result is essential for the work of this section and is proved as follows.

Suppose two single-valued and finite solutions φ_1 and φ_2 exist.

178

Therefore

$$\nabla^2 \varphi_1 = 0 \quad \text{and} \quad \nabla^2 \varphi_2 = 0$$

in the region V contained within the surface S. Now consider

$$\Phi = \varphi_1 - \varphi_2.$$
$$\nabla^2 \Phi = \nabla^2(\varphi_1 - \varphi_2) = \nabla^2 \varphi_1 - \nabla^2 \varphi_2 = 0.$$

At points on the boundary of the given region where φ is given, $\Phi = 0$ and at points on the boundary where $\partial \varphi / \partial n$ is specified, $\partial \Phi / \partial n = 0$. Thus $\Phi(\partial \Phi / \partial n) = 0$ at all points of the boundary.

Now in the region V

$$\text{div } (\Phi \text{ grad } \Phi) = \Phi \nabla^2 \Phi + (\text{grad } \Phi)^2$$
$$= (\text{grad } \Phi)^2 \quad \text{since} \quad \nabla^2 \Phi = 0 \text{ in V.}$$

Therefore

$$\int_V \text{div } (\Phi \text{ grad } \Phi) \, dv = \int_V (\text{grad } \Phi)^2 \, dv.$$

Now applying Gauss' theorem to the integral on the left hand side we obtain

$$\int_V \text{div } (\Phi \text{ grad } \Phi) \, dv = \int_S \Phi \text{ grad } \Phi \cdot ds = \int_S \Phi \frac{\partial \Phi}{\partial n} \, ds$$
$$= 0$$

since $\Phi(\partial \Phi / \partial n)$ is zero at each point of S.

Therefore

$$\int_V (\text{grad } \Phi)^2 \, dv = 0.$$

That is the integral of a strictly non-negative integrand is zero. Therefore the integrand $(\text{grad } \Phi)^2$ must itself be zero everywhere in the domain of integration. Hence grad $\Phi = 0$ everywhere in V.

Therefore Φ = constant everywhere in V and this constant must be zero if φ is specified anywhere on the boundary.

Thus $\varphi_1 = \varphi_2$ unless only $\partial \varphi / \partial n$ is specified on the boundary, and then $\varphi_1 - \varphi_2$ = constant.

The important consequence of this result is that if by any means whatever we can find a potential function φ satisfying Laplace's equation and all the required boundary conditions of a particular problem, then it is unique (apart perhaps from an additive constant). Also, since it is generally the associated vector field given by grad φ which is of primary interest, the arbitrary constant in the above is relatively unimportant.

The above theorem can be extended to infinite regions.

3.2 Electrostatic images

The type of problem to be dealt with here is essentially that of determining the effect of placing a conductor or dielectric in a given electrostatic field.

3.21 Images in conductors

For a conductor, on which there is no surface distribution of dipoles, in a given field, the potential will be constant inside the conductor and continuous at its surface and the main problem is to determine the form of the potential function in the region outside. This means obtaining, for this region, a solution of Laplace's equation which approaches a given constant value on the surface of the conductor in addition to satisfying any relevant boundary conditions associated with the undisturbed problem. This is done here by the method of images which consists of finding a system of point charges and dipoles (called 'image' charges and dipoles) all inside the conductor and such that the potential due to the given field and these images satisfies the above conditions. Now the potential distribution for the system of point charges and dipoles is easily obtained as in the previous chapter and by the uniqueness result proved above, this must also be the potential distribution outside the conductor for the given problem.

Readers familiar with the elementary theory of optics will recall analogous problems involving mirrors and light sources. For example, given a point light source a distance d in front of a plane mirror, then the paths of the light beams after reflection at the mirror may be determined by considering an 'image' source placed a distance d behind the mirror.

3.22 Point charge and an infinite, earthed, conducting plane

(The fact that the plane is earthed means that it is maintained at zero potential.)

For a point charge $+e$ at a point A, the equipotentials and lines of force are as shown in figure 69. Now suppose that an infinite, earthed (that is at zero potential) conducting plane is placed in this field at a perpendicular distance a from the point A. Choose an orthogonal cartesian coordinate system (x,y,z) such that the conducting plane coincides with the plane $x = 0$ and A is the point $(a,0,0)$ (figure 70). Since the plane is conducting, the potential φ at all points on it is constant and since it is earthed, this constant is zero. The problem is then to determine φ for the region $x \geqslant 0$ and excluding the point A such that $\nabla^2\varphi = 0$ in this region and $\varphi = 0$ on the surface $x = 0$.

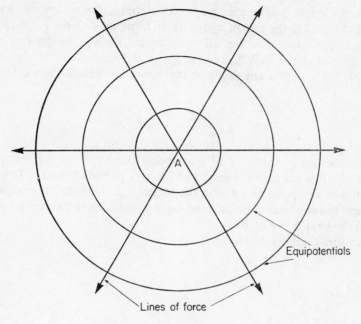

Equipotentials

Lines of force

FIGURE 69

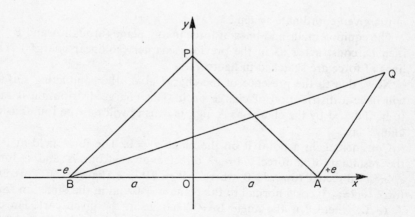

FIGURE 70

181

With the method of images we must determine the system of charges (image charges) in the region $x < 0$ which together with the given charge at A has the surface $x = 0$ as an equipotential surface. In this case, it is easily seen that the only image charge is the charge $-e$ at the point B$(-a,0,0)$. For if P is any point on the plane $x = 0$ then the potential at P is

$$\frac{1}{4\pi\varepsilon_0}\left(\frac{e}{AP} - \frac{e}{BP}\right)$$

where ε_0 is the permittivity of free space, but since by symmetry the length of AP is equal to the length of BP, this potential is zero. Thus the potential at any point Q(x,y,z) in the region $x \geqslant 0$ due to the charge $+e$ at A and the conducting plane is the same as that due to the charge $+e$ at A and the charge $-e$ at B.

Therefore the potential at Q is

$$\frac{1}{4\pi\varepsilon_0}\left(\frac{e}{AQ} - \frac{e}{BQ}\right)$$

$$= \frac{1}{4\pi\varepsilon_0}\left(\frac{e}{\sqrt{(x-a)^2 + y^2 + z^2}} - \frac{e}{\sqrt{(x+a)^2 + y^2 + z^2}}\right)$$

in the given coordinate system.

The equipotentials and lines of force in any plane through A and B can then be constructed as in the previous chapter (collinear charges). The lines of force are sketched in figure 71.

Now, due to the presence of the charge at A, the conducting surface will have a distribution of charge on it. This charge distribution is said to be induced by the charge at A and its density will now be found using Gauss' law.

Consider again a point P on the plane $x = 0$. The force field at P is the resultant of a force $e/4\pi\varepsilon_0 r^2$ directed away from A and a force $e/4\pi\varepsilon_0 r^2$ directed towards B where AP = BP = r (figure 72). This is a force $(e/2\pi\varepsilon_0 r^2)\cos\theta$ normal to the plane $x = 0$ and in the direction from A to B, where θ is the angle BÂP. That is, in the given cartesian coordinate system the force at P has components $(-e\cos\theta/2\pi\varepsilon_0 r^2, 0, 0)$. Thus, since the force field is zero in the region $x < 0$, it is seen that, on moving across the surface $x = 0$ in the direction of increasing x, the component of force normal to this surface is discontinuous by amount $-e\cos\theta/2\pi\varepsilon_0 r^2$.

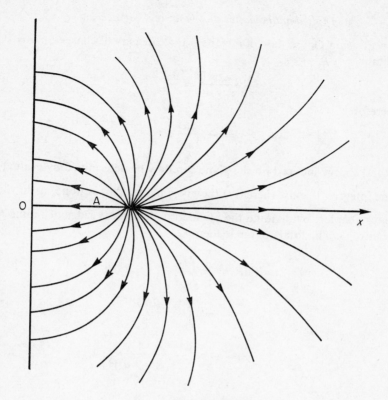

FIGURE 71

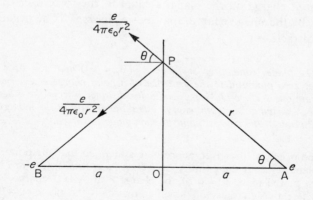

FIGURE 72

Hence, using Gauss' law, if σ is the surface density of charge on $x = 0$, we obtain

$$-\frac{e \cos \theta}{2\pi\varepsilon_0 r^2} = \frac{\sigma}{\varepsilon_0}.$$

Therefore

$$\sigma = -\frac{ea}{2\pi r^3}.$$

The total charge induced on the plane can now be determined by evaluating the integral $\int_S \sigma \, ds$ where S is the given plane. This is done, as in the previous chapter, by dividing the plane into elementary rings of radius R and centre O. The total charge is then

$$\int_0^\infty 2\pi\sigma R \, dR = 2\pi \int_0^\infty -\frac{ea}{2\pi r^3} R \, dR$$

$$= -ea \int_0^\infty \frac{R \, dR}{(R^2 + a^2)^{\frac{3}{2}}}$$

$$= -ea \left[-\frac{1}{(R^2 + a^2)^{\frac{1}{2}}} \right]_0^\infty$$

$$= -e.$$

Thus the total induced charge has the same magnitude and sign as the image charge. Now the total force exerted on the given point charge at A by the induced charge on the plane is equal to the force exerted on the charge at A by the image charge, that is $-e^2/16\pi\varepsilon_0 a^2$. (The negative sign implies an attraction.)

Example   The plane, conducting surfaces $x = 0$, $y \geqslant 0$ and $y = 0$, $x \geqslant 0$ are maintained at zero potential in the presence of a point charge e a perpendicular distance a from one surface and a perpendicular distance b from the other. Determine the force exerted on the point charge e and the density of the induced surface charge distribution at the nearest point of each plane to the charge e.

Choose the origin of the given coordinate system so that the point charge e lies in the plane $z = 0$ at the point A say. Then A $\equiv (a,b,0)$ (figure 73).

The image in the plane $y = 0$ of the charge e at A is a charge $-e$ at B $(a,-b,0)$. These two charges will together maintain the plane $y = 0$ at zero potential. The potential on the plane $x = 0$ will not however be zero. Now add a charge $-e$ at C $(-a,b,0)$, that is the image in the plane $x = 0$ of the charge e at A, and a charge e at D $(-a,-b,0)$, that is the image in the plane $x = 0$ of the

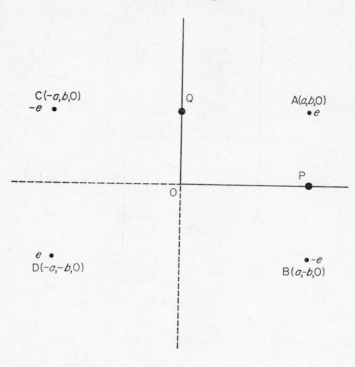

FIGURE 73

charge $-e$ at B. This system will now ensure that the plane $x = 0$ is at zero potential and since the additional charges at C and D are images of each other in the plane $y = 0$ they will not disturb the potential on this plane. Thus the complete image system for the given problem has been obtained. In other words, the potential and force fields in the region $x \geqslant 0$, $y \geqslant 0$ due to the charge e at A and the earthed, conducting planes $x = 0$, $y \geqslant 0$ and $y = 0$, $x \geqslant 0$ are the same as those due to the charges $+e$ at A and D together with the charges $-e$ at B and C.

Hence the forces on the charge at A are $e^2/16\pi\varepsilon_0 a^2$ in the negative x-direction due to the attraction of the charge at C, $e^2/16\pi\varepsilon_0 b^2$ in the negative y-direction due to the attraction of the charge at B and $e^2/16\pi\varepsilon_0(a^2 + b^2)$ in the direction OA due to the repulsion of the charge at D (figure 74).

Therefore the x-component of the resultant is

$$\frac{1}{4\pi\varepsilon_0}\left(-\frac{e^2}{4a^2} + \frac{e^2}{4(a^2 + b^2)}\cos\alpha\right)$$

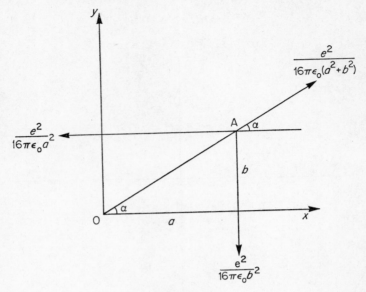

Figure 74

and the y-component of the resultant is

$$\frac{1}{4\pi\epsilon_0}\left(-\frac{e^2}{4b^2} + \frac{e^2}{4(a^2 + b^2)}\sin\alpha\right)$$

where $\tan\alpha = b/a$ and so $\cos\alpha = a/(a^2 + b^2)^{\frac{1}{2}}$ and $\sin\alpha = b/(a^2 + b^2)^{\frac{1}{2}}$.
  Hence the magnitude of the resultant is

$$\frac{e^2}{16\pi\varepsilon_0}\left\{\left(\frac{1}{a^2} - \frac{a}{(a^2 + b^2)^{\frac{3}{2}}}\right)^2 + \left(\frac{1}{b^2} - \frac{b}{(a^2 + b^2)^{\frac{3}{2}}}\right)^2\right\}^{\frac{1}{2}}.$$

  If $P \equiv (a,0,0)$, that is the point of the plane $y = 0$ nearest to the charge at A, then the forces at P due to the charges at A, B, C and D have magnitudes and directions as shown in figure 75. Hence the resultant force at P is in the direction Oy and has magnitude

$$\frac{1}{4\pi\varepsilon_0}\left(\frac{2e}{(PC)^2}\sin\beta - \frac{2e}{b^2}\right) \qquad \text{(since } PC = PD)$$

$$= \frac{1}{4\pi\varepsilon_0}\left(\frac{2e}{(b^2 + 4a^2)}\cdot\frac{b}{\sqrt{b^2 + 4a^2}} - \frac{2e}{b^2}\right)$$

$$\left(\text{since } PC = \sqrt{b^2 + 4a^2} \quad \text{and} \quad \sin\beta = \frac{b}{PC}\right).$$

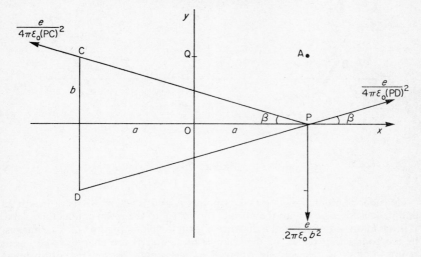

FIGURE 75

Hence, from Gauss law, if σ_P is the surface density of induced charge at P,

$$\frac{\sigma_P}{\varepsilon_0} = \frac{1}{4\pi\varepsilon_0} \left\{ \frac{2eb}{(b^2 + 4a^2)^{\frac{3}{2}}} - \frac{2e}{b^2} \right\} - 0 \qquad (\mathbf{E} = 0 \quad \text{for} \quad y < 0).$$

Therefore

$$\sigma_P = - \frac{e}{2\pi b^2} \left\{ 1 - \left(\frac{b}{\sqrt{b^2 + 4a^2}} \right)^3 \right\}.$$

Similarly, or by simply interchanging a and b in the above expression, it is seen that σ_Q, the surface density of induced charge at the point $Q(0,b,0)$, is

$$- \frac{e}{2\pi a^2} \left\{ 1 - \left(\frac{a}{\sqrt{a^2 + 4b^2}} \right)^3 \right\}.$$

Example   Show that the image in an infinite conducting plane of a dipole of moment m and axis making an angle θ with a normal to the plane held at a distance d from the plane is another dipole of moment m with axis making an angle −θ with the normal to the plane placed at a distance d on the other side of the plane.

Hence show that the dipole at P *is attracted towards the plane with a force*

$$\frac{3m^2}{64\pi\varepsilon_0 \, d^4} (1 + \cos^2 \theta)$$

if the medium containing the conducting plane and the dipole is free space.

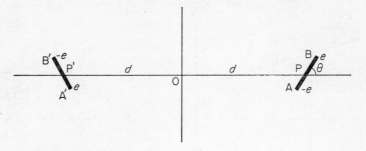

FIGURE 76

Consider point charges $-e$, $+e$ at points A and B where AB is of length $2a$ and makes an angle θ with the normal through P to the given conducting plane. P is the midpoint of AB and a distance d from the plane (figure 76). The images, in the conducting plane, of the charges at A and B are charges $+e$, $-e$ at points A$'$ and B$'$ respectively where B$'$A$'$ has length $2a$ and makes an angle $-\theta$ with the normal through P which also passes through P$'$ the midpoint of A$'$B$'$. P$'$ is a distance d behind the conducting plane and the points A, B, A$'$, B$'$ are all coplanar. Now let $a \to 0$ and $e \to \infty$ in such a way that $2ae \to m$. In the limit, the pair of charges at A and B becomes a dipole at P with moment m and axis making an angle θ with the normal to the plane. The image pair at A$'$ and B$'$ becomes a dipole at P$'$ with moment m and axis making an angle $-\theta$ with the normal to the plane.

Hence the potential energy W of the dipole at P in the presence of the conducting plane is equal to the potential of the dipole at P in the presence of the image dipole at P$'$.

Therefore, using the result proved on p. 171 in section 2.63, we obtain

$$W = \frac{m^2}{4\pi\varepsilon_0(2d)^3}(\cos 2\theta - 3\cos\theta\cos\theta)$$

$$= \frac{-m^2}{4\pi\varepsilon_0(2d)^3}(1 + \cos^2\theta).$$

In general, if the dipole is a distance $x/2$ from the plane and so a distance x from the image dipole,

$$W = \frac{-m^2}{4\pi\varepsilon_0 x^3}(1 + \cos^2\theta).$$

Then the force acting is

$$-\operatorname{grad} W = \frac{1}{4\pi\varepsilon_0}\left(-\frac{3m^2}{x^4}(1 + \cos^2\theta), 0, 0\right)$$

in terms of cartesian coordinates with origin at the image dipole and x-axis pointing towards the given dipole. Thus for the dipole at P the force is

$$\frac{1}{4\pi\varepsilon_0}\left(-\frac{3m^2}{(2d)^4}(1+\cos^2\theta),0,0\right),$$

that is an attraction of

$$\frac{3m^2}{64\pi\varepsilon_0\,d^4}(1+\cos^2\theta)$$

towards the conducting plane.

3.23 Point charge and a conducting sphere

The problem now to be considered is essentially that of determining the potential and force fields produced when a spherical conductor is placed in the field of a point charge e.

Firstly however it is necessary to define what is meant by the *inverse of a point* with respect to a sphere.

Consider a sphere with centre O and radius a and take a point P lying either inside or outside the sphere (figure 77).

Let P′ be a point on OP (or OP produced) such that if P is outside the sphere then P′ is inside and vice versa. Then P′ is called the *inverse* of the point P with respect to the sphere if

$$OP . OP' = a^2.$$

This implies that

$$\frac{OP}{a}=\frac{a}{OP'}$$

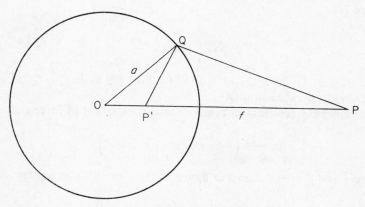

FIGURE 77

and hence that, if Q is any point on the surface of the sphere, the triangles OPQ and OP′Q are similar.

Therefore

$$\frac{PQ}{P'Q} = \frac{OP}{OQ} = \frac{f}{a}$$

where OP has length f.

Now suppose that a point charge $+e$ is placed at the point P and another point charge e' is placed at the point P′. Then the potential at any point Q on the surface of the sphere is

$$\frac{1}{4\pi\varepsilon_0}\left(\frac{e}{PQ} + \frac{e'}{P'Q}\right) = \frac{1}{4\pi\varepsilon_0}\left(\frac{e}{PQ} + \frac{fe'}{aPQ}\right)$$

$$= \frac{1}{4\pi\varepsilon_0}\left\{\frac{1}{PQ}\left(e + \frac{fe'}{a}\right)\right\}.$$

This can be made zero for all points Q on the sphere by choosing $e' = -ae/f$. Thus the image in the sphere of a point charge $+e$ at P is a point charge $-ae/f$ at P′ the inverse point of P with respect to the sphere.

Therefore the potential and force fields outside an earthed, conducting sphere of radius a placed with its centre a distance f $(> a)$ from a point charge e at P are the same as the potential and force fields due to the given point charge e at P and a point charge $-ae/f$ placed at the inverse point of P with respect to the sphere.

The sphere itself will have a total charge $-ae/f$ induced on its surface.

Once again the density σ of the charge induced on the surface of the sphere is determined using Gauss' law.

Consider any point Q on the surface of the sphere and let $PQ = r$ (figure 78).

Then

$$P'Q = \frac{a}{f}(PQ) = \frac{ar}{f}.$$

The forces at Q are $e/4\pi\varepsilon_0 r^2$ in the direction PQ and $ea/4\pi\varepsilon_0 f(P'Q)^2$ or $ef/4\pi\varepsilon_0 ar^2$ in the direction QP′.

Therefore the resultant outward normal component of force is

$$\frac{1}{4\pi\varepsilon_0}\left(-\frac{e}{r^2}\cos\alpha - \frac{ef}{ar^2}\cos\beta\right)$$

where α and β are as shown in figure 78. This can also be written as

$$-\frac{e}{4\pi\varepsilon_0 r^3}\left(r\cos\alpha + \frac{f}{a}r\cos\beta\right).$$

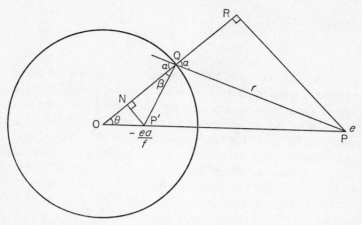

FIGURE 78

Now let R and N respectively be the feet of the perpendiculars from P and P' on to OQ (produced).

Then

$$r \cos \alpha = QR = OR - OQ$$
$$= f \cos \theta - a \quad \text{where} \quad P\hat{O}Q = \theta.$$

Also,

$$r \cos \beta = \frac{f}{a}(P'Q) \cos \beta = \frac{f}{a}(NQ)$$
$$= \frac{f}{a}(OQ - ON)$$
$$= \frac{f}{a}(a - OP' \cos \theta)$$
$$= \frac{f}{a}\left(a - \frac{a^2}{f} \cos \theta\right)$$

since OP . OP' $= a^2$ and OP $= f$. Therefore the resultant outward normal component of force at Q is

$$-\frac{e}{4\pi\varepsilon_0 r^3}\left\{f \cos \theta - a + \frac{f^2}{a^2}\left(a - \frac{a^2}{f} \cos \theta\right)\right\} = -\frac{e}{4\pi\varepsilon_0 r^3}\left(\frac{f^2 - a^2}{a}\right).$$

Hence, using Guass' law, if σ is the surface density of induced charge at Q, we obtain

$$\frac{\sigma}{\varepsilon_0} = -\frac{e}{4\pi\varepsilon_0 r^3}\left(\frac{f^2 - a^2}{a}\right).$$

Therefore

$$\sigma = -\frac{1}{4\pi}\frac{e}{r^3}\left(\frac{f^2 - a^2}{a}\right).$$

Suppose now that instead of being earthed the sphere is insulated (and so cannot gain or lose any charge) and initially uncharged. The total charge on the sphere must then remain zero and the potential will be constant, although not zero, on its surface. It is thus necessary to add a further image charge ea/f to make the total induced charge zero. This charge must of course be placed inside the sphere in such a position that the surface of the sphere still remains an equipotential surface. Thus this charge must be placed at the centre O of the sphere. The potential at all points on the surface is then e/f the potential due to the charge ea/f at O. (The potential on the surface due to the charges at P and P' is zero.)

Thus the potential and force fields outside an insulated, uncharged conducting sphere of radius a placed with its centre a distance f ($> a$) from a point charge e at P are the same as the potential and force fields due to the given point charge e at P together with a point charge $-ae/f$ at the inverse point of P with respect to the sphere and a point charge ae/f at the centre of the sphere.

If now, instead of being uncharged, the sphere has a total charge Q it is necessary to include a further point charge Q at the centre of the sphere. The surface of the sphere is still an equipotential surface and the total induced charge is Q.

The force exerted on the given point charge at P when the sphere is introduced is easily obtained by calculating the force due to the other point charges of the image system.

It is also easily shown that if a large slab of earthed, conducting material has in it a spherical hollow of radius a and if a point charge e is introduced at a point P' distant f' ($< a$) from the centre of the hollow, then the image system consists of the given charge at P' together with a point charge $-ea/f'$ at P the inverse point of P' with respect to the spherical surface of the hollow. This image system then enables the potential and force fields to be calculated at points within the hollow.

Example   Point charges e and $3e$ are held at points P and Q a distance $6a$ apart. An earthed, spherical conductor of radius a is introduced into the field with its centre O on the line PQ a distance $2a$ from P and a distance $4a$ from Q. Determine the density of the induced charge at the points of the conductor which are nearest to the points P and Q and the resulting force on the charge e at P.

The image, in the sphere, of the charge e at P is a charge $-ea/2a$ at P' where P' lies on OP and $OP' = a^2/2a = a/2$ (figure 79).

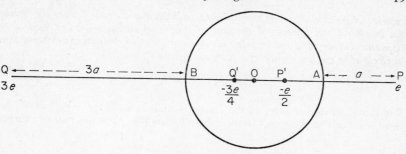

The image, in the sphere, of the charge $3e$ at Q is a charge $-3ea/4a$ at Q where Q' lies on OQ and $OQ' = a^2/4a = a/4$. Hence the complete image system consists of the given charges at P and Q together with the above charges at P' and Q'.

Therefore, considering these point charges it is seen that the force in the direction OP at A, the point of the sphere nearest to P, has magnitude

$$\frac{1}{4\pi\varepsilon_0}\left\{ -\frac{e}{a^2} - \frac{e}{2\left(\frac{a}{2}\right)^2} - \frac{3e}{4\left(\frac{5a}{4}\right)^2} + \frac{3e}{(5a)^2} \right\} = \frac{-21e}{25\pi\varepsilon_0 a^2}.$$

Then, from Gauss' law, the surface charge density σ_A at A is given by

$$\frac{\sigma_A}{\varepsilon_0} = \frac{-21e}{25\pi\varepsilon_0 a^2}$$

and so

$$\sigma_A = -\frac{21}{25\pi}\frac{e}{a^2}.$$

Similarly, the force in the direction OQ at B, the point of the sphere nearest to Q, has magnitude

$$\frac{1}{4\pi\varepsilon_0}\left\{ \frac{e}{(3a)^2} - \frac{e}{2(3a/2)^2} - \frac{3e}{4(a/4)^2} - \frac{3e}{(3a)^2} \right\} = -\frac{28e}{9\pi\varepsilon_0 a^2}.$$

Hence the surface charge density σ_B at B is $-\frac{28}{9\pi}\cdot\frac{e}{a^2}$. The force, in the direction OP, on the charge e at P is

$$\frac{1}{4\pi\varepsilon_0}\left\{ -\frac{e^2}{2(3a/2)^2} - \frac{3e^2}{4(9a/4)^2} + \frac{3e^2}{(6a)^2} \right\} = -\frac{31e^2}{432\pi\varepsilon_0 a^2}.$$

The negative sign indicates that the force is in the direction from P to O, that is it is an attraction towards the sphere.

Example *A point charge e is maintained at a point* P *a distance f from the centre of an insulated and uncharged conducting sphere of radius a ($< f$). Show that the positive and negative induced charges on the surface of the sphere are separated by a circle with circumference at a distance r from* P, *where*

$$r^3 = f(f^2 - a^2).$$

The image, in the sphere, of the charge e at P is a charge $-ea/f$ at P' the inverse point of P with respect to the sphere (figure 80). Now this pair of charges

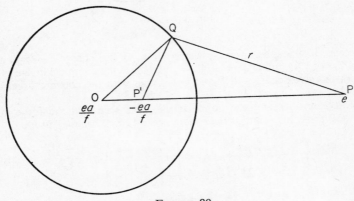

FIGURE 80

at P and P' results in an induced charge $-ea/f$ on the surface of the sphere. But the given sphere is insulated and uncharged and so a further charge ea/f must be placed at the centre O of the sphere to complete the image system. The complete image system then consists of the given charge e at P, a charge $-ea/f$ at P' and a charge ea/f at O.

Now, as shown previously (on p. 191), the resultant outward normal component of force at a point Q on the surface of the sphere distant r from P due to the charges at P and P' is

$$- \frac{e}{4\pi\varepsilon_0 r^3} \left(\frac{f^2 - a^2}{a} \right).$$

Hence, including also the charge at O, the resultant outward normal component of force at Q is

$$\frac{1}{4\pi\varepsilon_0} \left\{ \frac{ea}{fa^2} - \frac{e}{r^3} \left(\frac{f^2 - a^2}{a} \right) \right\}$$

Therefore if σ is the surface density of induced charge at Q,

$$\frac{\sigma}{\varepsilon_0} = \frac{1}{4\pi\varepsilon_0} \left\{ \frac{e}{fa} - \frac{e}{r^3} \left(\frac{f^2 - a^2}{a} \right) \right\}.$$

Thus $\sigma = 0$ when

$$\frac{e}{ra} - \frac{e}{r^3}\left(\frac{f^2 - a^2}{a}\right) = 0,$$

that is when

$$r^3 = f(f^2 - a^2).$$

If r is less than the real value given by this relation then it is seen that σ is negative while if r is greater than this value, σ is positive.

Hence the region of the sphere for which the induced charge is positive ($\sigma > 0$) is separated from that for which the induced charge is negative ($\sigma < 0$) by the circle with circumference a distance r from P where

$$r^3 = f(f^2 - a^2).$$

Example   An infinite earthed conducting plane plate has a hemispherical boss of radius a. A charge e is placed on the axis of the boss at a distance f ($> a$) from the plate and on the same side of the plate as the boss. Show that the charge is attracted towards the plate by a force

$$\frac{1}{4\pi\varepsilon_0}\left\{\frac{e^2}{4f^2} + \frac{4e^2 a^3 f^3}{(f^4 - a^4)^2}\right\}.$$

The charge e at P together with a charge $-e$ at Q the image point of P with respect to the plane will give constant potential on the plane without the boss. Potential will not however be constant on the boss. Now, adding charges $-ea/f$ and ea/f at the points P' and Q' the image points of P and Q with respect to the sphere of which the boss is half, will make the potential on the boss constant. These charges will not upset the potential on the plane itself as they are images of each other with respect to the plane.

Hence the complete image system consists of the point charge e at P together with a point charge $-e$ at Q, a point charge $-ea/f$ at P' and a point charge ea/f at Q' (figure 81). Then the resultant force attracting the charge at P towards the

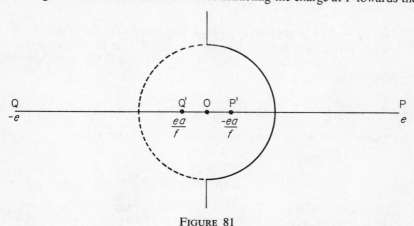

FIGURE 81

plate is

$$\frac{1}{4\pi\varepsilon_0}\left\{\frac{e^2}{4f^2} + \frac{e^2a}{f(f - a^2/f)^2} - \frac{e^2a}{f(f + a^2/f)^2}\right\} = \frac{1}{4\pi\varepsilon_0}\left\{\frac{e^2}{4f^2} + \frac{4e^2a^3f^3}{(f^4 - a^4)^2}\right\}.$$

Example   *A dipole of moment m is held at a point* P *distant f from the centre* O *of an earthed conducting sphere of radius a* (<f). *Determine the attraction of the sphere on the dipole*
(i) *when the axis of the dipole is perpendicular to* OP *and*
(ii) *when the axis of the dipole lies along* OP.

(i) Consider point charges $-e$ and e at the points A and B respectively where AB has length $2b$ and is perpendicular to OP. P is the midpoint of AB (figure 82). The image in the sphere of the charge at A is a charge

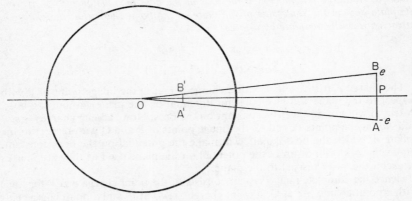

FIGURE 82

$ea/\sqrt{f^2 + b^2}$ at A′ the inverse point of A with respect to the sphere and the image of the charge at B is a charge $-ea\sqrt{f^2 + b^2}$ at B′ the inverse point of B with respect to the sphere. Now the triangles OA′B′ and OAB are similar and so

$$\frac{A'B'}{AB} = \frac{OA'}{OA} = \frac{a^2}{(OA)^2}$$

(since A and A′ are inverse points with respect to the sphere).
   Therefore

$$A'B' = \frac{2a^2b}{(OA)^2} = \frac{2a^2b}{b^2 + f^2}.$$

Now let $b \to 0$, $e \to \infty$ in such a way that $2be \to m$. In the limit the pair of charges at A and B becomes the dipole of moment m at P with axis perpendicular to OP. However, it is also seen that

$$\frac{2a^2b}{b^2 + f^2} \to 0 \quad \text{and} \quad \frac{ea}{\sqrt{b^2 + f^2}} \to \infty$$

in such a way that their product $\to ma^3/f^3$.

Thus, in the limit, the pair of image charges at A′ and B′ becomes the dipole of moment ma^3/f^3 at P′ the inverse point of P with respect to the sphere. This image dipole has the same direction as the dipole at P (that is perpendicular to OP) but its sense is opposite to that of the dipole at P. In vector notation, if the dipole at P has moment **m** perpendicular to OP, then the image dipole at P′ has moment $(-a^3/f^3)$**m**.

Now, using the result (proved in section 2.63) that the potential energy of a dipole **m** placed at the point with position vector **r** in the field of a dipole **m′** at the origin is given by

$$W = \frac{1}{4\pi\varepsilon_0}\left\{\frac{\mathbf{m}\cdot\mathbf{m'}}{r^3} - \frac{3(\mathbf{m}\cdot\mathbf{r})(\mathbf{m'}\cdot\mathbf{r})}{r^5}\right\},$$

we see that the potential energy of a dipole **m** at P a distance r from a dipole $-(a^3/f^3)$**m** at P′, where $\overline{PP'}$ is perpendicular to **m**, is given by

$$W = \frac{-m^2 a^3}{4\pi\varepsilon_0 f^3 r^3}.$$

Hence the force on the dipole **m** along the line P′P is

$$-\frac{\partial W}{\partial r} = -\frac{3m^2 a^3}{4\pi\varepsilon_0 f^3 r^4}.$$

Therefore putting $r = f - a^2/f$ we see that the dipole **m** is attracted towards the sphere by a force

$$\frac{3m^2 a^3}{4\pi\varepsilon_0 f^3 (f - a^2/f)^4} = \frac{3f m^2 a^3}{4\pi\varepsilon_0 (f^2 - a^2)^4}.$$

(ii) O, A, P, B are collinear. OA $= f - b$, OP $= f$, OB $= f + b$. A′, B′ and P′ are the inverse points with respect to the sphere of the points A, B and P respectively (figure 83). Therefore

$$\mathrm{OA'} = \frac{a^2}{\mathrm{OA}} = \frac{a^2}{f - b}, \qquad \mathrm{OB'} = \frac{a^2}{f + b} \quad \text{and} \quad \mathrm{OP'} = \frac{a^2}{f}.$$

Hence

$$\mathrm{B'A'} = \frac{a^2}{f - b} - \frac{a^2}{f + b} = \frac{2a^2 b}{f^2 - b^2}.$$

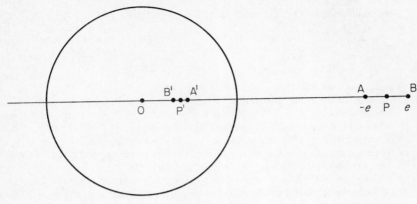

Charge $-e$ at A has image charge $ea/(f-b)$ at A′ and charge $+e$ at B has image charge $-ea/(f+b)$ at B′. Thus we have an image charge

$$\frac{ea}{f+b} + \left\{\left(\frac{ea}{f-b}\right) - \left(\frac{ea}{f+b}\right)\right\} \quad \text{or} \quad \left(\frac{ea}{f+b} + \frac{2eab}{f^2-b^2}\right)$$

at A′ and an image charge

$$\frac{-ea}{f+b}$$

at B′. The pair of charges $\pm ea/(f+b)$ at A′ and B′ becomes a dipole of moment ma^3/f^3 at P′ while the point charge $2eab/(f^2-b^2)$ at A′ becomes a point charge ma/f^2 at P′ in the limit as $b \to 0$, $e \to \infty$ and $2be \to m$, that is in the limit when the point charges at A and B become a dipole of moment m at P.

Hence the image in the sphere of a dipole of moment m at P with axis along OP is a parallel dipole of moment ma^3/f^3 at the inverse point of P together with a point charge ma/f^2 at the same point.

Now the potential energy W of a dipole **m** at P a distance r from a dipole (a^3/f^3)**m** and a point charge ma/f^2 at P′ where $\overline{\text{P′P}}$ is parallel to **m** is given by

$$W = \frac{1}{4\pi\varepsilon_0}\left\{-\frac{m}{r^2}\left(\frac{ma}{f^2}\right) + \frac{m^2a^3}{r^3f^3} - \frac{3}{r^5}(mr)\left(\frac{ma^3}{f^3}r\right)\right\}$$

$$= \frac{1}{4\pi\varepsilon_0}\left(-\frac{m^2a}{r^2f^2} - \frac{2m^2a^3}{r^3f^3}\right).$$

Hence the force on the dipole **m** along the line P′P is

$$-\frac{\partial W}{\partial r} = \frac{1}{4\pi\varepsilon_0}\left\{-\frac{2m^2a}{r^3f^2} - \frac{6m^2a^3}{r^4f^3}\right\}.$$

Therefore, putting $r = f - a^2/f$ we see that the dipole **m** is attracted towards the sphere by a force

$$\frac{1}{4\pi\varepsilon_0}\left\{\frac{2m^2 a}{f^2(f - a^2/f)^3} + \frac{6m^2 a^3}{f^3(f - a^2/f)^4}\right\} = \frac{fm^2 a(f^2 + 2a^2)}{2\pi\varepsilon_0(f^2 - a^2)^4}.$$

3.24 *Images in dielectrics*

Consider now the effect of introducing a dielectric into a given electrostatic field. The problem is to find a system of image charges or dipoles all within the region occupied by the dielectric which, together with the potential of the given field, determine a potential field φ_0 outside the dielectric and another system of image charges or dipoles all outside the region occupied by the dielectric which determine a potential field φ_i inside the region occupied by the dielectric such that φ_0 and φ_i satisfy the necessary boundary conditions for the problem on the surface of the dielectric. Then, as a consequence of the uniqueness result proved earlier, φ_0 and φ_i must represent the potential distributions due to the presence of the dielectric in the given field. Note that when calculating φ_i the medium is regarded as entirely free space.

Example A point charge e is placed in free space at a distance d from the plane face, $x = 0$, of a mass of uniform material of dielectric constant κ which occupies the whole of the space $x \leqslant 0$. Determine an image system which can be used to determine the potential field for this situation.

Let φ_i be the potential function inside the dielectric and let φ_0 be the potential function outside the dielectric.

The boundary conditions to be satisfied on the surface of the dielectric are (see section 2.29)

(i) potential is continuous, that is $\varphi_i = \varphi_0$,
(ii) the product of dielectric constant and normal component of force is continuous, that is $\kappa(\partial\varphi_i/\partial x) = \partial\varphi_0/\partial x$.

Suppose the potential outside the dielectric can be considered as due to the given charge at P together with an image charge e' at P' the image point of P and that the potential inside the dielectric can be considered as due to an image charge e'' at P (figure 84).

Then the potential φ_0 at any point A outside the dielectric will be given by

$$\varphi_0 = \frac{1}{4\pi\varepsilon_0}\left(\frac{e}{\text{PA}} + \frac{e'}{\text{P'A}}\right)$$

and the potential φ_i at any point B inside the dielectric will be given by

$$\varphi_i = \frac{e''}{4\pi\varepsilon_0 \text{PB}}.$$

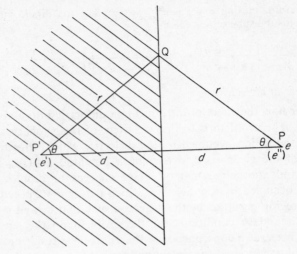

FIGURE 84

Hence, considering a point Q on the surface of the dielectric, it is seen that the first boundary condition implies

$$\frac{e}{r} + \frac{e'}{r} = \frac{e''}{r} \quad \text{or} \quad e + e' = e'' \quad (r = \text{P'Q} = \text{PQ}).$$

Also, the second boundary condition implies

$$\kappa \frac{e''}{r^2} \cos \theta = \frac{e}{r^2} \cos \theta - \frac{e'}{r^2} \cos \theta$$

or

$$\kappa e'' = e - e' \quad (\text{Q}\hat{\text{P}}'\text{P} = \text{Q}\hat{\text{P}}\text{P}' = \theta).$$

Hence

$$2e = (1 + \kappa)e'' \quad \text{and} \quad e' = \left(\frac{1 - \kappa}{1 + \kappa}\right)e.$$

Therefore the potential inside the dielectric is identical with that due to a charge $2e/(1 + \kappa)$ at P and the potential outside the dielectric is identical with that due to the charge e at P and a charge

$$\left(\frac{1 - \kappa}{1 + \kappa}\right)e$$

at P' the image point of P in the surface of the dielectric.

Note that when $\kappa = 1$ (that is the dielectric is replaced by free space), the potential in both regions reduces to that due solely to the charge e at P as expected.

Unfortunately, there is not a simple image solution for the problem of the point charge in the presence of a dielectric sphere. The solution of this problem requires spherical harmonics (see section 6.4) and is done as an example in section 6.4 (page 316).

Exercises

1. Point charges e, $2e$ are held a distance $2b$ apart and each a distance a from an infinite, earthed, conducting plane. Determine the surface density of induced charge at the point on the plane distant $\sqrt{a^2 + b^2}$ from each of the above charges.

2. Determine the image system for a point charge e at the point $(a,b,0)$ in the presence of the plane, earthed, conducting surfaces $x = 0$, $y \geqslant 0$ and $y = x$, $x \geqslant 0$. (Assume that $b > a > 0$.)

3. A dipole of moment **m** is equidistant from two mutually perpendicular, earthed, conducting, infinite planes. Its axis is in a plane perpendicular to the conducting planes and makes an angle $\frac{1}{4}\pi$ with each of these planes. Determine the resultant force on the dipole.

4. Charges $-e$, $2e$ are held at distances $3a$ and $4a$ respectively from the centre of an earthed conducting sphere of radius $2a$. If the line joining the charges passes through the centre of the sphere (and the charges are on opposite sides of the centre), calculate the density of the induced charge at the points where this line cuts the surface of the sphere.

5. A solid spherical conductor of radius a is held at the constant potential V in the presence of a point charge e at a distance f ($f > a$) from its centre. If the surface density of charge on the sphere is nowhere negative, show that

$$V \geqslant \frac{e(f + a)}{(f - a)^2}.$$

6. A solid spherical conductor of radius a is insulated and carries a total charge Q. If a point charge e is held at a distance f ($f > a$) from the centre of the sphere, show that the surface density of charge at the point of the sphere closest to the charge e will be zero if

$$Q = \frac{ea^2(3f - a)}{f(f - a)^2}.$$

7. An infinite earthed conducting plane plate has a hemispherical boss of radius a. On the same side of the plate as the boss there is a point charge e a distance a from both the plate and the axis of the boss. Find the components of the force acting on the point charge e.

8. A point charge e is placed in free space at a distance d from the plane face, $x = 0$, of a mass of uniform material of dielectric constant κ which occupies the whole of the space $x \leqslant 0$. Show that the energy of the system is

$$\left(\frac{1 - \kappa}{1 + \kappa}\right) \frac{e^2}{16\pi\varepsilon_0 d}.$$

9. Using cartesian coordinates, the plane sheet given by $x = 0$, $y^2 + z^2 \geqslant a^2$ together with the hemispherical surface $x^2 + y^2 + z^2 = a^2$, $x \geqslant 0$ form a conducting surface which is earthed. Apart from the above conductor, the region below the plane $z = 0$ is occupied by material of dielectric constant κ while the region above this plane is free space. If a point charge e is held at the point $(a,0,a)$, what is the image system?

10. A mass of uniform material of dielectric constant κ occupies the whole of the space $x \leqslant 0$. Obtain the force necessary to hold a dipole of moment $\mathbf{m}$ with its axis parallel to the plane $x = 0$ at a distance d from it in free space.

3.3 Hydrodynamic images

The type of problem to be dealt with here is essentially that of determining the effect of fixing a solid body somewhere in an inviscid, incompressible fluid moving irrotationally. This means obtaining for the fluid a potential function φ satisfying Laplace's equation (except at sources, sinks or doublets) and such that the component of the velocity vector $\mathbf{v}$ ($= \operatorname{grad} \varphi$) normal to the bounding surface of the given solid is zero. Any other relevant boundary conditions associated with the undisturbed problem must also be satisfied. Using the method of images, this is done by finding a system of sources (or sinks) and doublets (called images) all inside the solid and such that the potential due to the given field and these images satisfies the above conditions. Now the potential distribution for the system of point sources and doublets is again easily obtained and then by the uniqueness result proved earlier this must also be the potential distribution for the fluid in the given problem.

3.31 Point source and an infinite, plane boundary

The region on one side of an infinite plane wall is occupied by an inviscid, incompressible fluid. The problem is to determine the streamlines and equipotentials for the flow due to a point source of strength m held in the fluid at a perpendicular distance d from the wall. Choose an orthogonal coordinate system (x,y,z) such that the plane boundary coincides with the plane $x = 0$ and the point source is at the point A(d,0,0). Using the method of images we must determine the system of point sources and doublets in the region $x < 0$ which, together with the given point source

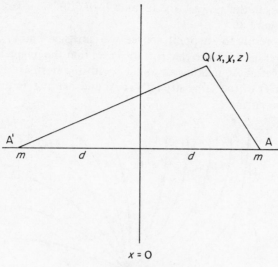

$x = 0$

FIGURE 85

at A, define a potential function φ such that the x-component of grad φ is zero on the plane $x = 0$.

Consider the potential due to point sources of equal strength m at the points A(d,0,0) and A'($-d$,0,0) (figure 85).

The potential φ at the point Q(x,y,z) is

$$\frac{-m}{\text{AQ}} - \frac{m}{\text{A'Q}} = \frac{-m}{\sqrt{(x-d)^2 + y^2 + z^2}} - \frac{m}{\sqrt{(x+d)^2 + y^2 + z^2}}.$$

Therefore the x-component of $\mathbf{v}$ (= grad φ) is

$$\frac{m(x-d)}{\{(x-d)^2 + y^2 + z^2\}^{\frac{3}{2}}} + \frac{m(x+d)}{\{(x+d)^2 + y^2 + z^2\}^{\frac{3}{2}}}.$$

Now if Q lies on the plane $x = 0$, this component is equal to

$$\frac{-md}{(d^2 + y^2 + z^2)^{\frac{3}{2}}} + \frac{md}{(d^2 + y^2 + z^2)^{\frac{3}{2}}}$$

which is zero for all values of y and z.

Hence there is no normal component of velocity anywhere on the plane $x = 0$. The point source m at A' is thus the image in the plane of the point source m at A.

The streamlines and equipotentials correspond respectively to the lines of force and equipotentials in the region $x \geqslant 0$ due to two equal positive point charges at the points $(d,0,0)$ and $(-d,0,0)$. The streamlines are sketched in figure 86.

Using this result for a point source we can then show, in a similar manner to that used for the electrostatic case, that the image in a plane of a point doublet at A, making an angle θ with the normal to the plane is another doublet of the same strength at A' and making an angle $(\pi - \theta)$ with the same normal.

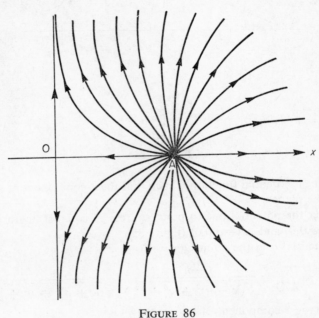

FIGURE 86

Hydrodynamic images with respect to a spherical surface are not so easily obtained as the corresponding electrostatic images and will not be considered here.

Further examples on images, using complex variables, are contained in Chapter 9.

Exercise

1. The region on one side of an infinite plane wall is occupied by an inviscid, incompressible fluid in which there is a source of strength m. If the fluid is at rest at infinity where the pressure is P, and the source is a distance d from the wall, obtain an expression for the pressure on the wall at a point distant r from the source and find for what value of r this pressure has its least value.

4

Fourier series

In Chapter 2 it was shown that Laplace's equation is satisfied by the potential function associated with a variety of problems occurring in hydrodynamics, electrostatics, magnetostatics and gravitation and Chapter 6 is concerned with determining solutions of this important partial differential equation using the method of separation of variables. Using this method, we shall frequently require to express the potential φ (or its normal derivative) on a boundary of the given region as the sum of constant multiples of the elementary periodic functions $\cos nx$ and $\sin nx$ for integer values of n. This is the problem to be considered in this section.

4.1 Periodic functions

A function f of a variable x is said to have a *period* T in x if for all $x, f(x + T) = f(x)$, where T is a positive constant. Any integer multiple of T will of course also be a period of f. The least value of $T > 0$ satisfying the above condition is called the *least period* or simply *the period* of $f(x)$.

For example, since

$$\sin x = \sin (2\pi + x) = \sin (4\pi + x) = \ldots$$

and

$$\cos x = \cos (2\pi + x) = \cos (4\pi + x) = \ldots \text{ for all } x,$$

it is seen that the functions $\sin x$ and $\cos x$ have periods $2\pi, 4\pi, 6\pi, \ldots$. In both cases, the least period is 2π.

Similarly, the functions $\sin nx$, $\cos nx$, for constant n, have periods $2\pi/n$, $4\pi/n$, $6\pi/n, \ldots$ and the functions $\sin 2\pi x$, $\cos 2\pi x$ have periods $1, 2, 3, \ldots$.

Example *Show that if* $f(x) \equiv k$ *where* k *is a constant, then any positive number is a period of* $f(x)$.

Since $f(x + c) = k = f(x)$ for all x and any positive constant c the result is proved. (Note that $f(x) \equiv k$ does not have a *least* period.)

4.2 Fourier series

Given a function $f(x)$ which is periodic with period 2π, the problem is to express it, where possible, as the sum of constant multiples of the elementary periodic functions $\cos nx$ and $\sin nx$ for integer values of n. In other words, the problem is to determine (where possible) the coefficients a_0, a_n and b_n, for all positive integer values of n, such that

$$f(x) = a_0 + \sum_{n=1}^{\infty} (a_n \cos nx + b_n \sin nx).$$

Integrating both sides with respect to x over one period from $x = -\pi$ to $x = \pi$ gives

$$\int_{-\pi}^{\pi} f(x)\,\mathrm{d}x = \int_{-\pi}^{\pi} a_0\,\mathrm{d}x + \int_{-\pi}^{\pi} \left\{ \sum_{n=1}^{\infty} (a_n \cos nx + b_n \sin nx) \right\} \mathrm{d}x.$$

Then, assuming that term by term integration of the series is permissible, we obtain

$$\int_{-\pi}^{\pi} f(x)\,\mathrm{d}x = \int_{-\pi}^{\pi} a_0\,\mathrm{d}x + \sum_{n=1}^{\infty} \left(a_n \int_{-\pi}^{\pi} \cos nx\,\mathrm{d}x + b_n \int_{-\pi}^{\pi} \sin nx\,\mathrm{d}x \right).$$

Now $\int_{-\pi}^{\pi} a_0\,\mathrm{d}x = 2\pi a_0$,

$$\int_{-\pi}^{\pi} \cos nx\,\mathrm{d}x = \left[\frac{1}{n} \sin nx \right]_{-\pi}^{\pi} = 0 \quad \text{for} \quad n = 1, 2, 3, \ldots$$

and

$$\int_{-\pi}^{\pi} \sin nx\,\mathrm{d}x = \left[-\frac{1}{n} \cos nx \right]_{-\pi}^{\pi} = 0 \quad \text{for} \quad n = 1, 2, 3, \ldots$$

Hence

$$\int_{-\pi}^{\pi} f(x)\,\mathrm{d}x = 2\pi a_0$$

and so

$$a_0 = \frac{1}{2\pi} \int_{-\pi}^{\pi} f(x)\,\mathrm{d}x.$$

Also, we see that

$$\int_{-\pi}^{\pi} f(x) \cos mx\,\mathrm{d}x = \int_{-\pi}^{\pi} a_0 \cos mx\,\mathrm{d}x$$
$$+ \int_{-\pi}^{\pi} \left\{ \sum_{n=1}^{\infty} (a_n \cos nx + b_n \sin nx) \right\} \cos mx\,\mathrm{d}x$$

for positive integers m. Term by term integration of the right hand side gives

$$\int_{-\pi}^{\pi} a_0 \cos mx \, dx + \sum_{n=1}^{\infty} \left(a_n \int_{-\pi}^{\pi} \cos nx \cos mx \, dx + b_n \int_{-\pi}^{\pi} \sin nx \cos mx \, dx \right).$$

Now

$$\int_{-\pi}^{\pi} a_0 \cos mx \, dx = \frac{a_0}{m} \left[\sin mx \right]_{-\pi}^{\pi} = 0,$$

and

$$\int_{-\pi}^{\pi} \cos nx \cos mx \, dx$$

$$= \begin{cases} \frac{1}{2} \int_{-\pi}^{\pi} \cos (n+m)x \, dx + \frac{1}{2} \int_{-\pi}^{\pi} \cos (n-m)x \, dx, & n \neq m \\[2mm] \frac{1}{2} \int_{-\pi}^{\pi} (1 + \cos 2nx) \, dx, & n = m \end{cases}$$

$$= \begin{cases} \dfrac{1}{2(n+m)} \left[\sin (n+m)x \right]_{-\pi}^{\pi} + \dfrac{1}{2(n-m)} \left[\sin (n-m)x \right]_{-\pi}^{\pi}, \\ \hspace{9cm} n \neq m \\[2mm] \dfrac{1}{2} \left[x + \dfrac{1}{2n} \sin 2nx \right]_{-\pi}^{\pi}, \hspace{4cm} n = m. \end{cases}$$

$$= \begin{cases} 0, & n \neq m \\ \pi, & n = m. \end{cases}$$

Also

$$\int_{-\pi}^{\pi} \sin nx \cos mx \, dx$$

$$= \begin{cases} \frac{1}{2} \int_{-\pi}^{\pi} \sin (n+m)x \, dx + \frac{1}{2} \int_{-\pi}^{\pi} \sin (n-m)x \, dx, & n \neq m \\[2mm] \frac{1}{2} \int_{-\pi}^{\pi} \sin 2nx \, dx, & n = m \end{cases}$$

$$= \begin{cases} \dfrac{1}{2(n+m)} \left[-\cos (n+m)x \right]_{-\pi}^{\pi} + \dfrac{1}{2(n-m)} \left[-\cos (n-m)x \right]_{-\pi}^{\pi}, \\ \hspace{9cm} n \neq m \\[2mm] \dfrac{1}{4n} \left[-\cos 2nx \right]_{-\pi}^{\pi}, \hspace{5cm} n = m \end{cases}$$

$$= 0.$$

Hence

$$\int_{-\pi}^{\pi} f(x) \cos mx \, dx = \begin{cases} 0, & n \neq m \\ a_n \pi, & n = m. \end{cases}$$

Therefore

$$a_n = \frac{1}{\pi} \int_{-\pi}^{\pi} f(x) \cos nx \, dx.$$

Finally,

$$\int_{-\pi}^{\pi} f(x) \sin mx \, dx$$

$$= \int_{-\pi}^{\pi} a_0 \sin mx \, dx + \int_{-\pi}^{\pi} \left\{ \sum_{n=1}^{\infty} (a_n \cos nx + b_n \sin nx) \right\} \sin mx \, dx$$

and hence (as above)

$$b_n = \frac{1}{\pi} \int_{-\pi}^{\pi} f(x) \sin nx \, dx.$$

The interval of integration used in the above is $(-\pi, \pi)$, but, because of the periodicity of the function $f(x)$, any interval $(\alpha, \alpha + 2\pi)$ for constant α may be used. Normally, in practice, the interval used is $(-\pi, \pi)$ or $(0, 2\pi)$.

The series

$$a_0 + \sum_{n=1}^{\infty} (a_n \cos nx + b_n \sin nx)$$

with a_n $(n = 0, 1, 2, \ldots)$ and b_n $(n = 1, 2, \ldots)$ defined as above is called the *Fourier series* of $f(x)$.

Now it is clear that for *any* integrable function defined in an interval of length 2π, constants a_n and b_n can be obtained as above, using this interval, and so a Fourier series can be obtained for the function. The sum of this series, for any particular value of x, in the given interval, need not however yield the corresponding value of the given function.

In order to state conditions for the convergence of Fourier series we introduce the following notation.

Let

$$f(x + 0) = \lim_{\varepsilon \to 0} f(x + \varepsilon) \quad \text{and} \quad f(x - 0) = \lim_{\varepsilon' \to 0} f(x - \varepsilon')$$

for positive ε and ε'.

It can be shown that if a function $f(x)$ is defined and bounded† in the

---

† $f(x)$ bounded in the interval $-\pi \leqslant x \leqslant \pi$ means that there exist constants m and M such that $m \leqslant f(x) \leqslant M$ for x in the interval $-\pi \leqslant x \leqslant \pi$.

interval $-\pi \leqslant x \leqslant \pi$ and if $f(x)$ has only a finite number of maxima and minima and only a finite number of discontinuities in the same interval, then its Fourier series is convergent for all x, $-\pi < x < \pi$, to the sum $\frac{1}{2}\{f(x + 0) + f(x - 0)\}$ and that at each of the end points $x = -\pi$, $x = \pi$ it is convergent to the sum $\frac{1}{2}\{f(-\pi + 0) + f(\pi - 0)\}$.

The above conditions are known as the *Dirichlet conditions*.

If the function $f(x)$ is continuous at $x = a$, then $f(a + 0) = f(a - 0) = f(a)$ and so when $x = a$ the Fourier series converges to the sum $f(a)$.

Since the Fourier series is periodic with period 2π, it will of course converge to the sum $f(a)$ when $x = a + 2n\pi$ for all integers n.

Example   Determine the Fourier series for $f(x) = x$ in the interval $-\pi \leqslant x \leqslant \pi$ and sketch the graph of the function represented by the sum of the series.

As above,

$$a_0 = \frac{1}{2\pi} \int_{-\pi}^{\pi} x \, dx = \frac{1}{2\pi} \left[\frac{x^2}{2} \right]_{-\pi}^{\pi} = 0.$$

$$a_n = \frac{1}{\pi} \int_{-\pi}^{\pi} x \cos nx \, dx$$

$$= \frac{1}{\pi} \left[x \frac{\sin nx}{n} \right]_{-\pi}^{\pi} - \frac{1}{\pi} \int_{-\pi}^{\pi} \frac{\sin nx}{n} \, dx$$

$$= \frac{1}{\pi n} \left[\frac{\cos nx}{n} \right]_{-\pi}^{\pi}$$

$$= 0.$$

$$b_n = \frac{1}{\pi} \int_{-\pi}^{\pi} x \sin nx \, dx$$

$$= \frac{1}{\pi} \left[-x \frac{\cos nx}{n} \right]_{-\pi}^{\pi} + \frac{1}{\pi} \int_{-\pi}^{\pi} \frac{\cos nx}{n} \, dx$$

$$= -\frac{2}{n} \cos n\pi + \frac{1}{\pi n} \left[\frac{\sin nx}{n} \right]_{-\pi}^{\pi}$$

$$= (-1)^{n+1} \frac{2}{n}.$$

Hence the Fourier series is

$$\sum_{n=1}^{\infty} (-1)^{n+1} \frac{2}{n} \sin nx$$

or

$$2\left(\sin x - \frac{\sin 2x}{2} + \frac{\sin 3x}{3} - \frac{\sin 4x}{4} + \cdots\right).$$

Since $f(x) = x$ is continuous and satisfies the Dirichlet conditions in the intervals $-\pi \leqslant x \leqslant \pi$, for any value of x in the interval $-\pi < x < \pi$ the sum of this Fourier series is the corresponding value of x itself. Also, when $x = \pm\pi$ the sum of the Fourier series converges to $\frac{1}{2}\{f(-\pi + 0) + f(\pi - 0)\} = \frac{1}{2}(-\pi + \pi) = 0$. This latter result is easily verified by putting $x = \pm\pi$ and so $\sin nx = 0$ in the above series. The graph of the function represented by the sum of the series is shown in figure 87. The value for $x = (2n + 1)\pi$ for integer values of

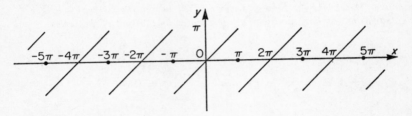

FIGURE 87

n is zero. Note that this graph is identical, in the interval $-\pi < x < \pi$, with the graph of $y = x$. Since the given function $f(x)$ is continuous at $x = \frac{1}{2}\pi$, it is seen that

$$f\left(\frac{\pi}{2}\right) = \sum_{n=1}^{\infty}(-1)^{n+1}\frac{2}{n}\sin\frac{n\pi}{2}$$

$$= 2\sum_{m=1}^{\infty}\frac{(-1)^{m-1}}{2m - 1}.$$

Therefore

$$\frac{\pi}{2} = 2(1 - \tfrac{1}{3} + \tfrac{1}{5} - \tfrac{1}{7} + \cdots)$$

or

$$\sum_{m=1}^{\infty}(-1)^{m-1}\frac{1}{2m - 1} = (1 - \tfrac{1}{3} + \tfrac{1}{5} - \tfrac{1}{7} + \cdots) = \frac{\pi}{4}.$$

Example Expand

$$f(x) = \begin{cases} 0 & \text{for} \quad -\pi \leqslant x < 0, \\ c & \text{for} \quad 0 \leqslant x \leqslant \pi, \end{cases}$$

where c is a positive constant, as a Fourier series and sketch the graph of the sum of the series.

$$a_0 = \frac{1}{2\pi} \int_{-\pi}^{\pi} f(x)\,dx = \frac{1}{2\pi} \int_0^{\pi} c\,dx$$

$$= \tfrac{1}{2}c.$$

$$a_n = \frac{1}{\pi} \int_{-\pi}^{\pi} f(x)\cos nx\,dx$$

$$= \frac{1}{\pi} \int_0^{\pi} c\cos nx\,dx$$

$$= 0.$$

$$b_n = \frac{1}{\pi} \int_{-\pi}^{\pi} f(x)\sin nx\,dx$$

$$= \frac{1}{\pi} \int_0^{\pi} c\sin nx\,dx$$

$$= \frac{c}{n\pi}(1 - \cos n\pi)$$

$$= \begin{cases} 0 & \text{when } n \text{ is even,} \\ \dfrac{2c}{n\pi} & \text{when } n \text{ is odd.} \end{cases}$$

Hence the Fourier series is

$$\tfrac{1}{2}c + \frac{2c}{\pi}\left(\sin x + \frac{\sin 3x}{3} + \frac{\sin 5x}{5} + \cdots\right).$$

Its sum at $x = \pm\pi$ is

$$\tfrac{1}{2}\{f(-\pi + 0) + f(\pi - 0)\} = \tfrac{1}{2}c$$

and its sum at $x = 0$ is

$$\tfrac{1}{2}\{f(0 - 0) + f(0 + 0)\} = \tfrac{1}{2}c.$$

The graph of the function represented by the sum of the series is shown in figure 88. When $x = n\pi$ $(n = 0, \pm 1, \pm 2, \ldots)$ the value of this function is $\tfrac{1}{2}c$.

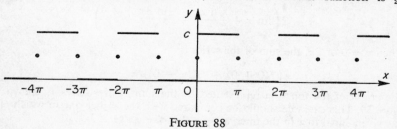

FIGURE 88

8

Since the given function $f(x)$ is continuous at $x = \frac{1}{2}\pi$, $f(\frac{1}{2}\pi)$ is equal to the sum of the Fourier series at $x = \frac{1}{2}\pi$. This once more leads to the result that

$$\sum_{m=1}^{\infty} (-1)^{m-1} \frac{1}{2m-1} = \frac{\pi}{4}.$$

Example   Determine the Fourier series for $f(x) = x$ in the interval $0 \leqslant x \leqslant 2\pi$ and sketch the graph of the function represented by the sum of the series.

$$a_0 = \frac{1}{2\pi} \int_0^{2\pi} x \, dx = \frac{1}{2\pi} \left[\frac{x^2}{2} \right]_0^{2\pi} = \pi.$$

$$a_n = \frac{1}{\pi} \int_0^{2\pi} x \cos nx \, dx$$

$$= \frac{1}{\pi} \left[\frac{x \sin nx}{n} \right]_0^{2\pi} - \frac{1}{\pi} \int_0^{2\pi} \frac{\sin nx}{n} \, dx$$

$$= \frac{1}{n\pi} \left[\frac{\cos nx}{n} \right]_0^{2\pi}$$

$$= 0.$$

$$b_n = \frac{1}{\pi} \int_0^{2\pi} x \sin nx \, dx$$

$$= \frac{1}{\pi} \left[-x \frac{\cos nx}{n} \right]_0^{2\pi} + \frac{1}{\pi} \int_0^{2\pi} \frac{\cos nx}{n} \, dx$$

$$= -\frac{2}{n} + \frac{1}{n\pi} \left[\frac{\sin nx}{n} \right]_0^{2\pi}$$

$$= -\frac{2}{n}.$$

Hence the Fourier series is

$$\pi - 2 \sum_1^{\infty} \frac{\sin nx}{n}$$

or

$$\pi - 2 \left(\sin x + \frac{\sin 2x}{2} + \frac{\sin 3x}{3} + \dots \right).$$

When $x = 0$ or 2π, the sum of the series is

$$\tfrac{1}{2}\{f(0+0) + f(2\pi-0)\} = \pi.$$

The graph of the function represented by the sum of the series is shown in figure 89. This example should be compared with the earlier one in which $f(x)$ was given equal to x in the interval $-\pi \leqslant x \leqslant \pi$ (page 209).

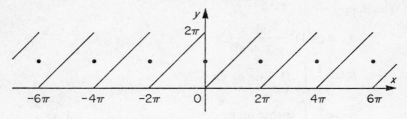

FIGURE 89

Example *Find the Fourier series for*

$$f(x) = \begin{cases} 0 & \text{for} \quad -\pi < x < 0, \\ \cosh x & \text{for} \quad 0 \leqslant x < \pi \end{cases}$$

and sketch the graph of the sum of this series.
  Deduce that

$$\sum_{n=1}^{\infty} \frac{(-1)^n}{1+n^2} = \frac{1}{2}\left(\frac{\pi}{\sinh \pi} - 1\right)$$

and that

$$\sum_{n=1}^{\infty} \frac{1}{1+n^2} = \frac{1}{2}\left(\frac{\pi}{\tanh \pi} - 1\right).$$

$$a_0 = \frac{1}{2\pi} \int_{-\pi}^{\pi} f(x)\,dx = \frac{1}{2\pi} \int_{0}^{\pi} \cosh x\,dx$$

$$= \frac{1}{2\pi} \left[\sinh x\right]_0^{\pi}$$

$$= \frac{\sinh \pi}{2\pi}.$$

$$a_n = \frac{1}{\pi} \int_{-\pi}^{\pi} f(x) \cos nx\,dx = \frac{1}{\pi} \int_{0}^{\pi} \cosh x \cos nx\,dx$$

$$= \frac{1}{\pi} \left[\sinh x \cos nx\right]_0^{\pi} - \frac{1}{\pi} \int_{0}^{\pi} \sinh x(-n \sin nx)\,dx$$

$$= (-1)^n \frac{\sinh \pi}{\pi} + \frac{n}{\pi} \left[\cosh x \sin nx\right]_0^{\pi} - \frac{n}{\pi} \int_{0}^{\pi} \cosh x(n \cos nx)\,dx$$

$$= (-1)^n \frac{\sinh \pi}{\pi} - n^2 a_n.$$

Therefore

$$a_n = (-1)^n \frac{\sinh \pi}{\pi(1 + n^2)} .$$

$$b_n = \frac{1}{\pi} \int_{-\pi}^{\pi} f(x) \sin nx \, dx$$

$$= \frac{1}{\pi} \int_{0}^{\pi} \cosh x \sin nx \, dx$$

$$= \frac{1}{\pi} \left[\sinh x \sin nx \right]_0^\pi - \frac{1}{\pi} \int_{0}^{\pi} \sinh x(n \cos nx) \, dx$$

$$= -\frac{n}{\pi} \left[\cosh x \cos nx \right]_0^\pi + \frac{n}{\pi} \int_{0}^{\pi} \cosh x(-n \sin nx) \, dx$$

$$= -\frac{n}{\pi} \left((-1)^n \cosh \pi - 1 \right) - n^2 b_n.$$

Therefore

$$b_n = \frac{n\left(1 - (-1)^n \cosh \pi\right)}{\pi(1 + n^2)} .$$

Hence the Fourier series is

$$\frac{\sinh \pi}{2\pi} + \sum_{n=1}^{\infty} \left\{ (-1)^n \frac{\sinh \pi}{\pi(1 + n^2)} \cos nx + \frac{n\left(1 - (-1)^n \cosh \pi\right)}{\pi(1 + n^2)} \sin nx \right\}$$

$$= \frac{\sinh \pi}{\pi} \left\{ \frac{1}{2} - \frac{\cos x}{2} + \frac{\cos 2x}{5} - \frac{\cos 3x}{10} + \ldots \right\}$$

$$+ \frac{1}{\pi} \left\{ \frac{1 + \cosh \pi}{2} \sin x + \frac{2(1 - \cosh \pi)}{5} \sin 2x \right.$$

$$\left. + \frac{3(1 + \cosh \pi)}{10} \sin 3x + \ldots \right\}.$$

The graph of the function represented by the sum of the series is shown in figure 90. The sum of the series at $x = 0$ is

$$\tfrac{1}{2}\{f(0 - 0) + f(0 + 0)\} = \tfrac{1}{2}(0 + \cos 0) = \tfrac{1}{2}.$$

The sum of the series at $x = \pm\pi$ is

$$\tfrac{1}{2}\{f(-\pi + 0) + f(\pi - 0)\} = \tfrac{1}{2}(0 + \cosh \pi) = \tfrac{1}{2} \cosh \pi.$$

Hence, putting $x = 0$ in the series, we obtain

$$\frac{1}{2} = \frac{\sinh \pi}{2\pi} + \sum_{n=1}^{\infty} (-1)^n \frac{\sinh \pi}{\pi(1 + n^2)} .$$

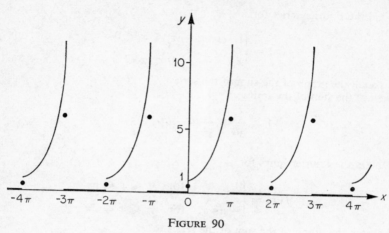

FIGURE 90

Therefore

$$\sum_{n=1}^{\infty} \frac{(-1)^n}{1+n^2} = \frac{\pi}{\sinh \pi}\left(\frac{1}{2} - \frac{\sinh \pi}{2\pi}\right)$$

$$= \frac{1}{2}\left(\frac{\pi}{\sinh \pi} - 1\right).$$

Then, putting $x = \pi$ in the series we obtain

$$\tfrac{1}{2}\cosh \pi = \frac{\sinh \pi}{2\pi} + \sum_{n=1}^{\infty}(-1)^n\frac{\sinh \pi}{\pi(1+n^2)}(-1)^n$$

$$= \frac{\sinh \pi}{2\pi} + \frac{\sinh \pi}{\pi}\sum_{n=1}^{\infty}\frac{1}{1+n^2}.$$

Therefore

$$\sum_{n=1}^{\infty}\frac{1}{1+n^2} = \frac{\pi}{\sinh \pi}\left(\tfrac{1}{2}\cosh \pi - \frac{\sinh \pi}{2\pi}\right)$$

$$= \frac{1}{2}\left(\frac{\pi}{\tanh \pi} - 1\right).$$

Exercises

1. Determine the Fourier series for the function

$$f(x) = \begin{cases} -\tfrac{1}{2}, & \text{for} \quad -\pi < x < 0 \\ \tfrac{1}{2}, & \text{for} \quad 0 < x < \pi. \end{cases}$$

What is the value of the sum of the series when $x = -\pi, 0, \pi$?

2. Find the Fourier series for

$$f(x) = \begin{cases} 1, & \text{for} \quad -\pi \leqslant x < 0 \\ x, & \text{for} \quad 0 \leqslant x \leqslant \pi \end{cases}$$

and sketch the graph of the sum of the series.
　Deduce the sum of the series

$$1 + \frac{1}{3^2} + \frac{1}{5^2} + \frac{1}{7^2} + \dots$$

3. Obtain the Fourier series for

$$f(x) = \begin{cases} 0, & \text{for} \quad 0 \leqslant x \leqslant \pi \\ -\sin x, & \text{for} \quad \pi \leqslant x \leqslant 2\pi \end{cases}$$

and sketch the graph of the sum of the series.

4. Find the Fourier series for $f(x) = e^x$ for $-\pi < x < \pi$.

5. Determine the Fourier series for

$$f(x) = \begin{cases} \sinh x & \text{for} \quad -\pi < x < 0, \\ 0 & \text{for} \quad 0 \leqslant x < \pi \end{cases}$$

and sketch the graph of the sum of the series.

6. Find the Fourier series expansion of

$$f(x) = \begin{cases} 0 & \text{for} \quad -\pi < x \leqslant 0, \\ x^2 & \text{for} \quad 0 \leqslant x < \pi \end{cases}$$

and sketch the graph of the sum of the series. Deduce that the sum of the series

$$1 + \frac{1}{2^2} + \frac{1}{3^2} + \frac{1}{4^2} + \dots$$

is equal to $\frac{1}{6}\pi^2$.

4.3 Odd and even functions

　A function $f(x)$ is said to be *even* if $f(-x) \equiv f(x)$ and *odd* if $f(-x) \equiv -f(x)$. For example x^2, $\cos x$, $x \sin x$, $x^3 \tan x$ are all even and x, $\sin x$, $x \cos x$, $x^2 \tan x$ are all odd.
　Many functions are of course neither even nor odd. For example e^x, $1 - x$, $(\cos x + \sin x)$ are neither even nor odd.

Now, if $F(x)$ is an odd function defined in the domain $-c < x < c$,

$$\int_{-c}^{c} F(x)\,dx = \int_{-c}^{0} F(x)\,dx + \int_{0}^{c} F(x)\,dx$$

$$= -\int_{c}^{0} F(-u)\,du + \int_{0}^{c} F(x)\,dx \quad \text{where} \quad u = -x$$

$$= \int_{c}^{0} F(u)\,du + \int_{0}^{c} F(x)\,dx \quad \text{since} \quad F(u) = -F(-u)$$

$$= 0.$$

Similarly, if $F(x)$ is an even function defined in $-c < x < c$,

$$\int_{-c}^{c} F(x)\,dx = 2\int_{0}^{c} F(x)\,dx.$$

Then, if $f(x)$ is an even function defined in $-\pi < x < \pi$, $F(x) = f(x) \sin nx$ is an odd function defined in this domain and so

$$\int_{-\pi}^{\pi} f(x) \sin nx\,dx = 0.$$

Therefore the coefficients b_n of the sine terms in the Fourier series for $f(x)$ in $-\pi < x < \pi$ are zero. Thus the Fourier series for an even function consists entirely of cosine terms. Further,

$$a_n = \frac{1}{\pi}\int_{-\pi}^{\pi} f(x) \cos nx\,dx$$

$$= \frac{2}{\pi}\int_{0}^{\pi} f(x) \cos nx\,dx \quad \text{since } f(x) \cos nx \text{ is an even function.}$$

Also,

$$a_0 = \frac{1}{2\pi}\int_{-\pi}^{\pi} f(x)\,dx = \frac{1}{\pi}\int_{0}^{\pi} f(x)\,dx.$$

Similarly it can be shown that if $f(x)$ is an odd function given in the interval $-\pi < x < \pi$, then its Fourier series consists entirely of sine terms and

$$b_n = \frac{2}{\pi}\int_{0}^{\pi} f(x) \sin nx\,dx.$$

Exercises

1. State which of the following functions are odd, which even and which are neither odd nor even

    (i) $\cos x - 3\cos 3x$,   (ii) $\sin x - 3\cos 3x$,   (iii) $1 + x$,

    (iv) $x/(x + 1)(x - 1)$,   (v) $x \tan x - x^2$,   (vi) $x \ln |x|$.

2. Show that if λ is a constant between 0 and 1 the Fourier series for the function $f(x) = \cos \lambda x$, $-\pi < x < \pi$ is

$$\frac{\sin \lambda \pi}{\pi}\left(\frac{1}{\lambda} + 2\lambda \sum_{n=1}^{\infty} \frac{(-1)^{n+1}}{n^2 - \lambda^2} \cos nx\right).$$

3. Determine the Fourier series for the function

$$f(x) = \begin{cases} 0 & \text{for} & -\pi < x < -\tfrac{1}{2}\pi, \\ \cos x & \text{for} & -\tfrac{1}{2}\pi \leqslant x < \tfrac{1}{2}\pi, \\ 0 & \text{for} & \tfrac{1}{2}\pi \leqslant x < \pi \end{cases}$$

and sketch the graph of the sum of the series.

4. Obtain the Fourier series for $f(x) = \sinh x$ in the interval $-\pi \leqslant x \leqslant \pi$. Sketch the graph of the sum of the series and deduce that

$$\sum_{n=0}^{\infty} \frac{(-1)^n(2n + 1)}{2n^2 + 2n + 1} = \tfrac{1}{2}\pi \operatorname{sech} \frac{\pi}{2}.$$

5. A function $f(x)$ has a period 2π and is defined for $-\pi \leqslant x \leqslant \pi$ as follows:

$$f(x) = \begin{cases} -\dfrac{3}{2\pi}(x + \dfrac{\pi}{3}), & -\pi \leqslant x < -\dfrac{\pi}{3} \\ 0, & -\dfrac{\pi}{3} \leqslant x \leqslant \dfrac{\pi}{3} \\ \dfrac{3}{2\pi}(x - \dfrac{\pi}{3}), & \dfrac{\pi}{3} < x \leqslant \pi. \end{cases}$$

Show that its Fourier series is

$$\frac{1}{3} - \frac{12}{\pi^2} \sum_{n=1}^{\infty} \frac{1}{n^2} \sin^2 \frac{n\pi}{3} \cos \frac{n\pi}{3} \cos nx$$

and deduce the sum of the series

$$\frac{1}{2^2} + \frac{1}{4^2} - \frac{1}{5^2} - \frac{1}{7^2} + \frac{1}{8^2} + \frac{1}{10^2} + \cdots.$$

4.4 Half-range series

Suppose now that it is desired to find a Fourier series for a function $f(x)$ which is defined only in the interval $0 < x < \pi$. This can be done of course by first defining a function $F(x)$ in $-\pi < x < \pi$ such that $F(x) \equiv f(x)$ in $0 < x < \pi$ and then determining the Fourier series for $F(x)$. Now the form of $F(x)$ in the interval $-\pi < x < 0$ can be chosen in an infinite variety of ways. In practice however only the following two ways are used.

1. Choose $F(x)$ to be an *even* function in $-\pi < x < \pi$, such that $F(x) \equiv f(x)$ in $0 < x < \pi$ (figure 91). This results in a Fourier cosine

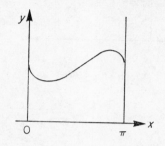

Graph of $y = f(x)$

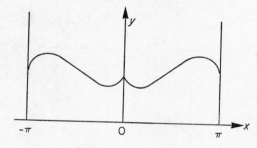

Graph of $y = F(x)$

FIGURE 91

series $a_0 + \sum\limits_{n=1}^{\infty} a_n \cos nx$ where

$$a_0 = \frac{1}{\pi} \int_0^{\pi} f(x)\, dx$$

and

$$a_n = \frac{2}{\pi} \int_0^{\pi} f(x) \cos nx\, dx.$$

2. Choose $F(x)$ to be an *odd* function in $-\pi < x < \pi$ such that $F(x) \equiv f(x)$ in $0 < x < \pi$ (figure 92). This results in a Fourier sine series $\sum\limits_{n=1}^{\infty} b_n \sin nx$ where

$$b_n = \frac{2}{\pi} \int_0^{\pi} f(x) \sin nx\, dx.$$

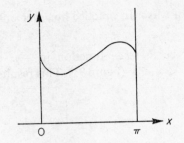

Graph of $y = f(x)$

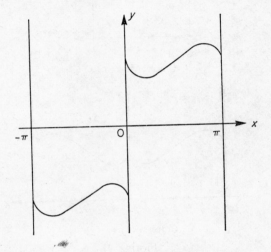

Graph of $y = F(x)$

FIGURE 92

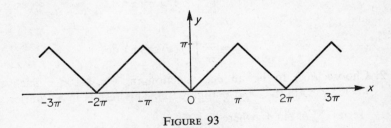

FIGURE 93

Thus given a function $f(x)$ defined in $0 < x < \pi$ it is possible, as above, to determine either a cosine series or a sine series associated with it.

Example   Obtain the half-range cosine series for the function $f(x) = x$ in the interval $0 \leqslant x \leqslant \pi$.

$$a_0 = \frac{1}{\pi} \int_0^\pi x \, dx = \frac{1}{\pi} \left[\frac{x^2}{2} \right]_0^\pi = \frac{\pi}{2}.$$

$$a_n = \frac{2}{\pi} \int_0^\pi x \cos nx \, dx$$

$$= \frac{2}{\pi} \left[x \frac{\sin nx}{n} \right]_0^\pi - \frac{2}{\pi} \int_0^\pi \frac{\sin nx}{n} \, dx$$

$$= \frac{2}{\pi n} \left[\frac{\cos nx}{n} \right]_0^\pi$$

$$= \frac{2}{\pi n^2} (\cos n\pi - 1)$$

$$= \begin{cases} - \dfrac{4}{\pi n^2} & \text{for } n \text{ odd} \\ 0 & \text{for } n \text{ even.} \end{cases}$$

Hence, for the interval $0 < x < \pi$, the Fourier cosine series for x is

$$\frac{\pi}{2} - \frac{4}{\pi} \sum_{m=1}^\infty \frac{1}{(2m-1)^2} \cos(2m-1)x$$

or

$$\frac{\pi}{2} - \frac{4}{\pi} \left(\cos x + \frac{\cos 3x}{3^2} + \frac{\cos 5x}{5^2} + \cdots \right).$$

The graph of the function represented by the sum of the series is shown in figure 93.

Now since the graph is continuous at $x = 0$, the sum of the series at $x = 0$ is equal to $f(0)$.

Therefore

$$0 = \frac{\pi}{2} - \frac{4}{\pi} \sum_{m=1}^\infty \frac{1}{(2m-1)^2} = \frac{\pi}{2} - \frac{4}{\pi} \left(1 + \frac{1}{3^2} + \frac{1}{5^2} + \cdots \right).$$

Hence

$$1 + \frac{1}{3^2} + \frac{1}{5^2} + \frac{1}{7^2} + \cdots = \frac{\pi^2}{8}.$$

(The same result is obtained by evaluating the sum of the series at $x = \pi$.) Note that the Fourier sine series for the above function is equivalent to the

Fourier series for $f(x) = x$ in the interval $-\pi < x < \pi$ and has already been obtained in an earlier example on p. 209. The graph is shown in figure 87.

Example   Obtain the half-range sine series for $f(x) = \cos x$ in $0 < x < \pi$.

$$b_n = \frac{2}{\pi} \int_0^{\pi} \cos x \sin nx \, dx$$

$$= \frac{1}{\pi} \int_0^{\pi} \{\sin (n + 1)x + \sin (n - 1)x\} \, dx$$

$$= \frac{1}{\pi} \left[-\frac{\cos (n + 1)x}{n + 1} - \frac{\cos (n - 1)x}{n - 1} \right]_0^{\pi} \quad \text{if} \quad n \neq 1$$

$$= \frac{1}{\pi} \left\{ \frac{1}{n + 1} + \frac{1}{n - 1} - \frac{\cos (n + 1)\pi}{n + 1} - \frac{\cos (n - 1)\pi}{n - 1} \right\}$$

$$= \frac{2n}{\pi(n^2 - 1)} (1 - (-1)^{n+1})$$

$$= \begin{cases} 0 & \text{if } n \text{ is odd and } > 1 \\ \dfrac{4n}{\pi(n^2 - 1)} & \text{if } n \text{ is even.} \end{cases}$$

Therefore

$$b_{2m} = \frac{8m}{\pi(4m^2 - 1)} \quad \text{for} \quad m = 1, 2, 3, \ldots$$

and $b_{2m+1} = 0$ for $m = 1, 2, 3, \ldots$ Also

$$b_1 = \frac{2}{\pi} \int_0^{\pi} \cos x \sin x \, dx$$

$$= \frac{1}{\pi} \left[\sin^2 x \right]_0^{\pi}$$

$$= 0.$$

Hence the half-range sine series for $\cos x$ in $0 < x < \pi$ is

$$\frac{8}{\pi} \sum_{m=1}^{\infty} \frac{m}{4m^2 - 1} \sin 2mx$$

$$= \frac{8}{\pi} \left(\frac{1}{1 \cdot 3} \sin 2x + \frac{2}{3 \cdot 5} \sin 4x + \frac{3}{5 \cdot 7} \sin 6x + \ldots \right).$$

The graph of the function represented by the sum of the series is shown in figure 94.

The half-range cosine series for $f(x) = \cos x$ in $0 < x < \pi$ is of course simply $\cos x$.

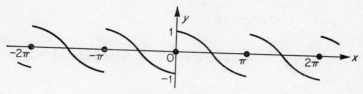

FIGURE 94

Example   Obtain a Fourier series for the function

$$f(x) = \begin{cases} \pi + x & \text{for} \quad -\pi \leqslant x < -\tfrac{1}{2}\pi, \\ \pi/2 & \text{for} \quad -\tfrac{1}{2}\pi \leqslant x < \tfrac{1}{2}\pi, \\ \pi - x & \text{for} \quad \tfrac{1}{2}\pi \leqslant x \leqslant \pi. \end{cases}$$

The graph of $y = f(x)$ is shown in figure 95.

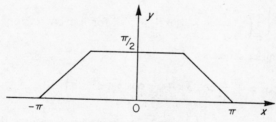

FIGURE 95

It is easily seen that $f(x)$ is an even function, and hence the Fourier series for $f(x)$ in $-\pi < x < \pi$ is the half-range cosine series for

$$f_1(x) = \begin{cases} \tfrac{1}{2}x & \text{in} \quad 0 < x < \tfrac{1}{2}\pi, \\ \pi - x & \text{in} \quad \tfrac{1}{2}\pi \leqslant x < \pi. \end{cases}$$

Therefore

$$a_0 = \frac{1}{\pi} \int_0^\pi f_1(x)\, dx$$

$$= \frac{1}{\pi} \left\{ \int_0^{\pi/2} \frac{\pi}{2}\, dx + \int_{\pi/2}^\pi (\pi - x)\, dx \right\}$$

$$= \frac{1}{\pi} \left\{ \left[\frac{\pi}{2} x \right]_0^{\pi/2} + \left[\pi x - \frac{x^2}{2} \right]_{\pi/2}^\pi \right\}$$

$$= \frac{3\pi}{8}.$$

$$a_n = \frac{2}{\pi} \int_0^\pi f_1(x) \cos nx \, dx$$

$$= \frac{2}{\pi} \left\{ \int_0^{\pi/2} \frac{\pi}{2} \cos nx \, dx + \int_{\pi/2}^\pi (\pi - x) \cos nx \, dx \right\}$$

$$= \frac{2}{\pi} \left\{ \frac{\pi}{2} \left[\frac{\sin nx}{n} \right]_0^{\pi/2} + \left[(\pi - x) \frac{\sin nx}{n} \right]_{\pi/2}^\pi + \frac{1}{n} \int_{\pi/2}^\pi \sin nx \, dx \right\}$$

$$= \frac{2}{\pi} \left\{ \frac{\pi}{2n} \sin \frac{n\pi}{2} - \frac{\pi}{2n} \sin \frac{n\pi}{2} + \frac{1}{n} \left[- \frac{\cos nx}{n} \right]_{\pi/2}^\pi \right\}$$

$$= \frac{2}{\pi} \left\{ - \frac{1}{n^2} (-1)^n + \frac{1}{n^2} \cos \frac{n\pi}{2} \right\}$$

$$= \begin{cases} \dfrac{2}{\pi} \dfrac{1}{n^2} & \text{for } n \text{ odd,} \\[3mm] \dfrac{2}{\pi} \left(- \dfrac{1}{n^2} + (-1)^{\frac{n}{2}} \dfrac{1}{n^2} \right) & \text{for } n \text{ even.} \end{cases}$$

Hence the Fourier series is

$$\frac{3\pi}{8} + \frac{2}{\pi} \left(\cos x + \frac{1}{3^2} \cos 3x + \frac{1}{5^2} \cos 5x + \cdots \right)$$

$$- \frac{4}{\pi} \left(\frac{1}{2^2} \cos 2x + \frac{1}{6^2} \cos 6x + \frac{1}{10^2} \cos 10x + \cdots \right).$$

Exercises

1. Obtain the half-range cosine series for $f(x) = \sin x$ in $0 \leqslant x \leqslant \pi$. Hence show that

$$\sum_{n=1}^\infty \frac{1}{4n^2 - 1} = \frac{1}{2}.$$

2. Obtain the half-range sine series for $f(x) = \pi - x$ in $0 < x < \pi$.

3. Expand

$$f(x) = \begin{cases} 1 - x/2h, & 0 \leqslant x \leqslant 2h \\ 0, & 2h < x \leqslant \pi \end{cases}$$

as a half-range cosine series. Hence find the sum of the series

$$\sum_{n=1}^\infty \left(\frac{\sin nh}{nh} \right)^2.$$

4.5 Change of scale

In the previous work only intervals of length 2π or π have been considered, but it is of course possible by a simple change of variable to map an interval of arbitrary length into one of length 2π or π and hence Fourier series can be determined for functions defined in intervals of arbitrary length. For example, if $f(x)$ is defined in $-l < x < l$, put $u = \pi x / l$ and then determine the Fourier series for $g(u) \equiv f(lu/\pi)$ in $-\pi < u < \pi$.

We obtain

$$
\begin{aligned}
a_0 &= \frac{1}{2\pi} \int_{-\pi}^{\pi} g(u)\,\mathrm{d}u \\
&= \frac{1}{2\pi} \int_{-\pi}^{\pi} f(lu/\pi)\,\mathrm{d}u \\
&= \frac{1}{2\pi} \int_{-l}^{l} f(x)\,\frac{\pi}{l}\,\mathrm{d}x \\
&= \frac{1}{2l} \int_{-l}^{l} f(x)\,\mathrm{d}x,
\end{aligned}
$$

$$
\begin{aligned}
a_n &= \frac{1}{\pi} \int_{-\pi}^{\pi} g(u) \cos nu\,\mathrm{d}u \\
&= \frac{1}{\pi} \int_{-\pi}^{\pi} f(lu/\pi) \cos nu\,\mathrm{d}u \\
&= \frac{1}{l} \int_{-l}^{l} f(x) \cos \frac{n\pi x}{l}\,\mathrm{d}x,
\end{aligned}
$$

$$
\begin{aligned}
b_n &= \frac{1}{\pi} \int_{-\pi}^{\pi} g(u) \sin nu\,\mathrm{d}u \\
&= \frac{1}{\pi} \int_{-\pi}^{\pi} f(lu/\pi) \sin nu\,\mathrm{d}u \\
&= \frac{1}{l} \int_{-l}^{l} f(x) \sin \frac{n\pi x}{l}\,\mathrm{d}x
\end{aligned}
$$

and hence the Fourier series

$$
a_0 = \sum_{n=1}^{\infty} (a_n \cos nu + b_n \sin nu)
$$

which, in terms of the variable x, is

$$
a_0 + \sum_{n=1}^{\infty} \left(a_n \cos \frac{n\pi x}{l} + b_n \sin \frac{n\pi x}{l} \right).
$$

Example   *Obtain the Fourier series for*

$$f(x) = \begin{cases} 0, & -\pi/\omega < x < 0 \\ A \sin \omega x, & 0 \leqslant x < \pi/\omega \end{cases}$$

where A and ω are constants, and deduce that

$$\sum_{m=1}^{\infty} \frac{1}{(2m-1)(2m+1)} = \frac{1}{2}.$$

The series will have the form

$$a_0 + \sum_{n=1}^{\infty} (a_n \cos n\omega x + b_n \sin n\omega x)$$

where

$$a_0 = \frac{\omega}{2\pi} \int_{-\pi/\omega}^{\pi/\omega} f(x) \, dx$$

$$= \frac{\omega}{2\pi} \int_{0}^{\pi/\omega} A \sin \omega x \, dx$$

$$= \frac{A}{\pi}.$$

$$a_n = \frac{\omega}{\pi} \int_{-\pi/\omega}^{\pi/\omega} f(x) \cos n\omega x \, dx$$

$$= \frac{\omega}{\pi} \int_{0}^{\pi/\omega} A \sin \omega x \cos n\omega x \, dx$$

$$= \frac{A\omega}{2\pi} \int_{0}^{\pi/\omega} \{\sin (n+1)\omega x - \sin (n-1)\omega x\} \, dx$$

$$= \frac{A\omega}{2\pi} \left[-\frac{\cos (n+1)\omega x}{(n+1)\omega} + \frac{\cos (n-1)\omega x}{(n-1)\omega} \right]_{0}^{\pi/\omega} \quad \text{if} \quad n \neq 1$$

$$= \frac{A}{2\pi} \left[-\frac{\cos (n+1)\pi}{n+1} + \frac{\cos (n-1)\pi}{n-1} + \frac{1}{n+1} - \frac{1}{n-1} \right]$$

$$= \begin{cases} 0 & \text{if } n \text{ is odd } (\neq 1), \\ \dfrac{-2A}{(n-1)(n+1)\pi} & \text{if } n \text{ is even.} \end{cases}$$

$$a_1 = \frac{\omega}{\pi} \int_{0}^{\pi/\omega} A \sin \omega x \cos \omega x \, dx$$

$$= \frac{\omega}{2\pi} A \int_{0}^{\pi/\omega} \sin 2\omega x \, dx$$

$$= \frac{\omega A}{2\pi}\left[-\frac{\cos 2\omega x}{2\omega} \right]_0^{\pi/\omega}$$

$$= 0.$$

$$b_n = \frac{\omega}{\pi} \int_{-\pi/\omega}^{\pi/\omega} f(x)\sin n\omega x\,dx$$

$$= \frac{\omega}{\pi} \int_0^{\pi/\omega} A\sin \omega x\sin n\omega x\,dx$$

$$= \frac{\omega A}{2\pi} \int_0^{\pi/\omega} \{\cos (n-1)\omega x - \cos (n+1)\omega x\}\,dx$$

$$= \frac{\omega A}{2\pi}\left[\frac{\sin (n-1)\omega x}{(n-1)\omega} - \frac{\sin (n+1)\omega x}{(n+1)\omega} \right]_0^{\pi/\omega} \quad \text{if } n \neq 1.$$

$$= 0.$$

$$b_1 = \frac{\omega}{\pi} \int_0^{\pi/\omega} A\sin^2 \omega x\,dx$$

$$= \frac{\omega A}{2\pi} \int_0^{\pi/\omega} (1 - \cos 2\omega x)\,dx$$

$$= \frac{\omega A}{2\pi}\left[x - \frac{\sin 2\omega x}{2\omega} \right]_0^{\pi/\omega}$$

$$= \frac{A}{2}.$$

Hence the Fourier series is

$$\frac{A}{\pi} + \frac{A}{2}\sin \omega x + \sum_{m=1}^{\infty} \frac{-2A}{(2m-1)(2m+1)\pi}\cos 2m\omega x$$

$$= \frac{A}{\pi} + \frac{A}{2}\sin \omega x - \frac{2A}{\pi}\left(\frac{1}{1.3}\cos 2\omega x + \frac{1}{3.5}\cos 4\omega x + \dots \right).$$

The graph of the function represented by the sum of this series is shown in figure 96.

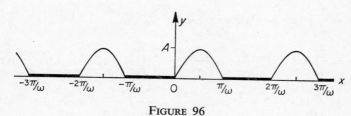

FIGURE 96

Since $f(x)$ is continuous at $x = 0$ we obtain, on putting $x = 0$ in the series and equating the resulting sum to $f(0)$, that

$$0 = \frac{A}{\pi} - \frac{2A}{\pi} \sum_{m=1}^{\infty} \frac{1}{(2m - 1)(2m + 1)}.$$

Hence

$$\frac{1}{1.3} + \frac{1}{3.5} = \frac{1}{5.7} + \ldots = \sum_{m=1}^{\infty} \frac{1}{(2m - 1)(2m + 1)} = \frac{1}{2}.$$

Example   Find the half-range cosine series for $f(x) = ax - x^2$ in $0 \leqslant x \leqslant a$ and deduce that

$$\sum_{m=1}^{\infty} \frac{1}{m^2} = \frac{\pi^2}{6}.$$

Put $u = \pi x/a$ and determine the cosine series for $f(au/\pi)$ in $0 \leqslant u \leqslant \pi$. If this series is

$$a_0 + \sum_{n=1}^{\infty} a_n \cos nu$$

then

$$a_0 = \frac{1}{\pi} \int_0^{\pi} f\left(\frac{au}{\pi}\right) du$$

$$= \frac{1}{a} \int_0^a f(x)\, dx$$

$$= \frac{1}{a} \int_0^a (ax - x^2)\, dx$$

$$= \frac{a^2}{6}$$

and

$$a_n = \frac{2}{\pi} \int_0^{\pi} f\left(\frac{au}{\pi}\right) \cos nu\, du$$

$$= \frac{2}{a} \int_0^a f(x) \cos \frac{n\pi x}{a}\, dx$$

$$= \frac{2}{a} \int_0^a (ax - x^2) \cos \frac{n\pi x}{a}\, dx$$

$$= \frac{2}{a}\left[(ax - x^2) \frac{a}{n\pi} \cos \frac{n\pi x}{a}\right]_0^a - \frac{2}{a} \int_0^a (a - 2x) \frac{a}{n\pi} \sin \frac{n\pi x}{a}\, dx$$

$$= \frac{2}{n\pi}\left[(a - 2x) \frac{a}{n\pi} \cos \frac{n\pi x}{a}\right]_0^a + \frac{2}{n\pi} \int_0^a \frac{2a}{n\pi} \cos \frac{n\pi x}{a}\, dx$$

$$= \frac{2a}{n^2\pi^2}\left(-a(-1)^n - a\right) + \frac{4a}{n^2\pi^2}\left[\frac{a}{n\pi}\sin\frac{n\pi x}{a}\right]_0^a$$

$$= -\frac{2a^2}{n^2\pi^2}\left(1 + (-1)^n\right)$$

$$= \begin{cases} 0 & \text{if } n \text{ is odd,} \\ -4a^2/n^2\pi^2 & \text{if } n \text{ is even.} \end{cases}$$

Hence the Fourier cosine series is

$$\frac{a^2}{6} - \frac{a^2}{\pi^2}\sum_{m=1}^{\infty}\frac{1}{m^2}\cos\frac{2m\pi x}{a}$$

$$= \frac{a^2}{6} - \frac{a^2}{\pi^2}\left(\cos\frac{2\pi x}{a} + \frac{1}{2^2}\cos\frac{4\pi x}{a} + \frac{1}{3^2}\cos\frac{6\pi x}{a} + \dots\right).$$

Putting $x = 0$, we obtain

$$\frac{a^2}{6} - \frac{a^2}{\pi^2}\sum_{m=1}^{\infty}\frac{1}{m^2} = \tfrac{1}{2}\{f(0+0) + f(0-0)\}.$$

Therefore

$$\sum_{m=1}^{\infty}\frac{1}{m^2} = \frac{\pi^2}{6}.$$

Exercises

1. Obtain the Fourier series for

$$f(x) = \begin{cases} x & \text{for} \quad -1 < x \leqslant 0, \\ x + 2 & \text{for} \quad 0 < x \leqslant 1. \end{cases}$$

By putting $x = \tfrac{1}{2}$ and $x = \tfrac{1}{4}$, deduce the sums of the series

$$1 - \tfrac{1}{3} + \tfrac{1}{5} - \tfrac{1}{7} + \dots$$

and

$$1 + \tfrac{1}{3} - \tfrac{1}{5} - \tfrac{1}{7} + \tfrac{1}{9} + \tfrac{1}{11} + \dots$$

2. Find the Fourier series for the function

$$f(x) = \begin{cases} -1 & \text{for} \quad -2 < x \leqslant -1, \\ x & \text{for} \quad -1 < x < 1, \\ 1 & \text{for} \quad 1 \leqslant x < 2. \end{cases}$$

3. Expand the function

$$f(x) = \begin{cases} 1 & \text{for } 0 < x \leqslant \tfrac{2}{3}, \\ 0 & \text{for } \tfrac{2}{3} < x < 1 \end{cases}$$

as a half range cosine series and sketch the graph of the sum of the series. Deduce the sum of the series

$$1 - \tfrac{1}{2} + \tfrac{1}{4} - \tfrac{1}{5} + \tfrac{1}{7} - \cdots.$$

4.6 Integration of Fourier series

Suppose that the function $f(x)$ defined in $-\pi \leqslant x \leqslant \pi$ satisfies the Dirichlet conditions so that its Fourier series

$$a_0 + \sum_{n=1}^{\infty} (a_n \cos nx + b_n \sin nx)$$

converges for all x in $-\pi < x < \pi$ to $\tfrac{1}{2}\{f(x - 0) + f(x + 0)\}$ and the sum of the series when $x = \pm\pi$ is $\tfrac{1}{2}\{f(-\pi + 0) + f(\pi - 0)\}$.

The function $F(x)$ given by

$$F(x) = \int_{-\pi}^{x} f(t)\, \mathrm{d}t - a_0 x, \qquad -\pi \leqslant x \leqslant \pi,$$

is then continuous everywhere in $-\pi < x < \pi$.

Also

$$F(\pi) = \int_{-\pi}^{\pi} f(t)\, \mathrm{d}t - a_0 \pi$$

$$= 2\pi a_0 - a_0 \pi = \pi a_0$$

and

$$F(-\pi) = \int_{-\pi}^{-\pi} f(t)\, \mathrm{d}t - a_0(-\pi) = \pi a_0$$

so that $F(\pi) = F(-\pi)$.

It can then be shown that the Fourier series for $F(x)$ converges to $F(x)$ for all x in $-\pi \leqslant x \leqslant \pi$.

Therefore $F(x) = A_0 + \sum_{n=1}^{\infty} (A_n \cos nx + B_n \sin nx)$ for $-\pi \leqslant x \leqslant \pi$, where

$$A_0 = \frac{1}{2\pi} \int_{-\pi}^{\pi} F(x)\, \mathrm{d}x, \qquad A_n = \frac{1}{\pi} \int_{-\pi}^{\pi} F(x) \cos nx\, \mathrm{d}x$$

and

$$B_n = \frac{1}{\pi} \int_{-\pi}^{\pi} F(x) \sin nx\, \mathrm{d}x.$$

Then, integrating the expression for A_n by parts, we obtain

$$A_n = \frac{1}{\pi}\left[F(x)\frac{\sin nx}{n}\right]_{-\pi}^{\pi} - \frac{1}{\pi}\int_{-\pi}^{\pi} F'(x)\frac{\sin nx}{n}\,dx$$

$$= -\frac{1}{\pi}\int_{-\pi}^{\pi} (f(x) - a_0)\frac{\sin nx}{n}\,dx$$

since $F'(x) = f(x) - a_0$.

Hence

$$A_n = -\frac{1}{n\pi}\int_{-\pi}^{\pi} f(x)\sin nx\,dx + \frac{a_0}{n\pi}\int_{-\pi}^{\pi}\sin nx\,dx$$

$$= -\frac{1}{n\pi}\int_{-\pi}^{\pi} f(x)\sin nx\,dx - \frac{a_0}{n\pi}\left[\frac{\cos nx}{n}\right]_{-\pi}^{\pi} = -\frac{b_n}{n}.$$

Similarly,

$$B_n = \frac{1}{\pi}\left[-F(x)\frac{\cos nx}{n}\right]_{-\pi}^{\pi} + \frac{1}{\pi}\int_{-\pi}^{\pi} F'(x)\frac{\cos nx}{n}\,dx$$

$$= \frac{1}{\pi}\int_{-\pi}^{\pi} (f(x) - a_0)\frac{\cos nx}{n}\,dx$$

$$= \frac{1}{n\pi}\int_{-\pi}^{\pi} f(x)\cos nx\,dx - \frac{a_0}{n\pi}\left[\frac{\sin nx}{n}\right]_{-\pi}^{\pi}$$

$$= \frac{a_n}{n}.$$

Therefore

$$F(x) = A_0 + \sum_{n=1}^{\infty}\left(-\frac{b_n}{n}\cos nx + \frac{a_n}{n}\sin nx\right).$$

Now putting $x = \pi$ we obtain

$$A_0 + \sum_{n=1}^{\infty}\left(-\frac{b_n}{n}\cos n\pi\right) = \tfrac{1}{2}\{F(-\pi + 0) + F(\pi - 0)\} = \pi a_0.$$

Hence

$$F(x) - \pi a_0 = \sum_{n=1}^{\infty}\left\{\frac{b_n}{n}(\cos n\pi - \cos nx) + \frac{a_n}{n}\sin nx\right\}.$$

Therefore

$$\int_{-\pi}^{x} f(t)\,dt = F(x) + a_0 x$$

$$= a_0(x + \pi) + \sum_{n=1}^{\infty}\left\{\frac{b_n}{n}(\cos n\pi - \cos nx) + \frac{a_n}{n}\sin nx\right\}.$$

Now it is readily seen that the right hand side of the above relation is obtained by integrating term by term the Fourier series for the function $f(x)$. Hence, if a function $f(x)$ satisfies the Dirichlet conditions in $-\pi \leqslant x \leqslant \pi$, then its Fourier series can be integrated term by term in this interval and the resulting series will converge to $\int_{-\pi}^{x} f(t)\,dt$ for all x in the interval. Note however that the integrated series will not be a Fourier series unless a_0 is zero. The above result also holds for intervals other than $-\pi \leqslant x \leqslant \pi$.

Example　Use the Fourier series for $f(x) = x$ in $-\pi \leqslant x \leqslant \pi$ (which was obtained in an earlier example on p. 209) to obtain the Fourier series for $g(x) = x^2$ in the same interval

We have

$$x = 2\left(\sin x - \frac{\sin 2x}{2} + \frac{\sin 3x}{3} - \frac{\sin 4x}{4} + \ldots\right).$$

Therefore, integrating term by term with respect to x from $-\pi$ to x we obtain

$$\frac{x^2}{2} - \frac{\pi^2}{2} = 2\Big\{-(\cos x - \cos \pi)$$

$$+ \frac{1}{2^2}(\cos 2x - \cos 2\pi) - \frac{1}{3^2}(\cos 3x - \cos 3\pi) + \ldots\Big\}.$$

Therefore

$$x^2 = \pi^2 - 4\left(1 + \frac{1}{2^2} + \frac{1}{3^2} + \ldots\right)$$

$$- 4\left(\cos x - \frac{1}{2^2}\cos 2x + \frac{1}{3^2}\cos 3x + \ldots\right)^{\dagger}$$

$$= a_0 + 4\sum_{n=1}^{\infty}(-1)^n \frac{\cos nx}{n^2}$$

where

$$a_0 = \pi^2 - 4\sum_{n=1}^{\infty}\frac{1}{n^2}.$$

Alternatively,

$$a_0 = \frac{1}{2\pi}\int_{-\pi}^{\pi} x^2\,dx = \frac{1}{2\pi}\left(\frac{2\pi^3}{3}\right) = \frac{\pi^2}{3}$$

† It is not always permissible to rearrange the terms in an infinite series in this way. However, for series obtained by term by term integration of the Fourier series for a function satisfying the Dirichlet conditions, such rearrangements are in fact permissible.

so that

$$x^2 = \frac{\pi^2}{3} + 4 \sum_{n=1}^{\infty} (-1)^n \frac{\cos nx}{n^2}.$$

The reader should verify for himself by evaluating

$$\frac{1}{\pi} \int_{-\pi}^{\pi} x^2 \cos nx \, dx$$

that the above is in fact the Fourier series for x^2 in $-\pi \leqslant x \leqslant \pi$.

Exercise

From the series obtained in Exercise 4 of section 4.3 deduce the Fourier series for $f(x) = \cosh x$ in the interval $-\pi < x < \pi$ and verify the result by direct evaluation.

4.7 Parseval's theorem

Suppose again that the function $f(x)$ satisfies the Dirichlet conditions in $-l \leqslant x \leqslant l$ so that

$$f(x) = a_0 + \sum_{n=1}^{\infty} \left(a_n \cos \frac{n\pi x}{l} + b_n \sin \frac{n\pi x}{l} \right) \quad \text{for} \quad -l < x < l,$$

where

$$a_0 = \frac{1}{2l} \int_{-l}^{l} f(x) \, dx, \quad a_n = \frac{1}{l} \int_{-l}^{l} f(x) \cos \frac{n\pi x}{l} dx, \quad b_n = \frac{1}{l} \int_{-\pi}^{\pi} f(x) \sin \frac{n\pi x}{l} dx$$

and where, if $x = \alpha$ is a point of discontinuity of f such that $-l < \alpha < l$, then $f(\alpha) = \frac{1}{2}\{f(\alpha - 0) + f(\alpha + 0)\}$. Then

$$[f(x)]^2 = a_0 f(x) + \sum_{n=1}^{\infty} f(x) \left(a_n \cos \frac{n\pi x}{l} + b_n \sin \frac{n\pi x}{l} \right).$$

Therefore

$$\int_{-l}^{l} [f(x)]^2 \, dx = \int_{-l}^{l} a_0 f(x) \, dx$$
$$+ \int_{-l}^{l} \sum_{n=1}^{\infty} f(x) \left(a_n \cos \frac{n\pi x}{l} + b_n \sin \frac{n\pi x}{l} \right) dx.$$

Hence, assuming that the infinite series on the right hand side can be integrated term by term, we obtain

$$\int_{-l}^{l} [f(x)]^2 \, dx = \int_{-l}^{l} a_0 f(x) \, dx + \sum_{n=1}^{\infty} \left\{ \int_{-l}^{l} a_n f(x) \cos \frac{n\pi x}{l} dx \right.$$
$$\left. + \int_{-l}^{l} b_n f(x) \sin \frac{n\pi x}{l} dx \right\}.$$

Therefore

$$\frac{1}{l} \int_{-l}^{l} [f(x)]^2 \, dx = \frac{a_0}{l} \int_{-l}^{l} f(x) \, dx$$

$$+ \sum_{n=1}^{\infty} \left\{ \frac{a_n}{l} \int_{-l}^{l} f(x) \cos \frac{n\pi x}{l} \, dx + \frac{b_n}{l} \int_{-l}^{l} f(x) \sin \frac{n\pi x}{l} \, dx \right\}$$

$$= 2a_0^2 + \sum_{n=1}^{\infty} (a_n^2 + b_n^2).$$

This result is known as Parseval's theorem.

Example *Obtain the Fourier series for $f(x) = |x|$ in $-\pi \leqslant x \leqslant \pi$ and hence evaluate*

$$\sum_{m=1}^{\infty} \frac{1}{m^4}$$

$$f(x) = |x| = \begin{cases} -x & -\pi \leqslant x \leqslant 0, \\ x & 0 \leqslant x \leqslant \pi. \end{cases}$$

$$a_0 = \frac{1}{2\pi} \int_{-\pi}^{\pi} f(x) \, dx = \frac{1}{2\pi} \left\{ \int_{-\pi}^{0} -x \, dx + \int_{0}^{\pi} x \, dx \right\}$$

$$= \frac{\pi}{2}.$$

$$a_n = \frac{1}{\pi} \int_{-\pi}^{\pi} f(x) \cos nx \, dx$$

$$= \frac{1}{\pi} \left\{ \int_{-\pi}^{0} -x \cos nx \, dx + \int_{0}^{\pi} x \cos nx \, dx \right\}$$

$$= \frac{2}{\pi} \int_{0}^{\pi} x \cos nx \, dx$$

$$= \frac{2}{\pi} \left\{ \left[x \frac{\sin nx}{n} \right]_{0}^{\pi} - \frac{1}{n} \int_{0}^{\pi} \sin nx \, dx \right\}$$

$$= \frac{2}{\pi n} \left[\frac{\cos nx}{n} \right]_{0}^{\pi} = \frac{2}{\pi n^2} (\cos n\pi - 1)$$

$$= \begin{cases} 0 & \text{for } n \text{ even} \\ -\dfrac{4}{\pi n^2} & \text{for } n \text{ odd.} \end{cases}$$

$b_n = 0$ since $f(x)$ is an even function.

Therefore

$$f(x) = \frac{\pi}{2} - \frac{4}{\pi} \sum_{m=1}^{\infty} \frac{1}{(2m-1)^2} \cos (2m-1)x.$$

Using Parseval's theorem, we obtain

$$2\left(\frac{\pi}{2}\right)^2 + \sum_{m=1}^{\infty}\left\{\frac{4}{\pi(2m-1)^2}\right\}^2 = \frac{1}{\pi}\int_{-\pi}^{\pi}[f(x)]^2\,\mathrm{d}x.$$

Therefore

$$\frac{\pi^2}{2} + \frac{16}{\pi^2}\sum_{m=1}^{\infty}\frac{1}{(2m-1)^4} = \frac{1}{\pi}\int_{-\pi}^{\pi}x^2\,\mathrm{d}x = \frac{2\pi^2}{3}.$$

Hence

$$\sum_{m=1}^{\infty}\frac{1}{(2m-1)^4} = \frac{\pi^4}{96}.$$

That is

$$\frac{1}{1^4} + \frac{1}{3^4} + \frac{1}{5^4} + \ldots = \frac{\pi^4}{96}.$$

Let

$$A_N = \sum_{n=1}^{N}\frac{1}{m^4}.$$

Therefore

$$A_{2N} = \sum_{m=1}^{2N}\frac{1}{m^4}$$

$$= \frac{1}{1^4} + \frac{1}{2^4} + \ldots + \frac{1}{(2N)^4}$$

$$= \left(\frac{1}{1^4} + \frac{1}{3^4} + \ldots + \frac{1}{(2N-1)^4}\right) + \left(\frac{1}{2^4} + \frac{1}{4^4} + \ldots + \frac{1}{(2N)^4}\right)$$

$$= \sum_{m=1}^{N}\frac{1}{(2m-1)^4} + \frac{1}{2^4}\left(1 + \frac{1}{2^4} + \frac{1}{3^4} + \ldots + \frac{1}{N^4}\right)$$

$$= B_N + \frac{1}{2^4}A_N \quad \text{where} \quad B_N = \sum_{m=1}^{N}\frac{1}{(2m-1)^4}.$$

Also,

$$A_{2N+1} = B_N + \frac{1}{2^4}A_N + \frac{1}{(2N+1)^4} = B_{N+1} + \frac{1}{2^4}A_N.$$

But it has been shown that the series

$$\sum\frac{1}{(2m-1)^4}$$

is convergent and that $B_N \to \pi^4/96$ as $N \to \infty$.

Also, it is known that the series $\Sigma(1/m^4)$ is convergent and so A_N tends to a limit A (say) as $N \to \infty$.

Hence, from the above, we obtain (on letting $N \to \infty$) that

$$A = \frac{\pi^4}{96} + \frac{1}{2^4} A.$$

Therefore

$$A\left(1 - \frac{1}{16}\right) = \frac{\pi^4}{96}$$

and so

$$\sum_{m=1}^{\infty} \frac{1}{m^4} = A = \frac{\pi^4}{90}.$$

Exercise

From the Fourier series for $f(x) = x$, $-\pi < x < \pi$, deduce, using Parseval's theorem, that

$$\sum_{n=1}^{\infty} \frac{1}{n^2} = \frac{\pi^2}{6}.$$

4.8 Differentiation of Fourier series

Suppose that $f(x)$ is continuous and satisfies the Dirichlet conditions in $-l \leqslant x \leqslant l$, then its Fourier series will converge for all x in $-l < x < l$ to $f(x)$.
Therefore

$$f(x) = a_0 + \sum_{n=1}^{\infty} \left(a_n \cos \frac{n\pi x}{l} + b_n \sin \frac{n\pi x}{l} \right)$$

for $-l < x < l$ where

$$a_0 = \frac{1}{2l} \int_{-l}^{l} f(x)\, dx, \qquad a_n = \frac{1}{l} \int_{-l}^{l} f(x) \cos \frac{n\pi x}{l}\, dx$$

and

$$b_n = \frac{1}{l} \int_{-l}^{l} f(x) \sin \frac{n\pi x}{l}\, dx.$$

Now the above conditions on $f(x)$ do not imply that $f'(x)$, the derivative of $f(x)$, is continuous nor that it satisfies the Dirichlet conditions. However if, in a particular case, $f'(x)$ does satisfy the Dirichlet conditions in $-l \leqslant x \leqslant l$ then its Fourier series

$$A_0 + \sum_{n=1}^{\infty} \left(A_n \cos \frac{n\pi x}{l} + B_n \sin \frac{n\pi x}{l} \right)$$

will converge for all x in $-l < x < l$ to $\frac{1}{2}\{f'(x + 0) + f'(x - 0)\}$.

$$A_0 = \frac{1}{2l} \int_{-l}^{l} f'(x)\, dx$$

$$= \frac{1}{2l}\left\{f(l) - f(-l)\right\}.$$

$$A_n = \frac{1}{l} \int_{-l}^{l} f'(x) \cos \frac{n\pi x}{l}\, dx$$

$$= \frac{1}{l}\left\{\left[f(x) \cos \frac{n\pi x}{l}\right]_{-l}^{l} + \frac{n\pi}{l} \int_{-l}^{l} f(x) \sin \frac{n\pi x}{l}\, dx\right\}$$

$$= \frac{1}{l}\{f(l) - f(-l)\} \cos n\pi + \frac{n\pi}{l} b_n.$$

$$B_n = \frac{1}{l} \int_{-l}^{l} f'(x) \sin \frac{n\pi x}{l}\, dx$$

$$= \frac{1}{l}\left\{\left[f(x) \sin \frac{n\pi x}{l}\right]_{-l}^{l} - \frac{n\pi}{l} \int_{-l}^{l} f(x) \cos \frac{n\pi x}{l}\, dx\right\}$$

$$= -\frac{n\pi}{l} a_n.$$

Hence, if $f'(x)$ satisfies the Dirichlet conditions, the Fourier series for $f'(x)$ will be

$$\sum_{n=1}^{\infty} \left(\frac{n\pi}{l} b_n \cos \frac{n\pi x}{l} - \frac{n\pi}{l} a_n \sin \frac{n\pi x}{l}\right),$$

the series obtained by differentiating term by term the Fourier series for $f(x)$, only if $f(l) = f(-l)$. For example the Fourier series for $f(x) = x$ in $-\pi < x < \pi$ is

$$\sum_{n=1}^{\infty} (-1)^{n+1} \frac{2}{n} \sin nx$$

but $f(-\pi) \neq f(\pi)$ and the series $\sum_{n=1}^{\infty} (-1)^{n+1} 2 \cos nx$ obtained by differentiating the above series term by term certainly does not converge to $f'(x) = 1$. Indeed the series does not even converge since the nth term does not tend to zero as n tends to infinity.

Example   The half-range cosine series for $f(x) = \sin x$ in $0 \leqslant x \leqslant \pi$ is

$$\frac{2}{\pi} - \frac{4}{\pi} \sum_{n=1}^{\infty} \frac{1}{(2n - 1)(2n + 1)} \cos 2nx.$$

Deduce the half-range sine series for $\cos x$ in $0 \leqslant x \leqslant \pi$.

$f'(x) = \cos x$ and so satisfies the Dirichlet conditions in $0 \leqslant x \leqslant \pi$. Also $f(\pi) = f(0) = 0$. Therefore the Fourier series for $\cos x$ can be obtained from that for $\sin x$ by differentiating term by term. Thus the required half-range series is

$$-\frac{4}{\pi} \sum_{n=1}^{\infty} \frac{1}{(2n-1)(2n+1)}(-2n \sin 2nx) = \frac{8}{\pi} \sum_{n=1}^{\infty} \frac{n}{(2n-1)(2n+1)} \sin 2nx.$$

which agrees with the result obtained in an earlier example on p. 222.

Exercise

From the series obtained in Exercise 3 of section 4.2, deduce the Fourier series for

$$f(x) = \begin{cases} 0 & \text{for} \quad 0 < x < \pi, \\ -\cos x & \text{for} \quad \pi < x < 2\pi, \end{cases}$$

and verify the result by direct evaluation.

5

Solution in series; Bessel functions; Legendre polynomials

5.1 Solution in series

In this chapter we shall be concerned with obtaining solutions of ordinary differential equations which can be written in the form

$$y'' + \frac{p(x)}{x} y' + \frac{q(x)}{x^2} y = 0$$

where $p(x)$ and $q(x)$ have power series expansions.

It can be shown that any differential equation of the above form has at least one solution which can be expressed in the form

$$y(x) = x^c \sum_{n=0}^{\infty} a_n x^n = x^c(a_0 + a_1 x + a_2 x^2 + \ldots) \qquad (a_0 \neq 0)$$

where the series is convergent for $|x|$ less than some constant R.

The method (which is also applicable to similar classes of higher order differential equations) essentially involves postulating such a series solution and then determining the values of $c, a_0, a_1, \ldots$ by substituting into the given differential equation. The following results will be assumed.*

(i) If $f(x) = \sum_{n=0}^{\infty} a_n x^{c+n}$ for $|x| < R \ (> 0)$, then

$$f'(x) = \sum_{n=0}^{\infty} (c + n) \times a_n x^{c+n-1} \text{ for } |x| < R.$$

That is term by term differentiation of the series is permissible.

---

* For proofs of these results see Jones and Jordan "Introductory Analysis for Scientists" (Wiley).

(ii) If also $g(x) = \sum\limits_{n=0}^{\infty} b_n x^{c+n}$ for $|x| < R_1 \, (> 0)$, then $f(x) + g(x) = \sum\limits_{n=0}^{\infty} (a_n + b_n)x^{c+n}$ for $|x|$ less than the smaller of R and R_1. That is the series can be added together term by term, and

(iii) $f(x)g(x) = \sum\limits_{n=0}^{\infty} (a_0 b_n + a_1 b_{n-1} + \ldots + a_{n-1}b_1 + a_n b_0)x^{2c+n}$

$$= a_0 b_0 x^{2c} + (a_0 b_1 + a_1 b_0)x^{2c+1} + \ldots$$

for $|x|$ less than the smaller of R and R_1. That is the series can be multiplied together term by term.

Example   Consider the differential equation

$$y'' + y = 0.$$

It is well known that the general solution of this equation can be written in the form

$$y(x) = A \cos x + B \sin x$$

for constants A and B. However this differential equation is of the form considered above, with $p(x) = 0$ and $q(x) = x^2$ and so we shall use the method indicated above to obtain a solution.

Assume

$$y(x) = \sum_{n=0}^{\infty} a_n x^{c+n} \text{ (with } a_0 \neq 0) \text{ is a solution.}$$

Therefore

$$y'(x) = \sum_{n=0}^{\infty} (c + n)a_n x^{c+n-1}$$

and

$$y''(x) = \sum_{n=0}^{\infty} (c + n)(c + n - 1)a_n x^{c+n-2}.$$

Hence, substituting in the differential equation we obtain

$$\sum_{n=0}^{\infty} (c + n)(c + n - 1)a_n x^{c+n-2} + \sum_{n=0}^{\infty} a_n x^{c+n} \equiv 0$$

and so all the coefficients in the resulting series on the left hand side must be zero.

Therefore, considering the term in x^{c-2} we obtain

$$c(c - 1)a_0 = 0.$$

This equation which determines c is called the *indicial equation*. Therefore $c = 0$ or $c = 1$ (since $a_0 \neq 0$). Also, considering the terms in x^{c-1} and x^{c+n-2} we obtain respectively

$$(c + 1)ca_1 = 0$$

and $(c + n)(c + n - 1)a_n + a_{n-2} = 0$ for $n \geqslant 2$.

Therefore when $c = 0$, a_1 is arbitrary and

$$a_n = -\frac{1}{n(n-1)} a_{n-2} \quad \text{for} \quad n \geqslant 2.$$

Thus we obtain the series

$$a_0 + a_1 x - \frac{1}{1 \cdot 2} a_0 x^2 - \frac{1}{3 \cdot 2} a_1 x^3 + \frac{1}{4 \cdot 3 \cdot 2 \cdot 1} a_0 x^4 + \frac{1}{5 \cdot 4 \cdot 3 \cdot 2} a_1 x^5 + \ldots$$

$$= a_0 \left(1 - \frac{1}{2!} x^2 + \frac{1}{4!} x^4 + \ldots \right) + a_1 \left(x - \frac{1}{3!} x^3 + \frac{1}{5!} x^5 + \ldots \right)$$

$$= a_0 \cos x + a_1 \sin x.$$

When $c = 1$, $a_1 = 0$ and we obtain again the solution $\cos x$.

Using this method to obtain the general solution of such a second order differential equation as is being considered, there arise three distinct cases requiring slightly different treatment. These are

1. The indicial equation has distinct roots which do not differ by an integer.
2. The indicial equation has equal roots.
3. The indicial equation has roots which differ by an integer.

Case 1   This is the simplest case. If the roots of the indicial equation are c_1 and c_2 where $c_1 - c_2 \neq$ an integer or zero, then a simple series solution will be obtained corresponding to each root. If these solutions are $y_1(x)$ and $y_2(x)$ respectively, then they will be linearly independent (that is one is not simply a constant multiple of the other) and the general solution of the differential equation may be written

$$y(x) = A y_1(x) + B y_2(x)$$

for arbitrary constants A and B.

Example   Obtain the general solution of the differential equation

$$3xy'' + 2y' + x^2 y = 0.$$

Assume that $y(x) = \sum_{n=0}^{\infty} a_n x^{n+c}$ with $a_0 \neq 0$ is a solution. Then

$$y'(x) = \sum_{n=0}^{\infty} (n + c) a_n x^{n+c-1}$$

and

$$y''(x) = \sum_{n=0}^{\infty} (n + c)(n + c - 1) a_n x^{n+c-2}.$$

Therefore, substituting into the given differential equation we obtain

$$3\sum_{n=0}^{\infty}(n+c)(n+c-1)a_n x^{n+c-1} + 2\sum_{n=0}^{\infty}(n+c)a_n x^{n+c-1} + \sum_{n=0}^{\infty}a_n x^{n+c+2} \equiv 0.$$

Therefore

$$\sum_{n=0}^{\infty}(c+n)(3c+3n-1)a_n x^{n+c-1} + \sum_{n=0}^{\infty}a_n x^{n+c+2} \equiv 0.$$

Hence, equating to zero the coefficient of x^{c-1} (this being the lowest power of x appearing) we obtain the indicial equation

$$c(3c-1)a_0 = 0$$

and so $c = 0$ or $c = \frac{1}{3}$ (since $a_0 \neq 0$). Then, considering the coefficient of x^c, we obtain

$$(c+1)(3c+2)a_1 = 0$$

and so $a_1 = 0$ since $c = 0$ or $c = \frac{1}{3}$. Similarly, picking out the coefficient of x^{c+1} gives

$$(c+2)(3c+5)a_2 = 0$$

and so $a_2 = 0$ since $c = 0$ or $c = \frac{1}{3}$. Then picking out the coefficient of x^{m+c-1} (for $m \geqslant 3$), we obtain

$$(c+m)(3c+3m-1)a_m + a_{m-3} = 0.$$

Therefore

$$a_m = -\frac{1}{(c+m)(3c+3m-1)}a_{m-3}$$

since $(c+m)$ and $(3c+3m-1)$ are not equal to zero for $c = 0$ or $c = \frac{1}{3}$ and $m \geqslant 3$.

Therefore

$$0 = a_1 = a_4 = a_7 = \ldots$$

and

$$0 = a_2 = a_5 = a_8 = \ldots.$$

When $c = 0$

$$a_3 = -\frac{1}{3 \cdot 8}a_0,$$

$$a_6 = -\frac{1}{6 \cdot 17}a_3 = \frac{1}{3 \cdot 6 \cdot 8 \cdot 17}a_0,$$

$$\cdots\cdots\cdots$$

$$\cdots\cdots\cdots$$

$$a_{3m} = -\frac{1}{3m(9m-1)}a_{3m-3}$$

$$= \ldots = (-1)^m \frac{1}{3 \cdot 6 \cdot 9 \ldots 3m \cdot 8 \cdot 17 \cdot 26 \ldots (9m-1)}a_0.$$

Hence the series solution is

$$a_0 \left(1 - \frac{x^3}{3 \cdot 8} + \frac{x^6}{3 \cdot 6 \cdot 8 \cdot 17} + \cdots \right).$$

When $c = \frac{1}{3}$

$$a_n = \frac{-1}{(n + \frac{1}{3})3n} a_{n-3} = \frac{-1}{n(3n + 1)} a_{n-3}.$$

Therefore

$$a_3 = \frac{-1}{3 \cdot 10} a_0,$$

$$a_6 = \frac{-1}{6 \cdot 19} a_3 = \frac{1}{3 \cdot 6 \cdot 10 \cdot 19} a_0,$$

. . . .

. . . .

$$a_{3m} = \frac{-1}{3m(9m + 1)} a_{3m-3}$$

$$= \cdots = (-1)^m \frac{1}{3 \cdot 6 \cdot 9 \cdots 3m \cdot 10 \cdot 19 \cdot 28 \cdots (9m + 1)} a_0.$$

Hence the series solution is

$$a_0 x^{\frac{1}{3}} \left(1 - \frac{x^3}{3 \cdot 10} + \frac{x^6}{3 \cdot 6 \cdot 10 \cdot 19} + \cdots \right)$$

which clearly is not simply a constant multiple of the series obtained above for $c = 0$. The general solution of the differential equation is then

$$y(x) = A \left(1 - \frac{x^3}{3 \cdot 8} + \frac{x^6}{3 \cdot 6 \cdot 8 \cdot 17} + \cdots \right)$$

$$+ B x^{\frac{1}{3}} \left(1 - \frac{x^3}{3 \cdot 10} + \frac{x^6}{3 \cdot 6 \cdot 10 \cdot 19} + \cdots \right)$$

for arbitrary constants A and B and for values of x for which the infinite series converge.

Case 2 When $c_1 = c_2$, the corresponding series solutions $y_1(x)$ and $y_2(x)$ will, of course, be identical and so $Ay_1(x) + By_2(x) = (A + B)y_1(x)$ does not give the general solution of the given differential equation. However it will be seen in the following example that if $y_1(x,c)$ denotes the series solution $y_1(x)$ before putting $c = c_1$, then $\{\partial y_1(x,c)/\partial c\}_{c=c_1}$ is also a solution of the given differential equation and is not merely a constant multiple

9

of $y_1(x)$. Hence the general solution of the given differential equation may be written

$$y(x) = Ay_1(x) + B \left\{ \frac{\partial y_1(x,c)}{\partial c} \right\}_{c=c_1}$$

for arbitrary constants A and B. This result is true in general.

Example   Obtain the general solution of the differential equation

$$x(x - 1)y'' + (5x - 1)y' + 4y = 0.$$

Assume that $y(x) = \sum_{n=0}^{\infty} a_n x^{n+c}$ with $a_0 \neq 0$ is a solution. Then

$$y'(x) = \sum_{n=0}^{\infty} (n + c)a_n x^{n+c-1}$$

and

$$y''(x) = \sum_{n=0}^{\infty} (n + c)(n + c - 1)a_n x^{n+c-2}.$$

Then substituting in the given differential equation we obtain

$$\sum_{n=0}^{\infty} (n + c)(n + c - 1)a_n(x^{n+c} - x^{n+c-1})$$

$$+ \sum_{n=0}^{\infty} (n + c)a_n(5x^{n+c} - x^{n+c-1}) + 4\sum_{n=0}^{\infty} a_n x^{n+c} \equiv 0.$$

Therefore

$$- \sum_{n=0}^{\infty} (n + c)^2 a_n x^{n+c-1} + \sum_{n=0}^{\infty} (n + c + 2)^2 a_n x^{n+c} \equiv 0.$$

Hence, equating to zero the coefficients of successive powers of x in this equation, we obtain firstly the indicial equation $-c^2 a_0 = 0$. The two roots of this equation are $c = 0$ (since $a_0 \neq 0$). Also $-(m + c)^2 a_m + (m + c + 1)^2 a_{m-1} = 0$ for $m \geqslant 1$ (picking out the coefficient of x^{m+c-1}).
   Therefore

$$a_m = \frac{(m + c + 1)^2}{(m + c)^2} a_{m-1}$$

since $(m + c)$ is not equal to zero when $c = 0$ and $m \geqslant 1$. Hence when $c = 0$,

$$a_m = \frac{(m + 1)^2}{m^2} a_{m-1}$$

and so

$$a_1 = \frac{2^2}{1^2}\, a_0,$$

$$a_2 = \frac{3^2}{2^2}\, a_1 = 3^2 a_0,$$

. . . .
. . . .

Therefore the series solution is

$$y_1(x) = a_0(1 + 2^2 x + 3^2 x^2 + \ldots + m^2 x^{m-1} + \ldots)$$

for arbitrary a_0. This only contains one arbitrary constant and so does not represent the general solution of the differential equation.

However, determining the coefficients $a_1, a_2, \ldots$ without first putting $c = 0$, we obtain

$$a_1 = \frac{(c+2)^2}{(c+1)^2}\, a_0,$$

$$a_2 = \frac{(c+3)^2}{(c+2)^2}\, a_1 = \frac{(c+3)^2}{(c+1)^2}\, a_0,$$

$$a_3 = \frac{(c+4)^2}{(c+1)^2}\, a_0,$$

. . . .

etc.

and so

$$y_1(x,c) = a_0 x^c \left\{ 1 + \left(\frac{c+2}{c+1}\right)^2 x + \left(\frac{c+3}{c+1}\right)^2 x^2 + \ldots \right\}.$$

Substituting this into the given differential equation then gives

$$x(x-1)y_1''(x,c) + (5x-1)y_1'(x,c) + 4y_1(x,c) = -c^2 a_0 x^{c-1}$$

since the coefficients of the other powers of x are zero because

$$a_m = [(m+c+1)/(m+c)]^2 a_0$$

as before.

Then, differentiating the above differential equation with respect to c we obtain

$$x(x-1)\frac{\partial}{\partial c}[y_1''(x,c)] + (5x-1)\frac{\partial}{\partial c}[y_1'(x,c)] + 4\frac{\partial}{\partial c}[y_1(x,c)]$$

$$= -2c a_0 x^{c-1} - c^2 a_0 x^{c-1}\ln x.$$

Now assuming that the order of differentiating with respect to x and with respect to c can be interchanged we obtain

$$x(x-1)\frac{d^2}{dx^2}\left\{\frac{\partial}{\partial c}[y_1(x,c)]\right\} + (5x-1)\frac{d}{dx}\left\{\frac{\partial}{\partial c}[y_1(x,c)]\right\} + 4\left\{\frac{\partial}{\partial c}[y_1(x,c)]\right\}$$

$$= -2ca_0x^{c-1} - c^2a_0x^{c-1}\ln x.$$

Then, putting $c = 0$, we obtain

$$x(x-1)\frac{d^2}{dx^2}\left\{\frac{\partial}{\partial c}[y_1(x,c)]\right\}_{c=0}$$

$$+ (5x-1)\frac{d}{dx}\left\{\frac{\partial}{\partial c}[y_1(x,c)]\right\}_{c=0} + 4\left\{\frac{\partial}{\partial c}[y_1(x,c)]\right\}_{c=0} = 0.$$

This states that

$$\left\{\frac{\partial}{\partial c}[y_1(x,c)]\right\}_{c=0}$$

is a solution of the given differential equation.

Now

$$y_1(x,c) = a_0x^c \sum_{n=0}^{\infty}\left(\frac{c+n+1}{c+1}\right)^2 x^n.$$

Therefore

$$\frac{\partial}{\partial c}[y_1(x,c)] = a_0\frac{\partial x^c}{\partial c}\sum_{n=0}^{\infty}\left(\frac{c+n+1}{c+1}\right)^2 x^n + a_0x^c\sum_{n=0}^{\infty}\frac{\partial}{\partial c}\left(\frac{c+n+1}{c+1}\right)^2 x^n$$

$$= a_0x^c\ln x\sum_{n=0}^{\infty}\left(\frac{c+n+1}{c+1}\right)^2 x^n + a_0x^c\sum_{n=0}^{\infty}\frac{-2n(c+n+1)}{(c+1)^3}x^n.$$

Therefore

$$\left\{\frac{\partial}{\partial c}[y_1(x,c)]\right\}_{c=0} = a_0\ln x\sum_{n=0}^{\infty}(n+1)^2x^n - 2a_0\sum_{n=1}^{\infty}n(n+1)x^n$$

$$= a_0\ln x\,(1 + 2^2x + 3^2x^2 + \ldots)$$

$$- 2a_0(1\,.\,2x + 2\,.\,3x^2 + 3\,.\,4x^3 + \ldots).$$

The presence of the factor $\ln x$ in the solution ensures that it is not simply a constant multiple of the solution $y_1(x)$. Hence the general solution of the given differential equation may be written

$$y(x) = A(1 + 2^2x + 3^2x^2 + \ldots) + B\{\ln x(1 + 2^2x + 3^2x^2 + \ldots)$$

$$- 2(1\,.\,2x + 2\,.\,3x^2 + 3\,.\,4x^3 + \ldots)\}$$

for arbitrary constants A and B and for values of x for which the above series converge.

Case 3 When the roots of the indicial equation differ by an integer, it may happen that either

(i) the series corresponding to one of the roots involves two arbitrary constants and so gives the general solution immediately, or

(ii) the recurrence relation for the coefficients in the series corresponding to one of the roots c_1 say leads to 'infinite' coefficients. The method of avoiding this difficulty involves putting $a_0 = k(c - c_1)$ $(k \neq 0)$ and then proceeding as in Case 2, and will be illustrated in an example.

Example *Obtain the general solution of the differential equation*

$$(x - x^2)y'' - (1 + 2x)y' + 2y = 0.$$

Assume that $y(x) = \sum_{n=0}^{\infty} a_n x^{n+c}$ with $a_0 \neq 0$ is a solution. Then

$$y'(x) = \sum_{n=0}^{\infty} (n + c)a_n x^{n+c-1}$$

and

$$y''(x) = \sum_{n=0}^{\infty} (n + c)(n + c - 1)a_n x^{n+c-2}.$$

Therefore substituting into the given differential equation we obtain

$$\sum_{n=0}^{\infty} (n + c)(n + c - 1)a_n(x^{n+c-1} - x^{n+c})$$

$$- \sum_{n=0}^{\infty} (n + c)a_n(x^{n+c-1} + 2x^{n+c}) + 2\sum_{n=0}^{\infty} a_n x^{n+c} \equiv 0.$$

Therefore

$$\sum_{n=0}^{\infty} (n + c)(n + c - 2)a_n x^{n+c-1}$$

$$- \sum_{n=0}^{\infty} \{(n + c)(n + c + 1) - 2\}a_n x^{n+c} \equiv 0.$$

Hence equating to zero the coefficients of successive powers of x we obtain the indicial equation

$$c(c - 2)a_0 = 0$$

and so $c = 0, 2$ (since $a_0 \neq 0$). Also

$$(m + c)(m + c - 2)a_m - \{(m + c - 1)(m + c) - 2\}a_{m-1} = 0 \text{ for } m \geqslant 1.$$

When $c = 0$

$$m(m - 2)a_m = (m + 1)(m - 2)a_{m-1}.$$

Therefore

$$a_1 = \frac{2(-1)}{1 \cdot (-1)} a_0 = 2a_0.$$

Putting $m = 2$ in the above recurrence relation supplies no information about a_2 as both sides of the relation vanish identically regardless of the value taken by a_2. Thus a_2 can take any arbitrary value, and

$$a_m = \frac{m+1}{m} a_{m-1} \quad \text{for} \quad m \geqslant 3.$$

Therefore

$$a_3 = \tfrac{4}{3} a_2,$$

$$a_4 = \tfrac{5}{4} a_3 = \tfrac{5}{3} a_2$$

etc. Thus we obtain the series solution

$$a_0(1 + 2x) + a_2 x^2(1 + \tfrac{4}{3}x + \tfrac{5}{3}x^2 + \ldots).$$

This contains two arbitrary constants and so represents the general solution of the given differential equation. Thus the general solution may be written

$$y(x) = A(1 + 2x) + Bx^2(3 + 4x + 5x^2 + \ldots)$$

for arbitrary constants A and B.

When $c = 2$

$$(m + 2)m a_m = m(m + 3)a_{m-1} \quad \text{for} \quad m \geqslant 1$$

and so

$$a_m = \frac{m+3}{m+2} a_{m-1}.$$

Therefore

$$a_1 = \tfrac{4}{3} a_0,$$

$$a_2 = \tfrac{5}{4} a_1 = \tfrac{5}{3} a_0$$

etc. and so the series $a_0 x^2(1 + \tfrac{4}{3}x + \tfrac{5}{3}x^2 + \ldots)$ which is contained in the above general solution is obtained. Note that it is the algebraically smaller root of the indicial equation which gives the general solution of the differential equation. This is true in general for Case 3(i).

Example　Obtain the general solution of the differential equation

$$x^2(x + 1)y'' + x(x + 1)y' - 4y = 0.$$

Putting $y(x) = \sum_{n=0}^{\infty} a_n x^{n+c}$ with $a_0 \neq 0$ and proceeding as above we obtain

$$\sum_{n=0}^{\infty} (n + c)(n + c - 1)a_n(x^{n+c+1} + x^{n+c})$$

$$+ \sum_{n=0}^{\infty} (n + c)a_n(x^{n+c+1} + x^{n+c}) - 4\sum_{n=0}^{\infty} a_n x^{n+c} \equiv 0.$$

Therefore

$$\sum_{n=0}^{\infty} (n + c)^2 a_n x^{n+c+1} + \sum_{n=0}^{\infty} \{(n + c)^2 - 4\}a_n x^{n+c} \equiv 0.$$

Hence equating to zero the coefficients of successive powers of x we obtain the indicial equation

$$(c^2 - 4)a_0 = 0$$

and so $c = \pm 2$ (since $a_0 \neq 0$). Also $(m + c - 1)^2 a_{m-1} + \{(m + c)^2 - 4\}a_m = 0$ for $m \geqslant 1$ (picking out the coefficient of x^{m+c}).

Therefore $(m + c + 2)(m + c - 2)a_m = -(m + c - 1)^2 a_{m-1}$.

When $c = -2$

$$m(m - 4)a_m = -(m - 3)^2 a_{m-1}.$$

Therefore

$$a_1 = \frac{-4}{1 \cdot (-3)} a_0,$$

$$a_2 = \frac{-1}{2(-2)} a_1 = \frac{1 \cdot 4}{(1 \cdot 2)(2 \cdot 3)} a_0,$$

$$a_3 = 0.$$

Now putting $m = 4$ in the above recurrence relation, each side vanishes regardless of the value taken by a_4. Thus a_4 cannot be determined uniquely but can take any arbitrary value.

Then

$$a_5 = \frac{-2^2}{5 \cdot 1} a_4,$$

$$a_6 = \frac{-3^2}{6 \cdot 2} a_5 = \frac{2^2 \cdot 3^2}{(5 \cdot 6)(1 \cdot 2)} a_4$$

etc.

Thus we obtain the series solution

$$x^{-2}\left\{a_0(1 + \tfrac{4}{3}x + \tfrac{1}{3}x^2) + a_4\left(x^4 - \frac{2^2}{5 \cdot 1} x^5 + \frac{2^2 \cdot 3^2}{(5 \cdot 6)(1 \cdot 2)} x^6 + \ldots\right)\right\}$$

or

$$\frac{a_0}{3}(3x^{-2} + 4x^{-1} + 1) + a_4 x^2\left(1 - \frac{2^2}{5 \cdot 1} x + \frac{2^2 \cdot 3^2}{(5 \cdot 6)(1 \cdot 2)} x^2 + \ldots\right).$$

This contains two arbitrary constants and so represents the general solution of the given differential equation. Thus the general solution may be written

$$y(x) = A(3x^{-2} + 4x^{-1} + 1) + Bx^2\left(1 - \frac{2^2}{5 \cdot 1} x + \frac{2^2 \cdot 3^2}{(5 \cdot 6)(1 \cdot 2)} x^2 + \ldots\right)$$

for arbitrary constants A and B.

Now when $c = 2$

$$a_m = \frac{-(m + 1)^2}{(m + 4)m} a_{m-1}$$

since m and $(m + 4)$ are not equal to zero for $m \geqslant 1$.

Hence

$$a_1 = \frac{-2^2}{5 \cdot 1} a_0,$$

$$a_2 = -\frac{3^2}{6 \cdot 2} a_1 = \frac{2^2 \cdot 3^2}{(5 \cdot 6)(1 \cdot 2)} a_0$$

etc. and so the series

$$a_0 x^2 \left(1 - \frac{2^2}{5 \cdot 1} x + \frac{2^2 \cdot 3^2}{(5 \cdot 6)(1 \cdot 2)} x^2 + \cdots \right)$$

which is contained in the above general solution is obtained.

Example   *Obtain the general solution of the differential equation*

$$x(1 - x)y'' - 3xy' - y = 0.$$

Putting $y(x) = \sum_{n=0}^{\infty} a_n x^{n+c}$ with $a_0 \neq 0$ and substituting in the differential equation we obtain

$$\sum_{n=0}^{\infty} (n + c)(n + c - 1)a_n(x^{n+c-1} - x^{n+c})$$

$$- \sum_{n=0}^{\infty} 3(n + c)a_n x^{n+c} - \sum_{n=0}^{\infty} a_n x^{n+c} \equiv 0.$$

Therefore

$$\sum_{n=0}^{\infty} (n + c)(n + c - 1)a_n x^{n+c-1}$$

$$- \sum_{n=0}^{\infty} \{(n + c)(n + c + 2) + 1\}a_n x^{n+c} \equiv 0.$$

Hence equating to zero the coefficients of successive powers of x we obtain the indicial equation

$$c(c - 1)a_0 = 0$$

and so $c = 0, 1$ (since $a_0 \neq 0$). Also $(m + c)(m + c - 1)a_m - (m + c)^2 a_{m-1} = 0$ for $m \geqslant 1$. Therefore $(m + c - 1)a_m - (m + c)a_{m-1} = 0$ since $m + c \neq 0$ for $m \geqslant 1$ and $c = 0, 1$.

When c = 1

$$ma_m = (m + 1)a_{m-1}.$$

Therefore

$$a_1 = 2a_0,$$
$$a_2 = \tfrac{3}{2}a_1 = 3a_0,$$
$$a_3 = \tfrac{4}{3}a_2 = 4a_0,$$

. . . .

etc.

Hence the series solution is

$$a_0 x(1 + 2x + 3x^2 + 4x^3 + \ldots) = a_0 x(1 - x)^{-2}.$$

When $c = 0$

$$(m - 1)a_m = ma_{m-1} \quad \text{for} \quad m \geqslant 1.$$

Now putting $m = 1$, this equation cannot be solved to give a finite value for a_1 if $a_0 \neq 0$.

To avoid this difficulty, put $a_0 = kc$ with $k \neq 0$ and do not put $c = 0$ until the end. Now of course the condition $a_0 \neq 0$ no longer holds, but this has been replaced by $k \neq 0$.

Then, proceeding as for Case 2, we see that

$$ca_1 = (c + 1)a_0 = c(c + 1)k.$$

Therefore

$$a_1 = k(c + 1),$$
$$(c + 1)a_2 = (c + 2)a_1.$$

Therefore

$$a_2 = \frac{c + 2}{c + 1} k(c + 1) = k(c + 2),$$

$$a_3 = \frac{c + 3}{c + 2} a_2 = k(c + 3),$$

$$\cdots \cdots$$

etc. Hence

$$y_1(x,c) = kx^c\{c + (c + 1)x + (c + 2)x^2 + (c + 3)x^3 + \ldots\}$$
$$= k \sum_{n=0}^{\infty} (c + n)x^{c+n}.$$

Substituting this into the given differential equation then gives

$$x(1 - x)y_1''(x,c) - 3xy_1'(x,c) - y_1(x,c) = kc^2(c - 1)x^{c-1}.$$

Putting $c = 0$, we see that $y_1(x,0)$ is a solution of the given differential equation. But

$$y_1(x,0) = k\{x + 2x^2 + 3x^3 + \ldots\} = kx(1 - x)^{-2}.$$

That is the solution already obtained for $c = 1$.

However, differentiating the above equation with respect to c and then putting $c = 0$ we obtain, as in Case 2, that

$$x(1 - x)\frac{d^2}{dx^2}\left\{\frac{\partial}{\partial c}[y_1(x,c)]\right\}_{c=0} - 3x\frac{d}{dx}\left\{\frac{\partial}{\partial c}[y_1(x,c)]\right\}_{c=0}$$

$$- \left\{\frac{\partial}{\partial c}[y_1(x,c)]\right\}_{c=0} = 0.$$

This illustrates that

$$\left\{ \frac{\partial}{\partial c} [y_1(x,c)] \right\}_{c=0}$$

is a solution of the given differential equation.

Now

$$\frac{\partial}{\partial c} [y_1(x,c)] = k \sum_{n=0}^{\infty} x^{c+n} + k \sum_{n=0}^{\infty} (c + n)x^n \frac{\partial}{\partial c} (x^c)$$

$$= k \sum_{n=0}^{\infty} x^{c+n} + k \sum_{n=0}^{\infty} (c + n)x^{c+n} \ln x.$$

Therefore

$$\left\{ \frac{\partial}{\partial c} [y_1(x,c)] \right\}_{c=0} = k \sum_{n=0}^{\infty} x^n + k \ln x \sum_{n=0}^{\infty} nx^n$$

$$= k(1 + x + x^2 + x^3 + \ldots)$$

$$+ k \ln x \, (x + 2x^2 + 3x^3 + \ldots)$$

$$= k(1 - x)^{-1} + kx(1 - x)^{-2} \ln x.$$

Once again, the factor $\ln x$ ensures that this is not merely a constant multiple of the solution already obtained.

Hence the general solution of the given differential equation may be written,

$$Ax(1 - x)^{-2} + B\{(1 - x)^{-1} + (1 - x)^{-2}x \ln x\}$$

for arbitrary constants A and B and for values of x for which the above series converge.

We have now seen that the two roots of the indicial equation either

1. lead directly to the general solution (containing two arbitrary constants) of the differential equation, or
2. lead only to a solution $\{y_1(x,c)\}_{c=c_1}$ with only one arbitrary constant in which case a second linear independent solution is given by $\{\partial y_1/\partial c\}_{c=c_1}$.

The work of this section has been concerned with obtaining, in terms of series expansions about $x = 0$, solutions of differential equations of the form

$$x^2 y'' + xp(x)y' + q(x)y = 0$$

in which $p(x)$ and $q(x)$ have power series expansions about $x = 0$.

More generally, it can be shown that any differential equation of the form

$$(x - x_0)^2 y'' + (x - x_0)p(x)y' + q(x)y = 0,$$

in which x_0 is a constant and $p(x)$ and $q(x)$ have power series expansions about $x = x_0$, has at least one solution which can be expressed in the form

$$y = (x - x_0)^c \sum_{n=0}^{\infty} a_n(x - x_0)^n \qquad (a_0 \neq 0)$$

where the series is convergent for $|x - x_0|$ less than some constant R. This solution can be obtained by first changing the independent variable in the differential equation from x to $u = x - x_0$ and then proceeding to find the series solution about the point $u = 0$ as above.

Exercises

Solve the following differential equations using the method of solution in series about $x = 0$ and, where possible, express the solutions in terms of elementary functions.

1. $4x \dfrac{d^2y}{dx^2} + 2 \dfrac{dy}{dx} + y = 0.$

2. $(x - x^2) \dfrac{d^2y}{dx^2} + (1 - 4x) \dfrac{dy}{dx} - 2y = 0.$

3. $(1 + x^2) \dfrac{d^2y}{dx^2} + x \dfrac{dy}{dx} - 4y = 0.$

4. $x^2 \dfrac{d^2y}{dx^2} + x(x + 1) \dfrac{dy}{dx} + (x - 1)y = 0.$

5. $x \dfrac{d^2y}{dx^2} + \dfrac{dy}{dx} - 2x^2y = 0.$

6. $x(1 - x) \dfrac{d^2y}{dx^2} - 2 \dfrac{dy}{dx} + 2y = 0.$

7. $2x(1 - x) \dfrac{d^2y}{dx^2} + (1 - 3x) \dfrac{dy}{dx} + 3y = 0.$

8. $x^2(x^2 - 1) \dfrac{d^2y}{dx^2} + 2x(x^2 + 1) \dfrac{dy}{dx} - 2y = 0.$

9. $x^2 \dfrac{d^2y}{dx^2} - 4x \dfrac{dy}{dx} + (6 - x^2)y = 0.$

10. $x^2(1 - x) \dfrac{d^2y}{dx^2} - x(2 + x) \dfrac{dy}{dx} + 2y = 0.$

5.2 The Wronskian

In the above examples it was easy to see in each particular case when we had two linearly independent series solutions $y_1(x)$ and $y_2(x)$, say, which could then be combined in the form $Ay_1(x) + By_2(x)$ for arbitrary constants A and B to give the general solution of the differential equation. In more complicated examples however it may not be so easy to see when the solutions $y_1(x)$ and $y_2(x)$ are linearly independent and in such cases the following result is useful.

If $y_1(x)$ and $y_2(x)$ are differentiable functions of x, and if $y_1(x)$ and $y_2(x)$ are linearly dependent, then

$$y_1 y_2' - y_1' y_2 = 0,$$

and conversely, if the above expression is zero and if either $y_1(x)$ or $y_2(x)$ is nowhere zero, then $y_1(x)$ and $y_2(x)$ are linearly dependent. The expression $y_1 y_2' - y_1' y_2$ is usually denoted by W and is called the *Wronskian* of the functions y_1 and y_2.

Proof: If $y_1(x)$ and $y_2(x)$ are linearly dependent then there exist non-zero constants a_1 and a_2 such that

$$a_1 y_1 + a_2 y_2 = 0.$$

Differentiating with respect to x we then obtain

$$a_1 y_1' + a_2 y_2' = 0.$$

Now eliminating a_1 and a_2, we obtain

$$y_1 y_2' - y_1' y_2 = 0.$$

That is $W = 0$.

Conversely, suppose that $y_1(x)$ is nowhere zero, then the equation

$$b_1(x) y_1(x) + y_2(x) = 0$$

will determine the function $b_1(x)$ which must be differentiable since both $y_1(x)$ and $y_2(x)$ are differentiable. Therefore

$$b_1'(x) y_1(x) + b_1(x) y_1'(x) + y_2'(x) = 0.$$

But, if $W = 0$, we also have

$$y_1(x) y_2'(x) - y_1'(x) y_2(x) = 0$$

or

$$y_1(x) y_2'(x) + y_1'(x) b_1(x) y_1(x) = 0$$

and so $b_1(x)y_1'(x) + y_2'(x) = 0$ since $y_1(x)$ is nowhere zero. Hence $b_1'(x)y_1(x) = 0$ and so $b_1'(x) = 0$. Therefore $b_1(x) = $ constant $= c$ say. Hence $cy_1(x) + y_2(x) = 0$. That is $y_1(x)$ and $y_2(x)$ are linearly dependent.

Further, if $y_1(x)$ and $y_2(x)$ are linearly independent solutions of the differential equation

$$y'' + f_1(x)y' + f_0(x)y = 0$$

then their Wronskian W is given by

$$W = y_1(x)y_2'(x) - y_1'(x)y_2(x)$$

and

$$
\begin{aligned}
W' &= y_1'(x)y_2'(x) + y_1(x)y_2''(x) - y_1''(x)y_2(x) - y_1'(x)y_2'(x) \\
&= y_1(x)\{-f_1(x)y_2'(x) - f_0(x)y_2(x)\} \\
&\quad - y_2(x)\{-f_1(x)y_1'(x) - f_0(x)y_1(x)\} \\
&= -f_1(x)\{y_1(x)y_2'(x) - y_1'(x)y_2(x)\} \\
&= -f_1(x)W.
\end{aligned}
$$

Hence $W = C \exp\{-\int f_1(x)\,\mathrm{d}x\}$ where C is an arbitrary constant. In particular, if there is a value of x for which $W \neq 0$, then $C \neq 0$ and so W does not vanish anywhere.

5.3 Bessel's equation—Bessel functions

The differential equation

$$x^2 y'' + xy' + (x^2 - v^2)y = 0$$

is known as *Bessel's equation of order v* and its solution for general v may be obtained using the method previously described. However, consider firstly the result of making the change of variable $u = x^{\frac{1}{2}}y$. Then

$$\frac{\mathrm{d}y}{\mathrm{d}x} = \frac{1}{x^{\frac{1}{2}}}\frac{\mathrm{d}u}{\mathrm{d}x} - \frac{u}{2x^{\frac{3}{2}}}$$

and

$$\frac{\mathrm{d}^2 y}{\mathrm{d}x^2} = \frac{1}{x^{\frac{1}{2}}}\frac{\mathrm{d}^2 u}{\mathrm{d}x^2} - \frac{1}{x^{\frac{3}{2}}}\frac{\mathrm{d}u}{\mathrm{d}x} + \frac{3}{4}\frac{u}{x^{\frac{5}{2}}}$$

and the differential equation becomes

$$u'' + ux^{-2}(\tfrac{1}{4} + x^2 - v^2) = 0$$

or

$$u'' + \left[1 - \frac{(v^2 - \tfrac{1}{4})}{x^2}\right]u = 0.$$

Now, in the special case of $v = \pm\frac{1}{2}$, we have,

$$u'' + u = 0$$

which has the general solution

$$u = A \sin x + B \cos x$$

for arbitrary constants A and B.

Hence the general solution of Bessel's equation of order $\frac{1}{2}$ is

$$y = \frac{A}{x^{\frac{1}{2}}} \sin x + \frac{B}{x^{\frac{1}{2}}} \cos x.$$

The graphs of $\sin x/x^{\frac{1}{2}}$ and $\cos x/x^{\frac{1}{2}}$ are sketched in figure 97.

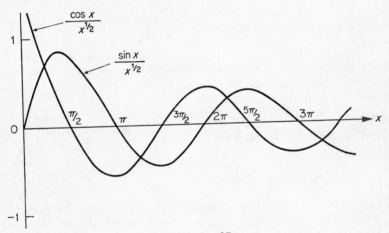

FIGURE 97

If $v \neq \pm\frac{1}{2}$ then, for large x, $(v^2 - \frac{1}{4})/x^2$ is small compared to 1 and so the solutions of the equation

$$u'' + \left[1 - \frac{(v^2 - \frac{1}{4})}{x^2} \right] u = 0$$

approximate to those of $u'' + u = 0$.*

Thus the solutions, for large x, approximate to decaying periodic functions. Now consider again the standard form of the equation of order v namely

$$x^2 y'' + x y' + (x^2 - v^2)y = 0.$$

---

* This argument is not so easy as it appears and must be applied with great care. For further details in this case see Watson, Theory of Bessel Functions.

Substituting $y(x) = \sum\limits_{n=0}^{\infty} a_n x^{n+c}$ with $a_0 \neq 0$ into this equation we obtain

$$\sum_{n=0}^{\infty} (n + c)(n + c - 1)a_n x^{n+c} + \sum_{n=0}^{\infty} (n + c)a_n x^{n+c}$$

$$+ \sum_{n=0}^{\infty} a_n(x^{n+c+2} - \nu^2 x^{n+c}) \equiv 0.$$

Hence equating to zero the coefficients of successive powers of x we obtain the indicial equation

$$(c^2 - \nu^2)a_0 = 0$$

and so $c = \nu \ (\geqslant 0)$ or $-\nu$ (since $a_0 \neq 0$). Also $\{(c + 1)^2 - \nu^2\}a_1 = 0$ and

$$\{(m + c)^2 - \nu^2\}a_m + a_{m-2} = 0 \quad \text{for} \quad m \geqslant 2.$$

When $2\nu \neq$ an integer or 0 (that is the roots of the indicial equation are not equal and do not differ by an integer).

From $\{(c + 1)^2 - \nu^2\}a_1 = 0$ we see that $(2c + 1)a_1 = 0$ and so $a_1 = 0$. Also $(m + c)^2 - \nu^2 = m(m + 2c) \neq 0$ for $m \geqslant 2$ and $c = \pm\nu$ (2ν not an integer). Therefore

$$a_m = \frac{-a_{m-2}}{m(m + 2c)}.$$

Hence

$$a_1 = a_3 = a_5 = \ldots = 0.$$

$$a_2 = \frac{-a_0}{2 \cdot 2(1 + c)} = \frac{-a_0}{2^2 \cdot 1 \cdot (1 + c)},$$

$$a_4 = \frac{-a_2}{4 \cdot 2(2 + c)} = \frac{a_0}{2^4 \cdot 2 \cdot 1(c + 2)(c + 1)},$$

$$a_6 = \frac{-a_4}{6 \cdot 2(3 + c)} = \frac{-a_0}{2^6 \cdot 3!(c + 3)(c + 2)(c + 1)},$$

$$\cdots\cdots\cdots\cdots\cdots\cdots$$

$$\cdots\cdots\cdots\cdots\cdots\cdots$$

$$a_{2m} = \frac{a_{2m-2}}{2m \cdot 2(m + c)} = \frac{(-1)^m a_0}{2^{2m} m!(c + m)(c + m - 1) \ldots (c + 2)(c + 1)}$$

$$= \frac{(-1)^m a_0 \Gamma(c + 1)}{2^{2m} m!(c + m)(c + m - 1) \ldots (c + 2)(c + 1)\Gamma(c + 1)}$$

$$= \frac{(-1)^m \Gamma(c + 1)a_0}{2^{2m} m!\Gamma(c + m + 1)}.^*$$

* For a definition of the Gamma or Factorial Function see Appendix 1.

Therefore, if $c = v$ the solution is

$$\sum_{m=0}^{\infty} a_{2m} x^{2m+v} = a_0 \Gamma(v+1) \sum_{m=0}^{\infty} \frac{(-1)^m}{2^{2m} m! \Gamma(v+m+1)} x^{2m+v}$$

$$= k_1 \sum_{m=0}^{\infty} \frac{(-1)^m}{m! \Gamma(v+m+1)} \left(\frac{x}{2}\right)^{2m+v}$$

$$\text{where} \quad k_1 = a_0 \Gamma(v+1) 2^v$$

$$= k_1 J_v(x)$$

where

$$J_v(x) = \sum_{m=0}^{\infty} \frac{(-1)^m}{m! \Gamma(v+m+1)} \left(\frac{x}{2}\right)^{2m+v}$$

is called *Bessel's function of the first kind of order v*. By the ratio test the series is convergent for all finite values of x.

Similarly, if $c = -v$ the solution obtained is $k_2 J_{-v}(x)$ where k_2 is a constant and

$$J_{-v}(x) = \sum_{m=0}^{\infty} \frac{(-1)^m}{m! \Gamma(m-v+1)} \left(\frac{x}{2}\right)^{2m-v}$$

is called Bessel's function of the first kind of order $-v$. (Note that for $v > 0$, $J_{-v}(x) \to \infty$ as $x \to 0$.)

Now the Wronskian W of these two solutions is (see section 5.2, p. 255)

$$C \exp\left\{-\int \frac{1}{x} \, dx\right\} = \frac{C}{x}$$

where C is a constant.

Hence

$$J_v(x) J'_{-v}(x) - J'_v(x) J_{-v}(x) = \frac{C}{x}.$$

The constant C may be determined by considering the first terms in each of the series $J_v(x)$, $J'_v(x)$, $J_{-v}(x)$ and $J'_{-v}(x)$.

Therefore

$$\frac{1}{\Gamma(v+1)} \left(\frac{-v}{\Gamma(-v+1)}\right) - \left(\frac{v}{\Gamma(v+1)}\right) \frac{1}{\Gamma(-v+1)} = C$$

and so

$$C = -\frac{2v}{\Gamma(v+1)\Gamma(-v+1)}.$$

Hence if v is not an integer or zero, $C \neq 0$ and so $W \neq 0$. Thus $J_v(x)$ and $J_{-v}(x)$ are linearly independent and the general solution of Bessel's equation of order v, in this case, is given by

$$y(x) = AJ_v(x) + BJ_{-v}(x)$$

for arbitrary constants A and B.

When $2v = 0$ or an integer. (That is the roots of the indicial equation are equal ($v = 0$) or differ by an integer.)

The solution $J_v(x)$ is again obtained as above, but to obtain a second independent solution the method of Case 2 or Case 3 for 'infinite coefficients' must be employed. This leads, after further manipulation, to

$$y(x) = k Y_v(x)$$

where k is a constant and $Y_v(x)*$ is called *Bessel's function of the second kind of order v*.

In this case the general solution of Bessel's equation of order v is

$$y(x) = AJ_v(x) + BY_v(x)$$

for arbitrary constants A and B.

When v is an integer or zero $Y_v(x)$ contains the term $J_{\pm v}(x) \ln \frac{1}{2}x$† and so tends to infinity as x tends to zero.

When v is equal to half an odd integer, it can be shown that

$$Y_v(x) = (-1)^{v+\frac{1}{2}} J_{-v}(x).$$

Hence the general solution of Bessel's equation of order v may be written

$$y(x) = AJ_v(x) + \begin{cases} BY_v(x), & v = \text{integer (or zero)} \\ BJ_{-v}(x), & v \neq \text{integer (or zero)} \end{cases}$$

for arbitrary constants A and B.

*

$$Y_v(x) = \frac{2}{\pi}\{\gamma + \ln(\tfrac{1}{2}x)\}J_v(x) - \frac{1}{\pi}\sum_{r=0}^{v-1}\frac{\Gamma(v-r)}{r!}\left(\frac{2}{x}\right)^{v-2r}$$

$$-\frac{1}{\pi}\sum_{r=0}^{\infty}\frac{(-1)^r(\tfrac{1}{2}x)^{v+2r}}{r!\Gamma(v+r+1)}\left\{\sum_{s=1}^{r+v}\frac{1}{s} + \sum_{s=1}^{r}\frac{1}{s}\right\}$$

where

$$\gamma = \lim_{n\to\infty}\left(1 + \tfrac{1}{2} + \tfrac{1}{3} + \ldots + \frac{1}{n} - \ln n\right) = 0\cdot5772\ldots$$

† See Appendix 2.

In most practical applications (as we shall see later) v is integral (equal to n say) and then we have

$$y(x) = AJ_n(x) + BY_n(x)$$

where $Y_n(x) \to \infty$ as $x \to 0$. If, as in many practical applications, the solution must be finite at the origin then we must take $B = 0$ and

$$y(x) = AJ_n(x).$$

Example   Show that

$$J_{\frac{1}{2}}(x) = \sqrt{\frac{2}{\pi}} \frac{\sin x}{x^{\frac{1}{2}}} \quad \text{and} \quad J_{-\frac{1}{2}}(x) = \sqrt{\frac{2}{\pi}} \frac{\cos x}{x^{\frac{1}{2}}}$$

$$J_{\frac{1}{2}}(x) = \sum_{m=0}^{\infty} \frac{(-1)^m}{\Gamma(m+1)\Gamma(m+\frac{3}{2})} \left(\frac{x}{2}\right)^{2m+\frac{1}{2}}$$

$$= \frac{x^{\frac{1}{2}}}{2^{\frac{1}{2}}\Gamma(\frac{3}{2})} \left(1 - \frac{x^2}{3!} + \frac{x^4}{5!} + \cdots\right)$$

$$= \sqrt{\frac{2}{\pi}} \, x^{-\frac{1}{2}} \left(x - \frac{x^3}{3!} + \frac{x^5}{5!} + \cdots\right)$$

$$= \sqrt{\frac{2}{\pi}} \frac{\sin x}{x^{\frac{1}{2}}}.$$

$$J_{-\frac{1}{2}}(x) = \sum_{m=0}^{\infty} \frac{(-1)^m}{\Gamma(m+1)\Gamma(m+\frac{1}{2})} \left(\frac{x}{2}\right)^{2m-\frac{1}{2}}$$

$$= \frac{x^{-\frac{1}{2}}}{2^{-\frac{1}{2}}\Gamma(\frac{1}{2})} \left(1 - \frac{x^2}{2!} + \frac{x^4}{4!} + \cdots\right)$$

$$= \sqrt{\frac{2}{\pi}} \frac{\cos x}{x^{\frac{1}{2}}}.$$

5.31. *Recurrence relations for Bessel functions*

In this sub-section we shall derive some recurrence relations for Bessel functions. These recurrence relations are useful for tabulating Bessel functions and will also be helpful in the manipulations of subsequent sections.

$$J_v(x) = \sum_{m=0}^{\infty} \frac{(-1)^m}{m!\,\Gamma(v+m+1)} \left(\frac{x}{2}\right)^{v+2m}.$$

Therefore

$$J_v'(x) = \frac{d}{dx}[J_v(x)] = \sum_{m=0}^{\infty} \frac{(-1)^m}{m!\,\Gamma(v+m+1)} \left(\frac{v+2m}{2}\right)\left(\frac{x}{2}\right)^{v+2m-1}.$$

Therefore

$$xJ'_\nu(x) = \sum_{m=0}^{\infty} \frac{(-1)^m}{m!\,\Gamma(\nu+m+1)} (\nu+2m)\left(\frac{x}{2}\right)^{\nu+2m}$$

$$= \nu J_\nu(x) + x \sum_{m=1}^{\infty} \frac{(-1)^m m}{m!\,\Gamma(\nu+m+1)}\left(\frac{x}{2}\right)^{\nu+2m-1}$$

and putting $m - 1 = r$ in the last summation we obtain

$$xJ'_\nu(x) = \nu J_\nu(x) + \sum_{r=0}^{\infty} \frac{(-1)^{r+1}}{r!\,\Gamma(\nu+1+r+1)}\left(\frac{x}{2}\right)^{\nu+1+2r}.$$

That is

$$xJ'_\nu(x) = \nu J_\nu(x) - xJ_{\nu+1}(x).$$

In particular, when $\nu = 0$, $J'_0(x) = -J_1(x)$. Also, from above, we have

$$xJ'_\nu(x) = \sum_{m=0}^{\infty} \frac{(-1)^m}{m!\,\Gamma(\nu+m+1)} (\nu+2m)\left(\frac{x}{2}\right)^{\nu+2m}$$

$$= \sum_{m=0}^{\infty} \frac{(-1)^m}{m!\,\Gamma(\nu+m+1)} [2(\nu+m)-\nu]\left(\frac{x}{2}\right)^{\nu+2m}$$

$$= -\nu J_\nu(x) + x \sum_{m=0}^{\infty} \frac{(-1)^m}{m!\,\Gamma(\nu+m+1)} (\nu+m)\left(\frac{x}{2}\right)^{\nu+2m-1}$$

$$= -\nu J_\nu(x) + x \sum_{m=0}^{\infty} \frac{(-1)^m}{m!\,\Gamma(\nu+m)}\left(\frac{x}{2}\right)^{\nu-1+2m}.$$

That is

$$xJ'_\nu(x) = -\nu J_\nu(x) + xJ_{\nu-1}(x).$$

Now adding and subtracting these two recurrence relations gives respectively

$$2J'_\nu(x) = J_{\nu-1}(x) - J_{\nu+1}(x)$$

and

$$\frac{1}{x}2\nu J_\nu(x) = J_{\nu-1}(x) + J_{\nu+1}(x).$$

Note that, using this last recurrence relation, we can generate in succession the values of the Bessel functions of all integral orders from the values of the functions of orders zero and one. This result is of course very useful when tabulating Bessel functions of integral orders.

Example   *Show that if ν is half an odd integer then $J_\nu(x)$ may be expressed wholly in terms of trigonometric functions and powers of x.*

Putting $v = \frac{1}{2}$ in the relation

$$\frac{1}{x} 2vJ_v(x) = J_{v-1}(x) + J_{v+1}(x)$$

we obtain

$$\frac{1}{x} J_{\frac{1}{2}}(x) = J_{-\frac{1}{2}}(x) + J_{\frac{3}{2}}(x).$$

Therefore

$$J_{\frac{3}{2}}(x) = \sqrt{\frac{2}{\pi x}} \left(\frac{\sin x}{x} - \cos x \right).$$

Similarly, when $v = \frac{3}{2}$ we obtain

$$\frac{3}{x} J_{\frac{3}{2}}(x) = J_{\frac{1}{2}}(x) + J_{\frac{5}{2}}(x).$$

Therefore

$$J_{\frac{5}{2}}(x) = \sqrt{\frac{2}{\pi x}} \left(\frac{3 \sin x}{x^2} - \frac{3 \cos x}{x} - \sin x \right).$$

This process can then be repeated as often as required.

5.32. Expansion in series of Bessel functions

When solving physical problems using cylindrical polar coordinates it will often be found desirable, in order to apply the boundary conditions, to be able to expand a given function in terms of Bessel functions in much the same way as functions were expanded in terms of sines and cosines in Fourier analysis. In order to do this we require to evaluate integrals involving the product of two Bessel functions and this is done in the following way.

Bessel's equation of order v is

$$u^2 \frac{d^2y}{du^2} + u \frac{dy}{du} + (u^2 - v^2)y = 0$$

and this has a solution $y_1 = J_v(u)$.

Now putting $u = ax$ where a is a constant, the differential equation becomes

$$x^2 y'' + xy' + (a^2 x^2 - v^2)y = 0$$

where the prime denotes differentiation with respect to x. A solution of this latter equation is then $y_1 = J_v(ax)$ and so we have

$$x^2 y_1'' + xy_1' + (a^2 x^2 - v^2)y_1 = 0.$$

Similarly

$$x^2 y_2'' + x y_2' + (b^2 x^2 - v^2) y_2 = 0$$

where $y_2 = J_v(bx)$ and b is a constant.

Therefore

$$x^2(y_1'' y_2 - y_2'' y_1) + x(y_1' y_2 - y_2' y_1) + x^2(a^2 - b^2) y_1 y_2 = 0.$$

Now integrate throughout this equation with respect to x from $x = 0$ to $x = 1$.

Therefore

$$(b^2 - a^2) \int_0^1 x y_1 y_2 \, dx = \int_0^1 (y_1' y_2 - y_2' y_1) \, dx + \int_0^1 x(y_1'' y_2 - y_2'' y_1) \, dx.$$

But

$$\frac{d}{dx} [x(y_1' y_2 - y_2' y_1)] = (y_1' y_2 - y_2' y_1) + x(y_1'' y_2 - y_2'' y_1)$$

and so

$$(b^2 - a^2) \int_0^1 x y_1 y_2 \, dx = \int_0^1 \frac{d}{dx} [x(y_1' y_2 - y_2' y_1)] \, dx$$

$$= \left[x(y_1' y_2 - y_2' y_1) \right]_0^1.$$

That is

$$(b^2 - a^2) \int_0^1 x J_v(ax) J_v(bx) \, dx = \left[x(a J_v'(ax) J_v(bx) - b J_v'(bx) J_v(ax)) \right]_0^1.$$

Therefore

$$(b^2 - a^2) \int_0^1 x J_v(ax) J_v(bx) \, dx = a J_v'(a) J_v(b) - b J_v'(b) J_v(a).$$

In the particular case when $a = \alpha_i$ and $b = \alpha_j$ are distinct roots of $J_v(\alpha) = 0$ (that is when $J_v(\alpha_i) = J_v(\alpha_j) = 0$ and $\alpha_i \neq \alpha_j$)

$$(\alpha_j^2 - \alpha_i^2) \int_0^1 x J_v(\alpha_i x) J_v(\alpha_j x) \, dx = 0$$

and so

$$\int_0^1 x J_v(\alpha_i x) J_v(\alpha_j x) \, dx = 0 \qquad (\alpha_i \neq \alpha_j)$$

Also

$$\int_0^1 x J_v(\alpha_i x) J_v(\alpha_j x) \, dx = \frac{\alpha_i J_v'(\alpha_i) J_v(\alpha_j) - \alpha_j J_v'(\alpha_j) J_v(\alpha_i)}{\alpha_j^2 - \alpha_i^2}$$

and so

$$\lim_{\alpha_j \to \alpha_i} \int_0^1 x J_\nu(\alpha_i x) J_\nu(\alpha_j x) \, dx$$

$$= \lim_{\alpha_j \to \alpha_i} \left\{ \frac{\alpha_i J_\nu'(\alpha_i) J_\nu(\alpha_j) - \alpha_j J_\nu'(\alpha_j) J_\nu(\alpha_i)}{\alpha_j^2 - \alpha_i^2} \right\}$$

$$= \lim_{\alpha_j \to \alpha_i} \left\{ \frac{\alpha_i J_\nu'(\alpha_i) J_\nu'(\alpha_j) - J_\nu'(\alpha_j) J_\nu(\alpha_i) - \alpha_j J_\nu''(\alpha_j) J_\nu(\alpha_i)}{2\alpha_j} \right\}$$

on using L'Hôpital's rule.

Therefore

$$\int_0^1 x [J_\nu(\alpha_i x)]^2 \, dx = \tfrac{1}{2} [J_\nu'(\alpha_i)]^2.$$

But putting $x = \alpha_i$ and $J_\nu(\alpha_i) = 0$ in the relation

$$x J_\nu'(x) = \nu J_\nu(x) - x J_{\nu+1}(x)$$

we obtain

$$J_\nu'(\alpha_i) = -J_{\nu+1}(\alpha_i)$$

and hence we have

$$\int_0^1 x [J_\nu(\alpha_i x)]^2 \, dx = \tfrac{1}{2} [J_{\nu+1}(\alpha_i)]^2.$$

The functions $\sqrt{x}\, J_\nu(\alpha_k x)$ for fixed ν and α_k one of the (infinitely many) roots of $J_\nu(x) = 0$ are said to be orthogonal in the interval $0 \leqslant x \leqslant 1$. Now suppose it is desired to expand the function $f(x)$ in $0 \leqslant x \leqslant 1$ in a series of Bessel functions in the form

$$f(x) = c_1 J_\nu(\alpha_1 x) + c_2 J_\nu(\alpha_2 x) + \cdots$$

where $\alpha_1, \alpha_2, \ldots$ are the successive roots of $J_\nu(x) = 0$.

Multiply both sides of the equation by $x J_\nu(\alpha_k x)$ and integrate with respect to x from $x = 0$ to $x = 1$.

From the above results we obtain

$$\int_0^1 x J_\nu(\alpha_k x) f(x) \, dx = \tfrac{1}{2} [J_{\nu+1}(\alpha_k)]^2 c_k$$

or

$$c_k = \frac{2}{[J_{\nu+1}(\alpha_k)]^2} \int_0^1 x J_\nu(\alpha_k x) f(x) \, dx$$

when $J_{\nu+1}(\alpha_k) \neq 0$.

Now it can be shown that all the zeros of $J_\nu(\alpha)$ are simple so that if $J_\nu(\alpha_k) = 0$, then $J_\nu'(\alpha_k) \neq 0$. Hence, with this assumption, we see from the first recurrence relation that if $J_\nu(\alpha_k) = 0$ then $J_{\nu+1}(\alpha_k) \neq 0$.

Example   *Express $f(x)$ in the form $c_1 J_0(\alpha_1 x) + c_2 J_0(\alpha_2 x) + \ldots$ in the interval $0 \leqslant x \leqslant 1$, where $\alpha_1, \alpha_2, \ldots$ are the successive roots of $J_0(x) = 0$ and (i) $f(x) = 1$ and (ii) $f(x) = x^2$.*

(i) If $1 = c_1 J_0(\alpha_1 x) + c_2 J_0(\alpha_2 x) + \ldots$ then multiplying both sides by $x J_0(\alpha_k x)$ and integrating with respect to x from $x = 0$ to $x = 1$ we obtain

$$c_k = \frac{2}{[J_1(\alpha_k)]^2} \int_0^1 x J_0(\alpha_k x) \, dx$$

$$= \frac{2}{[J_1(\alpha_k)]^2} \int_0^{\alpha_k} \frac{u}{\alpha_k} J_0(u) \frac{du}{\alpha_k}$$

on putting $u = \alpha_k x$. But one of the previously obtained recurrence relations (p. 261) gives

$$u J_0(u) = u J_1'(u) + J_1(u) = \frac{d}{du}(u J_1(u)).$$

Therefore

$$c_k = \frac{2}{\alpha_k^2 [J_1(\alpha_k)]^2} \int_0^{\alpha_k} \frac{d}{du}[u J_1(u)] \, du$$

$$= \frac{2}{\alpha_k^2 [J_1(\alpha_k)]^2} \left[u J_1(u) \right]_0^{\alpha_k} = \frac{2}{\alpha_k J_1(\alpha_k)}.$$

Hence

$$1 = 2 \left\{ \frac{J_0(\alpha_1 x)}{\alpha_1 J_1(\alpha_1)} + \frac{J_0(\alpha_2 x)}{\alpha_2 J_1(\alpha_2)} + \ldots \right\}.$$

(ii) As above, if $x^2 = c_1 J_0(\alpha_1 x) + c_2 J_0(\alpha_2 x) + \ldots$ then

$$c_k = \frac{2}{[J_1(\alpha_k)]^2} \int_0^1 x^3 J_0(\alpha_k x) \, dx$$

$$= \frac{2}{\alpha_k^4 [J_1(\alpha_k)]^2} \int_0^{\alpha_k} u^2 \frac{d}{du}[u J_1(u)] \, du$$

$$= \frac{2}{\alpha_k^4 [J_1(\alpha_k)]^2} \left\{ \left[u^3 J_1(u) \right]_0^{\alpha_k} - 2 \int_0^{\alpha_k} u^2 J_1(u) \, du \right\}.$$

Then since $J_1(u) = -J_0'(u)$,

$$c_k = \frac{2}{\alpha_k^4 [J_1(\alpha_k)]^2} \left\{ \alpha_k^3 J_1(\alpha_k) + 2 \int_0^{\alpha_k} u^2 J_0'(u) \, du \right\}$$

$$= \frac{2}{\alpha_k^4 [J_1(\alpha_k)]^2} \left\{ \alpha_k^3 J_1(\alpha_k) + 2 \left[u^2 J_0(u) \right]_0^{\alpha_k} - 4 \int_0^{\alpha_k} u J_0(u) \, du \right\}$$

$$= \frac{2}{\alpha_k^4 [J_1(\alpha_k)]^2} \left\{ \alpha_k^3 J_1(\alpha_k) + 2 \alpha_k^2 J_0(\alpha_k) - 4 \alpha_k J_1(\alpha_k) \right\}.$$

Now making use of the recurrence relation

$$\frac{2\nu}{x} J_\nu(x) = J_{\nu-1}(x) + J_{\nu+1}(x) \quad \text{with} \quad \nu = 1 \quad \text{and} \quad x = \alpha_k$$

we see that

$$c_k = \frac{2}{\alpha_k^4 [J_1(\alpha_k)]^2} \{\alpha_k^3 J_1(\alpha_k) - 2\alpha_k^2 J_2(\alpha_k)\}.$$

Therefore

$$c_k = \frac{2}{\alpha_k J_1(\alpha_k)} \left\{ 1 - \frac{2J_2(\alpha_k)}{\alpha_k J_1(\alpha_k)} \right\}.$$

Exercises

1. Use the substitution $y = zx^\nu$ where $\nu = \frac{1}{2}(1 - a) = $ constant to obtain the general solution of the differential equation

$$x\frac{d^2y}{dx^2} + a\frac{dy}{dx} + xy = 0.$$

2. Use the recurrence relations for Bessel functions to show that

(i) $$J_2(x) = J_0(x) + 2J_0''(x)$$

and (ii) $$J_3(x) = -3J_0'(x) - 4J_0'''(x).$$

3. Show that

$$x^2 J_1(x) = \frac{d}{dx}\left(x^2 J_2(x)\right)$$

and hence evaluate

$$\int_0^x x^3 J_0(x)\, dx.$$

4. Show that

$$\int x^\nu J_{\nu-1}(x)\, dx = x^\nu J_\nu(x) + \text{constant}$$

and that

$$\int x^{-\nu} J_{\nu+1}(x)\, dx = -x^{-\nu} J_{\nu+1}(x) + \text{constant}.$$

5.4 Legendre's equation—Legendre polynomials

The differential equation

$$(1 - x^2)y'' - 2xy' + m(m + 1)y = 0$$

is known as *Legendre's equation of degree m* and its solution can also be found by the method of solution in series.

Put

$$y(x) = \sum_{n=0}^{\infty} a_n x^{n+c} \quad \text{with} \quad a_0 \neq 0.$$

Then

$$\sum_{n=0}^{\infty} (n + c)(n + c - 1)a_n(x^{n+c-2} - x^{n+c})$$

$$- 2\sum_{n=0}^{\infty} (n + c)a_n x^{n+c} + m(m + 1)\sum_{n=0}^{\infty} a_n x^{n+c} = 0.$$

Hence equating to zero the coefficients of successive powers of x we obtain the indicial equation

$$c(c - 1)a_0 = 0,$$

$$(1 + c)ca_1 = 0$$

and

$$(r + c)(r + c - 1)a_r - (r + c - 2)(r + c - 3)a_{r-2}$$

$$- 2(r + c - 2)a_{r-2} + m(m + 1)a_{r-2} = 0 \quad (r \geqslant 2).$$

Therefore $c = 0, 1$ (since $a_0 \neq 0$),

$$c(1 + c)a_1 = 0$$

and

$$(r + c)(r + c - 1)a_r = (r + c - m - 2)(r + c + m - 1)a_{r-2} \text{ for } r \geqslant 2.$$

When c = 0

The equation $c(1 + c)a_1 = 0$ is now automatically satisfied for any arbitrary, finite value of a_1.

Also $r(r - 1)a_r = (r - m - 2)(r + m - 1)a_{r-2}$ $(r \geqslant 2)$.

Therefore

$$a_r = \frac{(r - m - 2)(r + m - 1)}{r(r - 1)} a_{r-2} \quad (r \geqslant 2).$$

Hence $a_2, a_4, a_6, \ldots$ can all be expressed in terms of a_0 and $a_3, a_5, a_7, \ldots$ can all be expressed in terms of a_1 as follows

$$a_2 = -\frac{m(m + 1)}{2} a_0,$$

$$a_4 = -\frac{(m - 2)(m + 3)}{4 . 3} a_2 = \frac{(m - 2)m(m + 1)(m + 3)}{4!} a_0,$$

$$a_6 = -\frac{(m - 4)(m - 2)m(m + 1)(m + 3)(m + 5)}{6!} a_0,$$

. . . .

$$a_3 = -\frac{(m - 1)(m + 2)}{3!} a_1,$$

$$a_5 = -\frac{(m - 3)(m + 4)}{5 . 4} a_3 = \frac{(m - 3)(m - 1)(m + 2)(m + 4)}{5!} a_1,$$

. . . .

Hence the general solution of Legendre's equation has been obtained in the form

$$y = a_0\left\{1 - \frac{m(m + 1)}{2!} x^2 + \frac{(m - 2)m(m + 1)(m + 3)}{4!} x^4 + \ldots\right\}$$

$$+ a_1\left\{x - \frac{(m - 1)(m + 2)}{3!} x^3\right.$$

$$+ \left.\frac{(m - 3)(m - 1)(m + 2)(m + 4)}{5!} x^5 + \ldots\right\}$$

for arbitrary constants a_0 and a_1. Using the ratio test it can be shown that both of the above series are convergent for $|x| < 1$.

When $c = 1$ it is easily shown that the series solution obtained is identical with the second series in the above solution.

Now if m is a positive integer, then whether it is even or odd, the expression for the coefficient a_{n+2} in terms of a_0 or a_1 will contain the factor $(m - n)$ in the numerator. Hence if, as happens in most cases of practical importance, m is a positive integer (equal to n) one or other of the above series will terminate after a finite number of terms (namely after the term $a_m x^m$). The terminated series will then be a polynomial of degree m in x which we shall denote by $p_m(x)$ and we now proceed to prove that $p_m(1) \neq 0$. The method of proof is by contradiction, that is we assume that $p_m(1) = 0$ and show that this leads to a contradiction. Since $p_m(x)$ satisfies Legendre's equation of degree m we have

$$(1 - x^2)p_m''(x) - 2xp_m'(x) + m(m + 1)p_m(x) = 0.$$

Then when $x = 1$ this yields

$$2p_m'(1) = m(m + 1)p_m(1).$$

Hence, if $p_m(1) = 0$, $p_m'(1)$ is also zero. But, differentiating the above equation in $p_m(x)$ with respect to x and then putting $x = 1$ we obtain

$$4p_m''(1) = \{m(m + 1) - 2\}p_m'(1)$$

and so if $p_m'(1) = 0$ then $p_m''(1)$ is also zero.

Continuing in this way we see that if $p_m(1) = 0$ then all the derivatives of $p_m(x)$ are zero when $x = 1$. But since $p_m(x)$ is a polynomial of degree m it can be expanded as a Taylor series about $x = 1$ in the form

$$p_m(x) = p_m(1) + (x - 1)p_m'(1) + \frac{(x - 1)^2}{2!}p_m''(1) + \ldots + \frac{(x - 1)^m}{m!}p_m^{(m)}(1)$$

from which it is seen that $p_m(x) \equiv 0$ if $p_m(1) = 0$. Hence $p_m(1)$ cannot be equal to zero. Thus we can divide $p_m(x)$ by the non-zero constant $p_m(1)$ to give a polynomial which has the value unity when $x = 1$. This polynomial is called the *Legendre polynomial of degree m* and is denoted by $P_m(x)$.

The other (infinite) series solution of Legendre's equation of degree m is called a *Legendre function of the second kind* and is denoted by $Q_m(x)$.

If m is even, say $m = 2r$, then the series having a finite number of terms is

$$1 - \frac{2r(2r + 1)}{2!}x^2$$

$$+ \ldots + \frac{(-1)^r 2 . 4 \ldots 2r . (2r + 1)(2r + 3) \ldots (4r - 1)}{(2r)!}x^{2r}$$

which after some manipulation may be written as

$$A_m \sum_{\lambda=0}^{\frac{1}{2}m} \frac{(-1)^\lambda (2m - 2\lambda)!}{2^m (m - \lambda)!(m - 2\lambda)!\lambda!} x^{m-2\lambda}$$

where

$$A_m = \frac{(-1)^{m/2} 2^m [(m/2)!]^2}{m!}.$$

Similarly, when m is odd, we arrive at

$$B_m \sum_{\lambda=0}^{\frac{1}{2}(m-1)} \frac{(-1)^\lambda (2m - 2\lambda)!}{2^m (m - \lambda)!(m - 2\lambda)!\lambda!} x^{m-2\lambda}$$

for the series having a finite number of terms.

Hence Legendre's polynomial of degree m, $P_m(x)$, may be written in the form

$$P_m(x) = \sum_{\lambda=0}^{[\frac{1}{2}m]} \frac{(-1)^\lambda (2m - 2\lambda)!}{2^m (m - \lambda)!(m - 2\lambda)!\lambda!} x^{m-2\lambda}$$

where

$$[\tfrac{1}{2}m] = \begin{cases} \tfrac{1}{2}m & \text{if } m \text{ is even} \\ \tfrac{1}{2}(m - 1) & \text{if } m \text{ is odd.} \end{cases}$$

(It will be proved later in an example on page 278 that this definition implies that $P_m(1) = 1$.)

In particular it is easily seen that

$$P_0(x) = 1,$$

$$P_1(x) = x,$$

$$P_2(x) = \tfrac{1}{2}(3x^2 - 1),$$

$$P_3(x) = \tfrac{1}{2}(5x^3 - 3x),$$

$$P_4(x) = \tfrac{1}{8}(35x^4 - 30x^2 + 3),$$

$$P_5(x) = \tfrac{1}{8}(63x^5 - 70x^3 + 15x)$$

etc.

5.41. *Rodrigue's formula for* $P_m(x)$

An alternative representation of Legendre polynomials is obtained as follows.

Using the binomial theorem we obtain

$$(x^2 - 1)^m = \sum_{\lambda=0}^{m} (-1)^\lambda \frac{m!}{\lambda!(m - \lambda)!} (x^2)^{m-\lambda}.$$

But, for $\lambda \leqslant \tfrac{1}{2}m$,

$$\frac{d^m}{dx^m} (x^{2m-2\lambda}) = \frac{(2m - 2\lambda)!}{(m - 2\lambda)!} x^{m-2\lambda}.$$

(For $\lambda > \tfrac{1}{2}m$ the derivatives are zero.)

Therefore

$$\frac{\mathrm{d}^m}{\mathrm{d}x^m}\{(x^2-1)^m\} = \sum_{\lambda=0}^{[\frac{1}{2}m]} \frac{(-1)^\lambda m!}{\lambda!(m-\lambda)!} \frac{(2m-2\lambda)!}{(m-2\lambda)!} x^{m-2\lambda}$$

$$= 2^m m! \sum_{\lambda=0}^{[\frac{1}{2}m]} \frac{(-1)^\lambda (2m-2\lambda)!}{2^m(m-\lambda)!\lambda!(m-2\lambda)!} x^{m-2\lambda}$$

$$= 2^m m! P_m(x).$$

Therefore

$$P_m(x) = \frac{1}{2^m m!} \frac{\mathrm{d}^m}{\mathrm{d}x^m}\{(x^2-1)^m\}.$$

This is known as *Rodrigue's formula.*

5.42 Recurrence relations for Legendre polynomials

Putting

$$w_m = \frac{1}{2^m m!}(x^2-1)^m$$

in Rodrigue's formula we obtain

$$P_m(x) = \frac{\mathrm{d}^m w_m}{\mathrm{d}x^m}.$$

Now

$$\frac{\mathrm{d}w_m}{\mathrm{d}x} = \frac{x(x^2-1)^{m-1}}{2^{m-1}(m-1)!} = xw_{m-1}$$

and so differentiating this m times using Leibniz's theorem* we obtain

$$\frac{\mathrm{d}^{m+1}w_m}{\mathrm{d}x^{m+1}} = x\frac{\mathrm{d}^m w_{m-1}}{\mathrm{d}x^m} + m\frac{\mathrm{d}^{m-1}w_{m-1}}{\mathrm{d}x^{m-1}}.$$

Therefore

$$\frac{\mathrm{d}P_m}{\mathrm{d}x} = x\frac{\mathrm{d}P_{m-1}}{\mathrm{d}x} + mP_{m-1}.$$

That is

$$P'_m - xP'_{m-1} = mP_{m-1}.$$

Also, since

$$\frac{\mathrm{d}w_m}{\mathrm{d}x} = xw_{m-1}$$

we have

$$x\frac{\mathrm{d}w_m}{\mathrm{d}x} = \{(x^2-1)+1\}w_{m-1}$$

* For the statement and proof of Leibniz's theorem see Appendix 3.

or

$$x \frac{\mathrm{d} w_m}{\mathrm{d} x} = 2m w_m + w_{m-1}.$$

Therefore again differentiating m times using Leibniz theorem we obtain

$$x \frac{\mathrm{d}^{m+1} w_m}{\mathrm{d} x^{m+1}} + m \frac{\mathrm{d}^m w_m}{\mathrm{d} x^m} = 2m \frac{\mathrm{d}^m w_m}{\mathrm{d} x^m} + \frac{\mathrm{d}^m w_{m-1}}{\mathrm{d} x^m}.$$

Hence

$$x \frac{\mathrm{d} P_m}{\mathrm{d} x} + m P_m = 2m P_m + \frac{\mathrm{d} P_{m-1}}{\mathrm{d} x}$$

or

$$x P'_m - P'_{m-1} = m P_m.$$

Now from these two results we obtain

$$(1 - x^2) P'_m = m P_{m-1} - m x P_m$$

and

$$(1 - x^2) P'_{m-1} = m x P_{m-1} - m P_m$$

and so

$$(1 - x^2) P'_m = (m + 1) x P_m - (m + 1) P_{m+1}.$$

Hence

$$m P_{m-1} - m x P_m = (m + 1) x P_m - (m + 1) P_{m+1}$$

and so

$$(m + 1) P_{m+1} - (2m + 1) x P_m + m P_{m-1} = 0.$$

Knowing $P_0(x)$ and $P_1(x)$ this last result can of course be used to generate in succession $P_2(x), P_3(x), \ldots$.

5.43. Expansion in series of Legendre polynomials

Proceeding as for Bessel functions, in order to be able to expand a given function as a series of Legendre polynomials for use in physical problems it is necessary to evaluate the integral of the product of two Legendre polynomials.

The Legendre polynomials P_n, P_m satisfy Legendre's equation of degree n and degree m respectively. That is

$$(1 - x^2) P''_n - 2x P'_n + n(n + 1) P_n = 0$$

and

$$(1 - x^2) P''_m - 2x P'_m + m(m + 1) P_m = 0$$

or

$$\frac{\mathrm{d}}{\mathrm{d} x} \{(1 - x^2) P'_n\} + n(n + 1) P_n = 0$$

and

$$\frac{d}{dx}\{(1 - x^2)P'_m\} + m(m + 1)P_m = 0.$$

Therefore

$$P_m \frac{d}{dx}\{(1 - x^2)P'_n\} - P_n \frac{d}{dx}\{(1 - x^2)P'_m\}$$

$$+ (n - m)(n + m + 1)P_m P_n = 0.$$

Then integrating this equation by parts we obtain

$$(m - n)(n + m + 1)\int_{-1}^{1} P_m P_n \, dx = \left[P_m(1 - x^2)P'_n \right]_{-1}^{1}$$

$$- \int_{-1}^{1} P'_m(1 - x^2)P'_n \, dx$$

$$- \left[P_n(1 - x^2)P'_m \right]_{-1}^{1}$$

$$+ \int_{-1}^{1} P'_n(1 - x^2)P'_m \, dx$$

$$= 0.$$

Therefore, if $m \neq n$,

$$\int_{-1}^{1} P_m(x)P_n(x) \, dx = 0 \qquad (n + m + 1 \neq 0 \text{ since } m, n \geqslant 0).$$

The functions $P_m(x)$ are said to be orthogonal in the interval $-1 \leqslant x \leqslant 1$.

Also, we have from a previously proved recurrence relation (on p. 272)

$$mP_m - (2m - 1)xP_{m-1} + (m - 1)P_{m-2} = 0.$$

Therefore multiplying this by P_m and integrating with respect to x from $x = -1$ to $x = 1$ we obtain

$$m\int_{-1}^{1} P_m^2 \, dx = (2m - 1)\int_{-1}^{1} xP_{m-1}P_m \, dx$$

since, as we have proved above,

$$\int_{-1}^{1} P_m P_{m-2} \, dx = 0.$$

Similarly, multiplying the relation

$$(m + 1)P_{m+1} - (2m + 1)xP_m + mP_{m-1} = 0$$

by P_{m-1} and integrating, we obtain

$$m \int_{-1}^{1} P_{m-1}^2 \, dx = (2m + 1) \int_{-1}^{1} x P_m P_{m-1} \, dx.$$

Therefore

$$(2m + 1) \int_{-1}^{1} P_m^2 \, dx = (2m - 1) \int_{-1}^{1} P_{m-1}^2 \, dx$$

$$= (2m - 3) \int_{-1}^{1} P_{m-2}^2 \, dx \quad \text{(similarly)}$$

$$= \cdots \cdots \cdots$$

$$= \int_{-1}^{1} P_0^2 \, dx = \int_{-1}^{1} dx \quad \text{(since } P_0(x) = 1)$$

$$= 2.$$

Therefore

$$\int_{-1}^{1} [P_m(x)]^2 \, dx = \frac{2}{2m + 1}.$$

Now if it is desired to expand the function $f(x)$ in $-1 \leqslant x \leqslant 1$ as a series of Legendre polynomials in the form

$$f(x) = c_0 P_0(x) + c_1 P_1(x) + c_2 P_2(x) + \cdots$$

multiply throughout by $P_m(x)$ and integrate with respect to x from $x = -1$ to $x = 1$. Using the above results, we obtain

$$\int_{-1}^{1} f(x) P_m(x) \, dx = \frac{2}{2m + 1} c_m$$

or

$$c_m = \frac{2m + 1}{2} \int_{-1}^{1} f(x) P_m(x) \, dx.$$

For polynomial functions of not too high degree however it will be found easier to determine the constants c_m by equating coefficients of corresponding powers of x on both sides of the equation

$$f(x) = c_0 P_0(x) + c_1 P_1(x) + c_2 P_2(x) + \cdots.$$

Example   Expand the function $f(x) = 5x^3 + 3x^2 + x + 1$ as a series of Legendre polynomials in the form

$$f(x) = c_0 P_0(x) + c_1 P_1(x) + \cdots.$$

Since $f(x)$ is a cubic polynomial it can be expanded in terms of P_0, P_1, P_2 and P_3 only. That is $c_m = 0$ for $m > 3$ and so

$$f(x) = c_0 P_0(x) + c_1 P_1(x) + c_2 P_2(x) + c_3 P_3(x).$$

Therefore $5x^3 + 3x^2 + x + 1 = c_0 + c_1 x + \frac{1}{2}c_2(3x^2 - 1) + \frac{1}{2}c_3(5x^3 - 3x)$.
Hence equating coefficients of corresponding powers of x we obtain

$$5 = \tfrac{5}{2}c_3 \quad \text{and so} \quad c_3 = 2,$$
$$3 = \tfrac{3}{2}c_2 \quad \text{and so} \quad c_2 = 2,$$
$$1 = c_1 - \tfrac{3}{2}c_3 \quad \text{and so} \quad c_1 = 4,$$
$$1 = c_0 - \tfrac{1}{2}c_2 \quad \text{and so} \quad c_0 = 2.$$

Thus $f(x) = 5x^3 + 3x^2 + x + 1 = 2P_0(x) + 4P_1(x) + 2P_2(x) + 2P_3(x)$.
Alternatively, making use of the results proved above, we have

$$c_0 = \tfrac{1}{2}\int_{-1}^{1} f(x)P_0(x)\,\mathrm{d}x$$

$$= \tfrac{1}{2}\int_{-1}^{1} (5x^3 + 3x^2 + x + 1)\,\mathrm{d}x$$

$$= 2.$$

$$c_1 = \tfrac{3}{2}\int_{-1}^{1} f(x)P_1(x)\,\mathrm{d}x$$

$$= \tfrac{3}{2}\int_{-1}^{1} (5x^3 + 3x^2 + x + 1)x\,\mathrm{d}x$$

$$= 4.$$

Similarly, after some manipulation, we obtain

$$c_2 = \tfrac{5}{2}\int_{-1}^{1} f(x)P_2(x)\,\mathrm{d}x = 2$$

and

$$c_3 = \tfrac{7}{2}\int_{-1}^{1} f(x)P_3(x)\,\mathrm{d}x = 2.$$

5.44. Generating function for Legendre polynomials

Consider the function $w = (1 - 2xu + u^2)^{-\frac{1}{2}}$

$$\frac{\partial w}{\partial x} = u(1 - 2xu + u^2)^{-\frac{3}{2}}$$

and

$$\frac{\partial w}{\partial u} = (x - u)(1 - 2xu + u^2)^{-\frac{3}{2}}$$

10

and so

$$u \frac{\partial w}{\partial u} = (x - u) \frac{\partial w}{\partial x}.$$

Now suppose that w is expanded in ascending powers of u as

$$c_0 + c_1 u + c_2 u^2 + \cdots$$

where the c_m's are of course functions of x.

Substituting into the last equation and assuming that term by term differentiation of the series both with respect to x and with respect to u is permissible, we obtain

$$\sum_{m=0}^{\infty} m c_m u^m = (x - u) \sum_{m=0}^{\infty} c'_m u^m \quad \text{where} \quad c'_m = \frac{d}{dx} \{c_m(x)\}.$$

Then, equating coefficients of u^m on each side of this equation,

$$m c_m = x c'_m - c'_{m-1}.$$

Also, since

$$\frac{\partial w}{\partial x} = u(1 - 2xu + u^2)^{-\frac{3}{2}},$$

$$\frac{\partial w}{\partial x} = u(1 - 2xu + u^2)^{-1} w$$

$$= \left(u - 2x + \frac{1}{u}\right)^{-1} w.$$

Therefore

$$w = \left(u - 2x + \frac{1}{u}\right) \frac{\partial w}{\partial x}$$

$$= (u - x) \frac{\partial w}{\partial x} + \left(\frac{1}{u} - x\right) \frac{\partial w}{\partial x}$$

$$= -u \frac{\partial w}{\partial u} + \left(\frac{1}{u} - x\right) \frac{\partial w}{\partial x}.$$

Therefore

$$\left(\frac{1}{u} - x\right) \frac{\partial w}{\partial x} = w + u \frac{\partial w}{\partial u}$$

$$= \frac{\partial}{\partial u} (uw).$$

Hence

$$\left(\frac{1}{u} - x\right) \sum_{m=0}^{\infty} c'_m u^m = \sum_{m=0}^{\infty} (m + 1) c_m u^m.$$

Then, equating coefficients of u^{m-1} on each side of this equation,

$$c'_m - xc'_{m-1} = mc_{m-1}.$$

Now, from this result and the one proved earlier that

$$mc_m = xc'_m - c'_{m-1},$$

we obtain

$$(1 - x^2)c'_m = mc_{m-1} - mxc_m$$

and

and so

$$(1 - x^2)c'_{m-1} = mxc_{m-1} - mc_m$$

Hence

$$(1 - x^2)c'_m = (m + 1)xc_m - (m + 1)c_{m+1}.$$

and so

$$mc_{m-1} - mxc_m = (m + 1)xc_m - (m + 1)c_{m+1}$$

$$(m + 1)c_{m+1} - (2m + 1)xc_m + mc_{m-1} = 0.$$

Thus the functions c_m satisfy the same recurrence relation as the Legendre polynomials (section 5.42).

But

$$w = (1 - 2xu + u^2)^{-\frac{1}{2}}$$
$$= \{1 - (2xu - u^2)\}^{-\frac{1}{2}}$$
$$= 1 + \tfrac{1}{2}(2xu - u^2) + \frac{1 \cdot 3}{2^2 2!}(2xu - u^2)^2 + \dots$$
$$+ \frac{1 \cdot 3 \dots (2m - 1)}{2^m m!}(2xu - u^2)^m + \dots$$

for $|x| \leqslant 1$ and $|u| < 1$.

Therefore picking out the coefficients of u^0 and u we see that

$$c_0 = 1 = P_0(x) \quad \text{and} \quad c_1 = x = P_1(x).$$

Then, it follows from the recurrence relation that $c_2 = P_2(x)$ and by induction that $c_m = P_m(x)$ for all positive integers m.

Thus $(1 - 2xu + u^2)^{-\frac{1}{2}} = P_0(x) + P_1(x)u + P_2(x)u^2 + \dots$ for $|x| \leqslant 1$ and $|u| < 1$. The function $(1 - 2xu + u^2)^{-\frac{1}{2}}$ is called the *generating function* for the Legendre polynomials and assumes some importance in potential theory.

If O, P and Q are three points such that OP has length r, OQ has length a and $Q\hat{O}P = \theta$ (figure 98), then the potential φ at P due to a unit point charge at Q may be expressed in the following ways.

$$\varphi = \frac{1}{QP} = \frac{1}{\sqrt{r^2 + a^2 - 2ar \cos \theta}}.$$

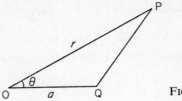

FIGURE 98

When $r < a$,

$$\varphi = \frac{1}{a}\left(1 - 2\left(\frac{r}{a}\right)\cos\theta + \left(\frac{r}{a}\right)^2\right)^{-\frac{1}{2}} \quad \text{where} \quad \frac{r}{a} < 1 \quad \text{and} \quad |\cos\theta| \leqslant 1$$

$$= \frac{1}{a}\left(1 + \frac{r}{a}P_1(\cos\theta) + \left(\frac{r}{a}\right)^2 P_2(\cos\theta) + \ldots\right).$$

(Putting $x = \cos\theta$ and $u = r/a$ in the above generating function.)
Similarly, when $r > a$,

$$\varphi = \frac{1}{r}\left(1 - 2\left(\frac{a}{r}\right)\cos\theta + \left(\frac{a}{r}\right)^2\right)^{-\frac{1}{2}}$$

$$= \frac{1}{r}\left(1 + \frac{a}{r}P_1(\cos\theta) + \left(\frac{a}{r}\right)^2 P_2(\cos\theta) + \ldots\right).$$

Example   Evaluate $P_m(1)$, $P_m(-1)$ and $P_m(0)$.

$$(1 - 2xu + u^2)^{-\frac{1}{2}} = P_0(x) + P_1(x)u + P_2(x)u^2 + \ldots \text{ for } |x| \leqslant 1 \text{ and } |u| < 1.$$

Put $x = 1$.
Therefore

$$(1 - 2u + u^2)^{-\frac{1}{2}} = P_0(1) + P_1(1)u + P_2(1)u^2 + \ldots + P_m(1)u^m + \ldots.$$

But

$$(1 - 2u + u^2)^{-\frac{1}{2}} = (1 - u)^{-1} = 1 + u + u^2 + \ldots + u^m + \ldots.$$

Hence

$$P_m(1) = 1.$$

Put $x = -1$.
Therefore

$$(1 + 2u + u^2)^{-\frac{1}{2}} = P_0(-1) + P_1(-1)u + P_2(-1)u^2 + \ldots + P_m(-1)u^m + \ldots.$$

But

$$(1 + 2u + u^2)^{-\frac{1}{2}} = (1 + u)^{-1} = 1 - u + u^2 + \ldots + (-1)^m u^m + \ldots.$$

Hence
$$P_m(-1) = (-1)^m.$$

Put $x = 0$.

Therefore
$$(1 + u^2)^{-\frac{1}{2}} = P_0(0) + P_1(0)u + P_2(0)u^2 + \ldots + P_m(0)u^m + \ldots.$$

But
$$(1 + u^2)^{-\frac{1}{2}} = 1 - \tfrac{1}{2}u^2 + \frac{1 \cdot 3}{2^2 2!} u^4 - \frac{1 \cdot 3 \cdot 5}{2^3 3!} u^6 + \ldots.$$

Hence $P_m(0) = 0$ if m is odd, and

$$P_m(0) = (-1)^{m/2} \frac{1 \cdot 3 \cdot 5 \ldots (m-1)}{2^{m/2} \left(\dfrac{m}{2}\right)!} \quad \text{if } m \text{ is even}$$

$$= (-1)^{m/2} \frac{m!}{2^{m/2} \left(\dfrac{m}{2}\right)! \, 2^{m/2} \left(\dfrac{m}{2}\right)!}$$

$$= \frac{m!}{\left\{\left(\dfrac{m}{2}\right)!\right\}^2} \left(-\frac{1}{4}\right)^{m/2}.$$

Exercises

1. Evaluate $\displaystyle\int_{-1}^{1} x P_n(x)\, \mathrm{d}x$ and $\displaystyle\int_{-1}^{1} x P_n(x) P_{n-1}(x)\, \mathrm{d}x$.

2. Deduce from the recurrence relations for Legendre polynomials that

(i)
$$x P_n'(x) = n P_n(x) + (2n - 3) P_{n-2}(x) + (2n - 7) P_{n-4}(x) + \ldots$$

and

(ii)
$$P_{n+1}'(x) - P_{n-1}'(x) = (2n + 1) P_n(x).$$

3. Prove that

$$\int_{t}^{1} P_n(x)\, \mathrm{d}x = \frac{1}{2n + 1} \{P_{n-1}(t) - P_{n+1}(t)\}$$

and deduce that

$$\int_{0}^{1} P_n(x)\, \mathrm{d}x = \begin{cases} 0 & \text{if } n \text{ is an even integer,} \\[2ex] \dfrac{(-1)^{\frac{1}{2}n - \frac{1}{2}}(n-1)!}{2^n (\frac{1}{2}n + \frac{1}{2})! (\frac{1}{2}n - \frac{1}{2})!} & \text{if } n \text{ is an odd integer.} \end{cases}$$

Further, deduce that

$$\int_0^1 xP_n(x)\,\mathrm{d}x = \begin{cases} 0 & \text{if } n \text{ is an odd integer,} \\ \dfrac{(-1)^{\frac{1}{2}n-1}(n-2)!}{2^n(\frac{1}{2}n+1)!(\frac{1}{2}n-1)!} & \text{if } n \text{ is an even integer.} \end{cases}$$

4. Express the function $f(x) = 35x^4 + 10x^2 - 2x - 3$ as a series of Legendre polynomials in the form

$$f(x) = c_0 P_0(x) + c_1 P_1(x) + \ldots + c_n P_n(x).$$

5.5 Numerical methods

The solutions found in this chapter are in general infinite power series which cannot be expressed in closed form and so for any value of the independent variable can only be evaluated approximately by evaluating the sum of a finite number of terms. The degree of approximation can however be increased for values of the independent variable within the circle of convergence of the series by including more and more terms in the partial sum. The convergence of such a series is however often too slow to make this method practicable. For example, although as stated the given series for Bessel's function of the first kind of order v, $J_v(x)$, is convergent for all finite values of x it is in fact useless for larger values of $|x|$. In this case (as in many others) we can derive an alternative expansion for $J_v(x)$ which is only valid for large values of x. Such an expansion is called an *asymptotic series expansion*.

Approximate solutions can however be obtained by other methods also. One such method is the method of finite differences. In this method the differential equation is replaced by a system of algebraic equations the solution of which tends to the solution of the differential equation as the 'mesh size' tends to zero. As a finite mesh size must always be used in practice it follows that the exact solution is never actually obtained.

6

Solutions of Laplace's equation

The method of solution to be used in this Chapter for Laplace's equation and in Chapter 7 for the heat conduction equation and in Chapter 8 for the wave equation is known as *separation of variables*. While it is a very important method of solution it must be emphasized strongly that it is certainly not the only method available. Indeed, while the method can be applied to equations of all of the above mentioned types, whether or not a solution to a particular problem can be obtained using the method depends also on the shape of the boundary and the form of the boundary conditions. For details of other methods, for example transform methods, Green's function methods and numerical methods the reader must refer to other texts dealing with these topics.

6.1 Introduction

We have already seen that a variety of physical variables under certain prescribed conditions satisfy Laplace's equation. The following is a list of some functions which satisfy this equation subject to the given conditions.

1. The gravitational potential in regions not occupied by attracting matter.
2. The electrostatic potential in a uniform dielectric (that is in regions not occupied by charges).
3. The magnetic potential in regions where there are no permanent magnets.
4. The electric potential in the theory of steady flow of electric currents in solid conductors.
5. The temperature of a solid when a steady (time-independent) state has been reached (see Chapter 7).
6. The velocity potential for a homogeneous liquid moving irrotationally.

7. The velocity potential for incompressible seepage flow through fine sands.

Now although physically the problems represented by the above situations appear vastly different, mathematically they are very similar. A solution of Laplace's equation, subject to given boundary conditions, can often be interpreted as a solution of a problem occurring in any one of the above seven different physical situations by interpreting the boundary conditions appropriately. Thus, for example, the problem of determining the steady state temperature distribution in a given solid when a given temperature distribution is maintained over its surface is the same as the problem of determining the electrostatic potential within the region occupied by the solid when a given potential distribution is maintained over the bounding surface.

Solutions of Laplace's equation are called *harmonic functions*. In addition to the uniqueness of such solutions subject to given boundary conditions (see Chapter 3) we have the following two general properties of harmonic functions.

1. A harmonic function φ cannot have a maximum or a minimum in a given region.
2. A harmonic function whose normal derivative is zero at all points of a closed boundary S must be constant throughout the region V enclosed by S.

Proof: (1) (The method of proof is by contradiction.) Suppose φ has a maximum at P. Then, moving radially away from P in any direction, φ must decrease and so there must exist a region surrounding P in which the radial derivative $\partial\varphi/\partial r$ is negative everywhere. Now construct a spherical surface S with centre P and lying wholly within this region (figure 99). At all points on S the normal derivative $\partial\varphi/\partial n$ is equal to the radial derivative $\partial\varphi/\partial r$ and so is always negative.

Therefore

$$\int_S \text{grad } \varphi \cdot ds \neq 0.$$

Then, applying Gauss' theorem to this result we obtain

$$\int_V \nabla^2\varphi \, dv \neq 0$$

where V is the region enclosed by the surface S.

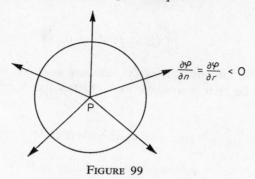

$$\frac{\partial \varphi}{\partial n} = \frac{\partial \varphi}{\partial r} < 0$$

FIGURE 99

But since φ is a harmonic function, $\nabla^2\varphi = 0$ everywhere within the region and so

$$\int_V \nabla^2\varphi \, dv = 0.$$

Thus we have obtained a contradiction and so our original assumption must be invalid.

Thus φ cannot have a maximum in the given region. Any maximum of φ must occur on the boundary of the region.

Similarly, φ cannot have a minimum in the given region.

It now follows that if φ is harmonic within a given region and has a constant value on the (closed) boundary of this region, then it must be constant throughout the region, for otherwise, if at any point φ had a value different from that on the boundary then a maximum or a minimum would occur.

(2)
$$\text{div} (\varphi \, \text{grad} \, \varphi) = \varphi \nabla^2\varphi + (\text{grad} \, \varphi)^2.$$

Therefore

$$\int_V \text{div} (\varphi \, \text{grad} \, \varphi) \, dv = \int_V (\text{grad} \, \varphi)^2 \, dv$$

since $\nabla^2\varphi = 0$ everywhere within V as φ is harmonic.

But applying Gauss' theorem to this, we have

$$\int_V \text{div} (\varphi \, \text{grad} \, \varphi) \, dv = \int_S \varphi \, \text{grad} \, \varphi \, . \, ds$$

$$= \int_S \varphi \left(\frac{\partial \varphi}{\partial n}\right) \, ds$$

$$= 0$$

since $\partial\varphi/\partial n$ is zero everywhere on S.

Therefore

$$\int_V (\text{grad } \varphi)^2 \, dv = 0.$$

Now, since the intergrand is everywhere non-negative, we deduce that it must in fact be zero everywhere in V, that is that $(\text{grad } \varphi)^2 = 0$ everywhere in V.

Hence

$$\text{grad } \varphi = 0 \text{ everywhere in V}$$

and so

$$\varphi = \text{constant everywhere in V.}$$

Finally, before proceeding to the determination of particular solutions of Laplace's equation it is important to note that the equation itself is linear so that if $\varphi_1, \varphi_2, \varphi_3, \ldots$ are all solutions then

$$\alpha_1 \varphi_1 + \alpha_2 \varphi_2 + \alpha_3 \varphi_3 + \cdots$$

where $\alpha_1, \alpha_2, \alpha_3, \ldots$ are arbitrary constants is also a solution.

6.2 Solution using Cartesian coordinates

In two dimensions Laplace's equation has the form

$$\frac{\partial^2 \varphi}{\partial x^2} + \frac{\partial^2 \varphi}{\partial y^2} = 0.$$

Using the method of solution called *separation of variables* we assume that a solution for φ can be found as the product of a function of x only, $X(x)$, and a function of y only, $Y(y)$; that is that the independent variables x and y can be completely separated out in the expression for φ. This assumption, which effects a considerable simplication would appear quite unjustified. It should be remembered however that if such a simplified solution satisfying the given boundary conditions can be found, then, as a consequence of the uniqueness theorem proved in Chapter 3, it is the unique solution of the given problem.

Thus put

$$\varphi = X(x)Y(y).$$

Then

$$Y(y)\frac{d^2 X}{dx^2} + X(y)\frac{d^2 Y}{dy^2} = 0$$

or, on dividing throughout by $X(x)Y(y)$,

$$\frac{1}{X}\frac{d^2X}{dx^2} + \frac{1}{Y}\frac{d^2Y}{dy^2} = 0.$$

Therefore

$$\frac{1}{X}\frac{d^2X}{dx^2} = -\frac{1}{Y}\frac{d^2Y}{dy^2}.$$

This equation states that a function of x only is equal to a function of y only. But, since x and y can vary independently, this implies that each of these functions must be identically equal to a constant (that is independent of x and y).

Therefore

$$\frac{1}{X}\frac{d^2X}{dx^2} = p^2$$

where p is a constant which may be real or complex. If p is wholly imaginary then, of course, p^2 will be negative. (The form p^2 is chosen rather than simply p to avoid square roots at a later stage.)

Therefore

$$\frac{d^2X}{dx^2} - p^2X = 0$$

and so

$$X = Ae^{-px} + Be^{px}$$

for $p \neq 0$ and arbitrary constants A and B, and

$$X = A_0 + B_0x$$

for $p = 0$ and arbitrary constants A_0 and B_0. Also,

$$-\frac{1}{Y}\frac{d^2Y}{dy^2} = p^2.$$

Therefore

$$\frac{d^2Y}{dy^2} + p^2Y = 0$$

and so

$$Y = C\cos py + D\sin py$$

for $p \neq 0$ and arbitrary constants C and D, and

$$Y = C_0 + D_0y$$

for $p = 0$ and arbitrary constants C_0 and D_0.

Hence

$$\varphi = \begin{cases} (Ae^{-px} + Be^{px})(C\cos py + D\sin py) & p \neq 0, \\ (A_0 + B_0 x)(C_0 + D_0 y) & p = 0 \end{cases}$$

satisfies Laplace's equation

$$\frac{\partial^2 \varphi}{\partial x^2} + \frac{\partial^2 \varphi}{\partial y^2} = 0.$$

(Note that, putting $\dfrac{1}{X}\dfrac{\mathrm{d}^2 X}{\mathrm{d}x^2} = -p'^2$ instead of p^2 results in the solution

$$\varphi = \begin{cases} (A'\cos p'x + B'\sin p'x)(C'e^{-p'y} + D'e^{p'y}) & p' \neq 0, \\ (A_0 + B_0 x)(C_0 + D_0 y) & p' = 0 \end{cases}$$

for arbitrary constants A', B', C', D', A_0, B_0, C_0 and D_0 being obtained and both forms of solution are equivalent.) Values for p (or p') and the other constants occurring are then determined by the boundary conditions of a particular problem as illustrated in the following example.

Obtain the solution of Laplace's equation which satisfies the following boundary conditions.

1. $\varphi = 0$ on $y = 0$ for $x \geqslant 0$.
2. $\varphi = 0$ on $y = l$ for $x \geqslant 0$.
3. φ remains finite as x tends to infinity for $0 < y < l$.
4. $\varphi = f(y)$ for $0 < y < l$ when $x = 0$.

As above, putting $\varphi = X(x)Y(y)$ we obtain

$$X(x) = Ae^{-px} + Be^{px} \quad \text{and} \quad Y(y) = C\cos py + D\sin py \quad \text{for} \quad p \neq 0,$$

$$X(x) = A_0 + B_0 x \quad \text{and} \quad Y(y) = C_0 + D_0 y \quad \text{for} \quad p = 0.$$

Boundary condition 1 implies that $Y(y) = 0$ when $y = 0$.

Therefore $C = 0$ and so $Y(y) = D\sin py$, $p \neq 0$, and $C_0 = 0$ when $p = 0$. Boundary condition 2 implies that $Y(y) = 0$ when $y = l$.

Therefore when $p \neq 0$, $0 = D\sin pl$ and so $\sin pl = 0$ ($D = 0$ is also a solution, but this simply leads to the trivial result $\varphi = 0$ for all x and all y and does not satisfy boundary condition 4 unless $f(y) \equiv 0$).

Therefore $pl = n\pi$ for integers n. When $p = 0$, we obtain $D_0 = 0$ and so the solution for φ corresponding to $p = 0$ is identically zero and so need not be considered further. Boundary condition 3 implies that $X(x)$ remains finite as $x \to \infty$ and so, taking $p > 0$ (without loss of generality), we see that $B = 0$.

Thus the solution for φ now has the form

$$A D e^{-(n\pi x/l)} \sin \frac{n\pi y}{l}$$

for positive integer values of n. That is for each positive integer value of n we have a solution of the form

$$b_n e^{-(n\pi x/l)} \sin \frac{n\pi y}{l}$$

where b_n is an arbitrary constant. However such a solution for a particular value of n will not in general satisfy the final boundary condition. Indeed it is clear that this condition can only be satisfied by a solution of the above form when $f(y) \equiv \alpha \sin (n\pi y/l)$ for some integer n and constant α. But, since Laplace's equation is linear, a more general solution (of separated form) satisfying the boundary conditions 1 to 3 can be written

$$\varphi = \sum_{n=1}^{\infty} b_n e^{-(n\pi x/l)} \sin \frac{n\pi y}{l} .$$

Now boundary condition 4 gives

$$f(y) = \sum_{n=1}^{\infty} b_n \sin \frac{n\pi y}{l} \quad \text{for} \quad 0 \leqslant y \leqslant l$$

from which we see that if $f(y)$ satisfies the Dirichlet conditions in $0 < y < l$ then the b_n's are the coefficients of its Fourier sine series. That is

$$b_n = \frac{2}{l} \int_0^l f(y) \sin \frac{n\pi y}{l} \, \mathrm{d}y.$$

Example   Obtain a solution φ of Laplace's equation in two dimensions subject to the following boundary conditions in which x and y are rectangular cartesian coordinates.

(i) $\varphi \to 0$ *as* $x \to \infty$.
(ii) $\varphi = 0$ *when* $y = 0$ *and when* $y = a$ *for all* $x > 0$.
(iii) $\varphi = Ky(a - y)$ *when* $x = 0$ *for* $0 \leqslant y \leqslant a$, *where* K *is a constant*.

The function φ satisfies the equation

$$\frac{\partial^2 \varphi}{\partial x^2} + \frac{\partial^2 \varphi}{\partial y^2} = 0.$$

Put $\varphi = X(x)Y(y)$ and proceed as above, to obtain the solution

$$\varphi_p = \begin{cases} (Ae^{-px} + Be^{px})(C\cos py + D\sin py) & p \neq 0, \\ (A_0 + B_0 x)(C_0 + D_0 y) & p = 0. \end{cases}$$

Since $\varphi \to 0$ as $x \to \infty$, $A_0 = 0$, $B_0 = 0$ and $B = 0$.
Since $\varphi = 0$ when $y = 0$ for all $x > 0$, $C_0 = 0$ and $C = 0$.
Since $\varphi = 0$ when $y = a$ for all $x > 0$, $D_0 = 0$ and $\sin pa = 0$ ($D = 0$ leads to the trivial solution $\varphi = 0$ for all x and y and this would not satisfy condition (iii).)
Therefore

$$pa = n\pi$$

for positive integer values of n. (The result, $B = 0$, of applying boundary condition (i) implies that p is non-negative.)
Hence the solution now has the form

$$Ae^{-(n\pi x/a)} D \sin\frac{n\pi y}{a}$$

for positive integer values of n.
Such a solution, by itself, does not however satisfy the final boundary condition. However a more general solution of separated form satisfying conditions (i) and (ii) is

$$\varphi = \sum_{n=1}^{\infty} b_n e^{-(n\pi x/a)} \sin\frac{n\pi y}{a}$$

where b_n is independent of x and y. Now $\varphi = Ky(a - y)$ when $x = 0$ for $0 \leqslant y \leqslant a$ and so

$$Ky(a - y) = \sum_{n=1}^{\infty} b_n \sin\frac{n\pi y}{a} \quad \text{for} \quad 0 \leqslant y \leqslant a.$$

Hence, expanding the left hand side as a Fourier sine series we obtain

$$\begin{aligned} b_n &= \frac{2}{a}\int_0^a Ky(a - y)\sin\frac{n\pi y}{a}\,dy \\ &= \frac{2}{a}\left\{\left[-\frac{a}{n\pi}Ky(a - y)\cos\frac{n\pi y}{a}\right]_0^a + \frac{aK}{n\pi}\int_0^a (a - 2y)\cos\frac{n\pi y}{a}\,dy\right\} \\ &= \frac{2Ka}{n^2\pi^2}\left\{\left[(a - 2y)\sin\frac{n\pi y}{a}\right]_0^a + 2\int_0^a \sin\frac{n\pi y}{a}\,dy\right\} \\ &= \frac{2Ka}{n^2\pi^2}\left\{-\frac{2a}{n\pi}(\cos n\pi - 1)\right\} \\ &= \begin{cases} \dfrac{8Ka^2}{n^3\pi^3} & \text{for } n \text{ odd}, \\ 0 & \text{for } n \text{ even}. \end{cases} \end{aligned}$$

Thus

$$\varphi_2 = \frac{2}{\sinh{(\pi b/a)}} \left\{\sinh{\frac{\pi b}{a}} \cosh{\frac{\pi y}{a}} + \left(1 - \cosh{\frac{\pi b}{a}}\right) \sinh{\frac{\pi y}{a}}\right\} \sin{\frac{\pi x}{a}}$$

$$= \frac{2}{\sinh{(\pi b/a)}} \left\{\sinh{\frac{\pi y}{a}} + \sinh{\frac{\pi (b - y)}{a}}\right\} \sin{\frac{\pi x}{a}}.$$

The complete solution of the given problem is then $(\varphi_1 + \varphi_2)$ where φ_1 and φ_2 are as given above.

Example   The function φ satisfies Laplace's equation in the rectangular region $0 < x < a, 0 < y < b$. On the boundary of the rectangle φ satisfies the following conditions

(i) $\varphi = 0$ when $y = 0$ and when $y = b$ and $0 < x < a$,
(ii) $\varphi = k$, where k is a constant, when $x = 0$ and $0 < y < b$ and
(iii) $\partial \varphi / \partial x = 0$ when $x = a$ and $0 < y < b$.
Find the form of the solution φ within the rectangle.

Take

$$\varphi = \begin{cases} (A \cosh{px} + B \sinh{px})(C \cos{py} + D \sin{py}) & p \neq 0, \\ (A_0 + B_0 x)(C_0 + D_0 y) & p = 0. \end{cases}$$

Now, since $\varphi = 0$ when $y = 0, b$, and $0 < x < a$, we obtain, as in the last example, that

$$\varphi = \left(A D \cosh{\frac{n \pi x}{b}} + B D \sinh{\frac{n \pi x}{b}}\right) \sin{\frac{n \pi y}{b}}$$

for $n \neq 0$.

A more general solution satisfying the boundary condition (i) is

$$\varphi = \sum_{n=1}^{\infty} \left(\alpha_n \cosh{\frac{n \pi x}{b}} + \beta_n \sinh{\frac{n \pi x}{b}}\right) \sin{\frac{n \pi y}{b}}.$$

(As in the last example it is not necessary to consider negative integer values for n). Now $\varphi = k$ when $x = 0$ and $0 < y < b$ and so

$$k = \sum_{n=1}^{\infty} \alpha_n \sin{\frac{n \pi y}{b}}.$$

Hence the α_n's are the coefficients of the Fourier sine series for k in $0 < y < b$. That is

$$\alpha_n = \frac{2}{b} \int_0^b k \sin{\frac{n \pi y}{b}} \, dy$$

$$= \begin{cases} 4k/n\pi & \text{for } n \text{ odd} \\ 0 & \text{for } n \text{ even}. \end{cases}$$

But $\partial\varphi/\partial x = 0$ when $x = a$ and $0 < y < b$ and so

$$0 = \sum_{n=1}^{\infty} \frac{n\pi}{b}\left(\alpha_n \sinh \frac{n\pi a}{b} + \beta_n \cosh \frac{n\pi a}{b}\right) \sin \frac{n\pi y}{b}.$$

Hence

$$\beta_n = -\alpha_n \tanh \frac{n\pi a}{b}$$

$$= \begin{cases} -\dfrac{4k}{n\pi} \tanh \dfrac{n\pi a}{b} & \text{for } n \text{ odd} \\ 0 & \text{for } n \text{ even.} \end{cases}$$

In three dimensions Laplace's equation has the form

$$\frac{\partial^2\varphi}{\partial x^2} + \frac{\partial^2\varphi}{\partial y^2} + \frac{\partial^2\varphi}{\partial z^2} = 0$$

and the process of solution is similar to that above.

Put

$$\varphi = X(x)\,Y(y)Z(z).$$

Therefore

$$YZ\frac{d^2 X}{dx^2} + XZ\frac{d^2 Y}{dy^2} + XY\frac{d^2 Z}{dz^2} = 0.$$

Therefore

$$\frac{1}{X}\frac{d^2 X}{dx^2} + \frac{1}{Y}\frac{d^2 Y}{dy^2} + \frac{1}{Z}\frac{d^2 Z}{dz^2} = 0.$$

That is a function of x only plus a function of y only plus a function of z only is equal to zero. Hence each of the terms

$$\frac{1}{X}\frac{d^2 X}{dx^2}, \quad \frac{1}{Y}\frac{d^2 Y}{dy^2} \quad \text{and} \quad \frac{1}{Z}\frac{d^2 Z}{dz^2}$$

must be equal to a constant.

Put

$$\frac{1}{X}\frac{d^2 X}{dx^2} = -p^2 \quad \text{and} \quad \frac{1}{Y}\frac{d^2 Y}{dy^2} = -q^2$$

and then

$$\frac{1}{Z}\frac{d^2 Z}{dz^2} = p^2 + q^2 = r^2 \quad \text{(say).}$$

Therefore we have

$$\frac{d^2 X}{dx^2} + p^2 X = 0,$$

$$\frac{d^2 Y}{dy^2} + q^2 Y = 0$$

and

$$\frac{d^2 Z}{dz^2} - r^2 Z = 0.$$

Hence

$$X = \begin{cases} A \cos px + B \sin px & p \neq 0, \\ A_0 + B_0 x & p = 0, \end{cases}$$

for arbitrary constants A, A_0, B and B_0.

$$Y = \begin{cases} C \cos qy + D \sin qy & q \neq 0, \\ C_0 + D_0 y & q = 0, \end{cases}$$

for arbitrary constants C, C_0, D and D_0. Also

$$Z = \begin{cases} E e^{-rz} + F e^{rz} & r \neq 0, \\ E_0 + F_0 z & r = 0 \end{cases}$$

for arbitrary constants E, E_0, F and F_0.

Values for p and q (and hence r) and the other constants occurring are then determined by the boundary values of a particular problem.

Exercises

1. Obtain a solution φ of Laplace's equation in two-dimensions subject to each of the following sets of boundary conditions in which x and y are rectangular cartesian coordinates.

   (a) (i) $\varphi \to 0$ as $y \to \infty$, (ii) $\varphi = 0$ when $x = 0$ and when $x = 1$ for all $y > 0$, and (iii) $\varphi = \sin^3 \pi x$ when $y = 0$ for $0 \leqslant x \leqslant 1$.

   (b) (i) φ remains finite as $y \to \infty$, (ii) $\partial \varphi / \partial x = 0$ when $x = 0$ and when $x = 1$ for all $y > 0$, and (iii) $\varphi = 1 - x$ when $y = 0$ for $0 \leqslant x \leqslant 1$.

2. Laplace's equation is satisfied in the rectangle $0 < x < a$, $0 < y < b$. Find a solution φ such that

   (i) $\varphi = 0$ when $x = 0$ and when $x = a$ and $0 < y < b$,

   (ii) $\varphi = 0$ when $y = 0$ and $0 < x < a$ and

   (iii) $\varphi = x(x - a)$ when $y = b$ and $0 < x < a$.

3. Find a function φ satisfying Laplace's equation in the rectangle $0 < x < a$, $0 < y < b$ and subject to the following boundary conditions
   (i) $\varphi = 0$ when $x = 0$ and $0 < y < b$,
   (ii) $\varphi = 0$ when $y = 0$ and $0 < x < a$,
   (iii) $\varphi = y \sin (\pi y/b)$ when $x = a$ and $0 < y < b$ and
   (iv) $\varphi = x \sin (\pi x/a)$ when $y = b$ and $0 < x < a$.

6.3 Solution using cylindrical polar coordinates (ρ, ψ, z)

6.31 Circular harmonics

Solutions of Laplace's equation independent of z are called *circular harmonics*.

Laplace's equation is

$$\nabla^2 \varphi = \frac{\partial^2 \varphi}{\partial \rho^2} + \frac{1}{\rho} \frac{\partial \varphi}{\partial \rho} + \frac{1}{\rho^2} \frac{\partial^2 \varphi}{\partial \psi^2} = 0 \quad \text{(see Chapter 1).}$$

Put $\varphi = R(\rho)F(\psi)$ where $R(\rho)$ is a function of ρ only and $F(\psi)$ is a function of ψ only.

Therefore

$$F \frac{d^2 R}{d\rho^2} + \frac{1}{\rho} F \frac{dR}{d\rho} + \frac{1}{\rho^2} R \frac{d^2 F}{d\psi^2} = 0.$$

Hence

$$\frac{\rho^2}{R} \frac{d^2 R}{dr^2} + \frac{\rho}{R} \frac{dR}{d\rho} + \frac{1}{F} \frac{d^2 F}{d\psi^2} = 0.$$

and so

$$\frac{\rho^2}{R} \left(\frac{d^2 R}{d\rho^2} + \frac{1}{\rho} \frac{dR}{d\rho} \right) = - \frac{1}{F} \frac{d^2 F}{d\psi^2}.$$

Now the left hand side is a function of ρ only while the right hand side is a function of ψ only and so each must be constant. Let this constant be v^2.*

Therefore

$$\frac{d^2 F}{d\psi^2} + v^2 F = 0,$$

and so

$$F = \begin{cases} A \cos v\psi + B \sin v\psi & \text{if} \quad v \neq 0, \\ A_0 \psi + B_0 & \text{if} \quad v = 0 \end{cases}$$

* A positive value is chosen for this constant since this will give a periodic (instead of exponential) type solution for F.

for arbitrary constants A, B, A_0 and B_0. Also

$$\frac{\rho}{R}\left(\rho\frac{d^2R}{d\rho^2} + \frac{dR}{d\rho}\right) = v^2.$$

Therefore

$$\frac{\rho}{R}\frac{d}{d\rho}\left(\rho\frac{dR}{d\rho}\right) - v^2 = 0.$$

Putting $u = \ln\rho$, we have $d/du = \rho(d/d\rho)$ and this latter equation becomes

$$\frac{d^2R}{du^2} - v^2R = 0.$$

Therefore when $v \neq 0$,

$$R = Ce^{vu} + De^{-vu}$$

$$= Ce^{v\ln\rho} + De^{-v\ln\rho}$$

$$= C\rho^v + D\rho^{-v}$$

for arbitrary constants C and D. When $v = 0$,

$$R = C_0u + D_0 = C_0\ln\rho + D_0.$$

Hence the solutions of Laplace's equation in cylindrical polar coordinates independent of z take the form

$$\varphi_v = (C_v\rho^v + D_v\rho^{-v})(A_v\cos v\psi + B_v\sin v\psi), \quad v \neq 0,$$

$$\varphi_0 = (C_0\ln\rho + D_0)(A_0\psi + B_0), \quad\quad\quad\quad v = 0.$$

A more general solution is obtained by adding any number of such solutions for different values of v.

In many applications to physical problems however the solution $\varphi(\rho,\psi)$ is required to be single valued in ψ in the sense that

$$\varphi(\rho,\psi + 2m\pi) = \varphi(\rho,\psi) \quad\text{for}\quad m = 1, 2, 3, \ldots.$$

Therefore $A_0 = 0$ and v must be an integer n say. Hence a general solution in which φ is single valued is

$$\varphi = \sum_{n=1}^{\infty}\rho^n(a_n\cos n\psi + b_n\sin n\psi)$$

$$+ \sum_{n=1}^{\infty}\rho^{-n}(c_n\cos n\psi + d_n\sin n\psi) + b_0\ln\rho + a_0$$

where a_0, a_n, b_0, b_n, c_n and d_n are arbitrary constants.

If in a particular problem φ remains finite as $\rho \to 0$, then b_0, c_n and d_n are all zero and the solution has the form

$$\varphi = a_0 + \sum_{n=1}^{\infty} \rho^n (a_n \cos n\psi + b_n \sin n\psi).$$

Further, if it is given that $\varphi = g(\psi)$ when $\rho = \rho_0$, then

$$g(\psi) = a_0 + \sum_{n=1}^{\infty} (a_n \rho_0^n \cos n\psi + b_n \rho_0^n \sin n\psi).$$

Hence, if $g(\psi)$ satisfies the Dirichlet conditions in $0 \leqslant \psi \leqslant 2\pi$, the constants a_0, $a_n \rho_0^n$ and $b_n \rho_0^n$ (and hence a_n and b_n) can be determined using Fourier's method.

Again, if a solution valid for $\rho \geqslant a$ and such that φ remains finite as $\rho \to \infty$ is required, then b_0, a_n, b_n in the above general solution must all be zero and so we have

$$\varphi = a_0 + \sum_{n=1}^{\infty} (c_n \rho^{-n} \cos n\psi + d_n \rho^{-n} \sin n\psi).$$

Once again, if it is given that $\varphi = g_1(\psi)$ when $\rho = \rho_1 (\geqslant a)$ then a_0, c_n, d_n can be determined using Fourier's method.

Example   An infinitely long, hollow, circular cylinder of conducting material has internal radius a. This cylinder is halved by two cuts parallel to its axis, and one half is maintained at potential φ_0 while the other half is earthed. Determine the form of the potential at all points within the region enclosed by the cylinder.

The potential φ in the region enclosed by the cylinder satisfies Laplace's equation $\nabla^2 \varphi = 0$.

Choose cylindrical polar coordinates (ρ, ψ, z) with z-axis along the axis of the cylinder and let the half-planes $\psi = 0$ and $\psi = \pi$ each contain one of the two cuts which divide the given cylinder into the halves which are maintained at different potentials.

Then φ is independent of z and satisfies the boundary conditions

$$\varphi = \varphi_0 \quad \text{when} \quad \rho = a \quad \text{and} \quad 0 \leqslant \psi < \pi$$

$$\varphi = 0 \quad \text{(earthed) when} \quad \rho = a \quad \text{and} \quad \pi \leqslant \psi < 2\pi.$$

Since φ is single valued and remains finite along the axis of the cylinder (that is when $\rho = 0$), we start with the solution

$$\varphi = A_0 + \sum_{n=1}^{\infty} \rho^n (A_n \cos n\psi + B_n \sin n\psi).$$

Then, on application of the above boundary conditions, we obtain

$$A_0 + \sum_{n=1}^{\infty} a^n (A_n \cos n\psi + B_n \sin n\psi) = g(\psi)$$

where

$$g(\psi) = \begin{cases} \varphi_0 & \text{for} \quad 0 \leqslant \psi < \pi, \\ 0 & \text{for} \quad \pi \leqslant \psi < 2\pi. \end{cases}$$

Using Fourier's method to obtain the constant coefficients, we have

$$A_0 = \frac{1}{2\pi} \int_0^{2\pi} g(\psi) \, d\psi = \frac{1}{2\pi} \int_0^{\pi} \varphi_0 \, d\psi$$

$$= \frac{\varphi_0}{2}.$$

$$A_n = \frac{1}{a^n \pi} \int_0^{2\pi} g(\psi) \cos n\psi \, d\psi = \frac{\varphi_0}{a^n \pi} \int_0^{\pi} \cos n\psi \, d\psi$$

$$= 0.$$

$$B_n = \frac{1}{a^n \pi} \int_0^{2\pi} g(\psi) \sin n\psi \, d\psi = \frac{\varphi_0}{a^n \pi} \int_0^{\pi} \sin n\psi \, d\psi$$

$$= \begin{cases} \dfrac{2\varphi_0}{a^n \pi n} & \text{for } n \text{ odd,} \\ 0 & \text{for } n \text{ even.} \end{cases}$$

Hence

$$\varphi(\rho,\psi) = \frac{\varphi_0}{2} + \frac{2\varphi_0}{\pi} \sum_{m=1}^{\infty} \left(\frac{\rho}{a}\right)^{2m-1} \frac{\sin (2m-1)\psi}{2m-1}.$$

Example   An infinitely long, uncharged, conducting, circular cylinder of radius a is placed in a uniform electric field of strength E_0 with its axis at right angles to the lines of force. Determine the resulting potential and force field outside the cylinder.

In terms of cylindrical polar coordinates (ρ,ψ,z) with z-axis along the axis of the cylinder it is seen that the potential φ is independent of z. If φ_0 is the potential of the uniform field E_0 before the cylinder is present, then

$$\mathbf{E}_0 = -\text{grad } \varphi_0$$

and so

$$(E_0 \cos \psi, -E_0 \sin \psi, 0) = -\left(\frac{\partial \varphi_0}{\partial \rho}, \frac{1}{\rho} \frac{\partial \varphi_0}{\partial \psi}, \frac{\partial \varphi_0}{\partial z}\right)$$

(using cylindrical polars and resolving $\mathbf{E}_0$ as illustrated in figure 100).

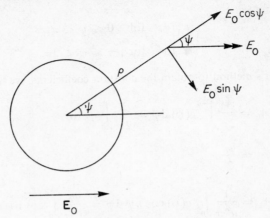

FIGURE 100

Therefore

$$\frac{\partial \varphi_0}{\partial \rho} = -E_0 \cos \psi$$

$$\frac{\partial \varphi_0}{\partial \psi} = E_0 \rho \sin \psi$$

$$\frac{\partial \varphi_0}{\partial z} = 0$$

Hence $\varphi_0 = -E_0 \rho \cos \psi$ apart from an arbitrary additive constant which can be taken to be zero.

Then

$$\varphi = \varphi_0 + \varphi_i$$

where φ_i is the potential induced by the presence of the cylinder and tends to zero as ρ tends to infinity. Also

$$\nabla^2 \varphi = 0.$$

Therefore

$$0 = \nabla^2 \varphi_0 + \nabla^2 \varphi_i = \text{div (grad } \varphi_0) + \nabla^2 \varphi_i$$

$$= -\text{div } E_0 + \nabla^2 \varphi_i = \nabla^2 \varphi_i.$$

Since the cylinder is a perfect conductor, its surface is an equipotential, say $\varphi = 0$. That is, the boundary condition to be satisfied is $\varphi = 0$ when $\rho = a$. Therefore $\varphi_i = -\varphi_0$ when $\rho = a$. Hence $\varphi_i = E_0 a \cos \psi$ when $\rho = a$.

Thus we must solve $\nabla^2 \varphi_i \; 0$ where = (i). $\varphi_i \to 0$ as $\rho \to \infty$ and (ii). $\varphi_i = E_0 a \cos \psi$ when $\rho = a$.

Since φ_i is single valued we have, using condition (i), that φ_i has the general form

$$\varphi_i = \sum_{n=1}^{\infty} (C_n \rho^{-n} \cos n\psi + D_n \rho^{-n} \sin n\psi)$$

for arbitrary constants C_n and D_n.

But clearly the induced potential φ_i must be symmetric about $\psi = 0$ since both the initial field and the cylinder itself are symmetric. Hence

$$\varphi_i(\rho,\psi) = \varphi_i(\rho,-\psi).$$

Therefore $D_n = 0$ for $n \geqslant 1$, and

$$\varphi_i = \sum_{n=1}^{\infty} C_n \rho^{-n} \cos n\psi.$$

Now condition (ii). leads to the result

$$E_0 a \cos \psi = \sum_{n=1}^{\infty} C_n a^{-n} \cos n\psi.$$

Therefore, equating coefficients of $\cos n\psi$, we obtain

$$E_0 a = C_1 a^{-1},$$

$$C_n = 0, \qquad n \neq 1.$$

Hence

$$\varphi_i = \frac{E_0 a^2}{\rho} \cos \psi.$$

Therefore

$$\varphi = \varphi_0 + \varphi_i = -E_0 \rho \cos \psi \left(1 - \frac{a^2}{\rho^2}\right).$$

Then

$$\mathbf{E} = -\text{grad } \varphi = \left(-\frac{\partial \varphi}{\partial \rho}, -\frac{1}{\rho}\frac{\partial \varphi}{\partial \psi}, \frac{\partial \varphi}{\partial z}\right)$$

$$= \left(E_0 \cos \psi \left(1 + \frac{a^2}{\rho^2}\right), -E_0 \sin \psi \left(1 - \frac{a^2}{\rho^2}\right), 0\right),$$

Example *An infinitely long, uncharged, cylindrical shell of circular cross-section has internal radius a and external radius b and is placed in a uniform electric field of strength E_0 with its axis at right angles to the lines of force. If the dielectric constant for the material of the shell is κ, determine the electric field in the region enclosed by the inner surface of the shell.*

Let φ_0 be the potential of the uniform field E_0. Then, as in the previous example, $\varphi_0 = -E_0 \rho \cos \psi$ in terms of cylindrical polar coordinates (ρ,ψ,z) with z-axis along the axis of the cylinder.

Let φ_1, φ_2, φ_3 be the potentials inside, within and outside the shell respectively (figure 101).

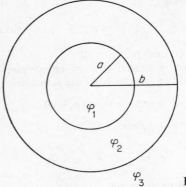

FIGURE 101

Then $\nabla^2\varphi_1 = 0$, $\nabla^2\varphi_2 = 0$ and $\nabla^2\varphi_3 = 0$, and φ_1, φ_2, φ_3 are all functions of ψ and ρ only. Also φ_1, φ_2, φ_3 must all be symmetric about $\psi = 0$.

Since φ_1 remains finite as $\rho \to 0$ the general form of solution for φ_1 is

$$\varphi_1 = A_0 + \sum_{n=1}^{\infty} \rho^n A_n \cos n\psi.$$

The general form of the solution for φ_2 is

$$\varphi_2 = \sum_{n=1}^{\infty} \rho^n B_n \cos n\psi + \sum_{n=1}^{\infty} \rho^{-n} C_n \cos n\psi + k \ln \rho + l.$$

As in the previous example, φ_3 may be expressed in the form $\varphi_0 + \varphi_i$ where φ_i satisfies Laplace's equation and tends to zero as ρ tends to infinity.

Therefore

$$\varphi_3 = -E_0\rho \cos \psi + \sum_{n=1}^{\infty} \rho^{-n} D_n \cos n\psi.$$

The boundary conditions to be satisfied are

1. $\varphi_1 = \varphi_2$ when $\rho = a$.
2. $\varphi_2 = \varphi_3$ when $\rho = b$.
3. $\partial\varphi_1/\partial\rho = \kappa(\partial\varphi_2/\partial\rho)$ when $\rho = a$.
4. $\kappa(\partial\varphi_2/\partial\rho) = \partial\varphi_3/\partial\rho$ when $\rho = b$.

Hence

$$\sum_{n=1}^{\infty} a^n A_n \cos n\psi + A_0 = \sum_{n=1}^{\infty} a^n B_n \cos n\psi$$

$$+ \sum_{n=1}^{\infty} a^{-n} C_n \cos n\psi + k \ln a + l.$$

Therefore

$$A_0 = k \ln a + l$$

and

Also

$$a^n A_n = a^n B_n + a^{-n} C_n.$$

$$\sum_{n=1}^{\infty} b^n B_n \cos n\psi + \sum_{n=1}^{\infty} b^{-n} C_n \cos n\psi + k \ln b + l$$

$$= -E_0 b \cos \psi + \sum_{n=1}^{\infty} b^{-n} D_n \cos n\psi.$$

Therefore

$$k \ln b + l = 0,$$

and

$$bB_1 + b^{-1}C_1 = -E_0 b + b^{-1} D_1$$

Also

$$b^n B_n + b^{-n} C_n = b^{-n} D_n \qquad (n > 1).$$

$$\sum_{n=1}^{\infty} n a^{n-1} A_n \cos n\psi$$

$$= \kappa \left\{ \sum_{n=1}^{\infty} n a^{n-1} B_n \cos n\psi - \sum_{n=1}^{\infty} n a^{-(n+1)} C_n \cos n\psi + \frac{k}{a} \right\}.$$

Therefore

$$\frac{k}{a} = 0$$

(and hence $k = 0$, $l = 0$ and $A_0 = 0$) and

Also

$$a^{n-1} A_n = \kappa(a^{n-1} B_n - a^{-(n+1)} C_n).$$

$$\kappa \left\{ \sum_{n=1}^{\infty} n b^{n-1} B_n \cos n\psi - \sum_{n=1}^{\infty} n b^{-(n+1)} C_n \cos n\psi \right\}$$

$$= -E_0 \cos \psi - \sum_{n=1}^{\infty} n b^{-(n+1)} D_n \cos n\psi.$$

Therefore

$$\kappa \left(B_1 - \frac{1}{b^2} C_1 \right) = -\frac{D_1}{b^2} - E_0$$

and

$$\kappa(b^{n-1} B_n - b^{-(n+1)} C_n) = -b^{-(n+1)} D_n \quad (n > 1).$$

Thus we have the following two systems of equations to solve

$$\left. \begin{array}{l} a^n A_n \quad\quad - a^n B_n \quad\quad - a^{-n} C_n \quad\quad\quad\quad = 0 \\ \quad\quad\quad\quad\quad b^n B_n \quad + b^{-n} C_n \quad - b^{-n} D_n = 0 \\ a^{n-1} A_n - \kappa a^{n-1} B_n + \kappa a^{-(n+1)} C_n \quad\quad\quad = 0 \\ \kappa b^{n-1} A_n \quad\quad\quad\quad - \kappa b^{-(n+1)} C_n + b^{-(n+1)} D_n = 0 \end{array} \right\} \quad \text{for} \quad n > 1,$$

and

$$aA_1 - aB_1 - \frac{1}{a} C_1 \hspace{2cm} = 0$$

$$bB_1 + \frac{1}{b} C_1 - \frac{1}{b} D_1 = -E_0 b$$

$$A_1 - \kappa B_1 + \frac{\kappa}{a^2} C_1 \hspace{1.5cm} = 0$$

$$\kappa B_1 - \frac{\kappa}{b^2} C_1 + \frac{1}{b^2} D_1 = -E_0$$

It can be shown that the only solution of the first system is

$$A_n = B_n = C_n = D_n = 0 \qquad (n > 1).$$

(This can be done by simply solving the equations or by showing that the determinant of the coefficient matrix is non-zero.)

Solving the second system we obtain

$$A_1 = \frac{-4\kappa E_0}{(\kappa + 1)^2 - (\kappa - 1)^2 (a/b)^2}.$$

Therefore

$$\varphi_1 = A_1 \rho \cos \psi = \frac{-4\kappa E_0}{(\kappa + 1)^2 - (\kappa - 1)^2 (a/b)^2} \rho \cos \psi.$$

(Note that, on putting $\kappa = 1$, this reduces to $\varphi_1 = -E_0 \rho \cos \psi$ the potential of the uniform field.) Hence the electric field in the region enclosed by the inner surface of the shell is

$$\mathbf{E}_1 = -\text{grad } \varphi_1 = \frac{4\kappa}{(\kappa + 1)^2 - (\kappa - 1)^2 (a/b)^2} \mathbf{E}_0.$$

That is the field strength is

$$\frac{4\kappa E_0}{(\kappa + 1)^2 - (\kappa - 1)^2 (a/b)^2}$$

and the lines of force are parallel to the lines of force of the initial field $\mathbf{E}_0$.

6.32 *General cylindrical harmonics*

Solutions of Laplace's equation in cylindrical polar coordinates are called *cylindrical harmonics*.

Laplace's equation is

$$\nabla^2 \varphi = \frac{\partial^2 \varphi}{\partial \rho^2} + \frac{1}{\rho} \frac{\partial \varphi}{\partial \rho} + \frac{1}{\rho^2} \frac{\partial^2 \varphi}{\partial \psi^2} + \frac{\partial^2 \varphi}{\partial z^2} = 0 \quad \text{(see Chapter 1)}.$$

Put

$$\varphi = R(\rho)F(\psi)Z(z).$$

Therefore

$$FZ\frac{d^2R}{d\rho^2} + \frac{1}{\rho}FZ\frac{dR}{d\rho} + \frac{1}{\rho^2}RZ\frac{d^2F}{d\psi^2} + RF\frac{d^2Z}{dz^2} = 0.$$

Then

$$\frac{1}{R}\frac{d^2R}{d\rho^2} + \frac{1}{\rho R}\frac{dR}{d\rho} + \frac{1}{\rho^2 F}\frac{d^2F}{d\psi^2} + \frac{1}{Z}\frac{d^2Z}{dz^2} = 0.$$

Hence

$$\frac{1}{\rho R}\frac{1}{d\rho}\left(\rho\frac{dR}{d\rho}\right) + \frac{1}{\rho^2 F}\frac{d^2F}{d\psi^2} + \frac{1}{Z}\frac{d^2Z}{dz^2} = 0.$$

The first two terms are independent of z while the third term is a function of z only and so must be equal to a constant μ^2 (say). That is

$$\frac{1}{Z}\frac{d^2Z}{dz^2} = \mu^2.$$

Therefore

$$\frac{d^2Z}{dz^2} - \mu^2 Z = 0$$

and hence

$$Z = \begin{cases} Ae^{\mu z} + Be^{-\mu z} & \mu \neq 0, \\ A_0 z + B_0 & \mu = 0 \end{cases}$$

for arbitrary constants A, A_0, B, B_0.

Then

$$\frac{1}{\rho R}\frac{d}{d\rho}\left(\rho\frac{dR}{d\rho}\right) + \frac{1}{\rho^2 F}\frac{d^2F}{d\psi^2} + \mu^2 = 0.$$

Therefore

$$\frac{\rho}{R}\frac{d}{d\rho}\left(\rho\frac{dR}{d\rho}\right) + \rho^2\mu^2 = -\frac{1}{F}\frac{d^2F}{d\psi^2}.$$

Now the left hand side is a function of ρ only while the right hand side is a function of ψ only and so each must be equal to a constant ν^2 (say). Therefore

$$\frac{1}{F}\frac{d^2F}{d\psi^2} = -\nu^2.$$

and so

$$\frac{d^2F}{d\psi^2} + \nu^2 F = 0.$$

Hence

$$F = \begin{cases} C \cos \nu\psi + D \sin \nu\psi & \nu \neq 0, \\ C_0 + D_0\psi & \nu = 0 \end{cases}$$

for arbitrary constants C, C_0, D, D_0.

If F is single valued, then ν must be an integer and D_0 must be zero. Then

$$\frac{\rho}{R}\frac{d}{d\rho}\left(\rho\frac{dR}{d\rho}\right) - \nu^2 + \rho^2\mu^2 = 0.$$

Therefore

$$\rho\frac{d}{d\rho}\left(\rho\frac{dR}{d\rho}\right) + (\mu^2\rho^2 - \nu^2)R = 0.$$

Now if $\mu \neq 0$ put $x = \mu\rho$ and this equation becomes

$$x^2\frac{d^2R}{dx^2} + x\frac{dR}{dx} + (x^2 - \nu^2)R = 0$$

which is Bessel's equation of order ν and has the general solution

$$R = EJ_\nu(x) + GY_\nu(x)$$
$$= EJ_\nu(\mu\rho) + GY_\nu(\mu\rho)$$

for arbitrary constants E and G.

However, since in practical problems φ must be single valued (and so F must be single valued) ν is an integer. But $Y_\nu(\mu\rho)$ becomes infinite when $\rho \to 0$ if ν is an integer and so for practical applications in which R remains finite at $\rho = 0$, $G = 0$ and

$$R = EJ_n(\mu\rho)$$

for integer n.

When $\mu = 0$ the equation for R becomes

$$\rho\frac{d}{d\rho}\left(\rho\frac{dR}{d\rho}\right) - \nu^2R = 0$$

and when $\nu = 0$ this has solution

$$E_0 + G_0 \ln \rho$$

for arbitrary constants E_0 and G_0. Since $\ln \rho$ becomes infinite as $\rho \to 0$ or as $\rho \to \infty$, for practical applications requiring a solution which remains finite as $\rho \to 0$ or as $\rho \to \infty$, $G_0 = 0$.

When $\nu \neq 0$, put $\rho = e^t$ and the differential equation becomes

$$\frac{d^2R}{dt^2} - \nu^2R = 0$$

with general solution

$$R = E_0' e^{vt} + G_0' e^{-vt}$$
$$= E_0' \rho^v + G_0' \rho^{-v}$$

for arbitrary constants E_0' and G_0'.

For practical applications in which R remains finite as $\rho \to 0$ then, if v is positive, $G_0' = 0$. Similarly, for practical applications in which R remains finite as $\rho \to \infty$, $E_0' = 0$. Hence, for given $\mu(\neq 0)$ the general solution (of separated form) for single-valued φ finite at $\rho = 0$ has the form

$$\varphi = (a_0 e^{\mu z} + c_0 e^{-\mu z}) J_0(\mu z)$$
$$+ \sum_{n=1}^{\infty} \{e^{\mu z}(a_n \cos n\psi + b_n \sin n\psi) + e^{-\mu z}(c_n \cos n\psi + d_n \sin n\psi)\} J_n(\mu \rho)$$

which can also be written

$$\varphi = \sum_{n=0}^{\infty} \{e^{\mu z}(a_n \cos n\psi + b_n \sin n\psi) + e^{-\mu z}(c_n \cos n\psi + d_n \sin n\psi)\} J_n(\mu \rho).$$

When $\mu = 0$, the general solution (of separated form) for single-valued φ finite at $\rho = 0$ has the form

$$\varphi = (a_0 + c_0 z) + \sum_{n=1}^{\infty} \{(a_n \cos n\psi + b_n \sin n\psi) + z(c_n \cos n\psi + d_n \sin n\psi)\} \rho^n$$

which can also be written

$$\varphi = \sum_{n=0}^{\infty} \{(a_n \cos n\psi + b_n \sin n\psi) + z(c_n \cos n\psi + d_n \sin n\psi)\} \rho^n.$$

The reader should be able to write down the corresponding expressions for φ when the condition φ finite at $\rho = 0$ is removed.

The arbitrary constants are then determined by the boundary conditions for any particular problem.

Example A hollow right-circular cylinder of height h and radius a has its ends closed by thin discs of radius a. The curved surface and one of the plane ends are earthed while the other plane end is maintained at the constant potential φ_0. Show that the potential φ inside the cylinder is given by

$$\varphi = 2\varphi_0 \sum_{n=1}^{\infty} \frac{\sinh\left(\dfrac{\alpha_n z}{a}\right) J_0\left(\dfrac{\alpha_n \rho}{a}\right)}{\alpha_1 J_1(\alpha_n) \sinh \dfrac{\alpha_n h}{a}}$$

11

where ρ is measured from the axis of the cylinder and z is measured along the axis from the earthed end and α_n *(n = 1,2,3, . . .) are the zeros of the Bessel function* $J_0(x)$.

Choose cylindrical polar coordinates (ρ,ψ,z) with origin at the centre of the earthed end (figure 102).

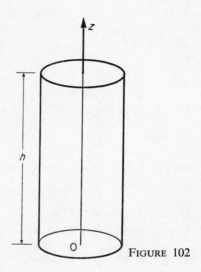

FIGURE 102

The potential φ is independent of ψ and satisfies Laplace's equation in the form

$$\frac{\partial^2 \varphi}{\partial \rho^2} + \frac{1}{\rho}\frac{\partial \varphi}{\partial \rho} + \frac{\partial^2 \varphi}{\partial z^2} = 0.$$

Put $\varphi = R(\rho)Z(z)$. Then we obtain

$$\frac{1}{R}\frac{\mathrm{d}^2 R}{\mathrm{d}\rho^2} + \frac{1}{R\rho}\frac{\mathrm{d}R}{\mathrm{d}\rho} + \frac{1}{Z}\frac{\mathrm{d}^2 Z}{\mathrm{d}z^2} = 0$$

or

$$\frac{1}{R}\frac{\mathrm{d}^2 R}{\mathrm{d}\rho^2} + \frac{1}{R\rho}\frac{\mathrm{d}R}{\mathrm{d}\rho} = -\frac{1}{Z}\frac{\mathrm{d}^2 Z}{\mathrm{d}z^2}.$$

The left hand side is a function of ρ only while the right hand side is a function of z only and so each must be constant.

Therefore put

$$\frac{1}{Z}\frac{\mathrm{d}^2 Z}{\mathrm{d}z^2} = p^2.$$

Hence

$$Z = \begin{cases} Ae^{pz} + Be^{-pz} & p \neq 0, \\ A_0 z + B_0 & p = 0 \end{cases}$$

for arbitrary constants A, A_0, B, B_0. Therefore

$$Z = \begin{cases} (A + B)\left(\dfrac{e^{pz} + e^{-pz}}{2}\right) + (A - B)\left(\dfrac{e^{pz} - e^{-pz}}{2}\right) & p \neq 0, \\ A_0 z + B_0 & p = 0 \end{cases}$$

$$= \begin{cases} C \cosh pz + D \sinh pz & p \neq 0, \\ A_0 z + B_0 & p = 0 \end{cases}$$

where $C = (A + B)$ and $D = (A - B)$ are arbitrary constants. Also,

$$\rho^2 \frac{d^2 R}{d\rho^2} + \rho \frac{dR}{d\rho} + p^2 \rho^2 R = 0.$$

When $p \neq 0$ this is Bessel's equation of order zero and has solution

$$R = EJ_0(p\rho) + FY_0(p\rho)$$

for arbitrary constants E and F. But since we require a solution φ which remains finite when $\rho = 0$, F must be zero. When $p = 0$, the equation for R is

$$\rho^2 \frac{d^2 R}{d\rho^2} + \rho \frac{dR}{d\rho} = 0$$

which has general solution

$$R = E_0 + F_0 \ln \rho.$$

Since φ must remain finite when $\rho = 0$, $F_0 = 0$.
Therefore when $p \neq 0$,

$$\varphi = (C \cosh pz + D \sinh pz)EJ_0(p\rho)$$
$$= (c \cosh pz + d \sinh pz)J_0(p\rho)$$

for constants c and d. When $p = 0$,

$$\varphi = E_0(A_0 z + B_0)$$
$$= c_0 + d_0 z$$

for constants c_0 and d_0. Now $\varphi = 0$ when $z = 0$ and so, when $p \neq 0$,

$$cJ_0(p\rho) = 0 \quad \text{for} \quad 0 < \rho < a.$$

Therefore $c = 0$ and $\varphi = d \sinh pz J_0(p\rho)$ for $p \neq 0$. Also, when $p = 0$, $c_0 = 0$ and so $\varphi = d_0 z$ for $p = 0$. Then $\varphi = 0$ when $\rho = a$ and so, when $p \neq 0$,

$$d \sinh pz \, J_0(pa) = 0 \quad \text{for} \quad 0 < z < h.$$

Therefore $J_0(pa) = 0$ ($d = 0$ gives the trivial solution $\varphi = 0$). Hence $p = \alpha_n/a$ ($n = 1,2,3,\ldots$) where the α_n are the roots of $J_0(x) = 0$. Also, $d_0 = 0$ and so $\varphi = 0$ for $p = 0$. Therefore a general solution (of separated form) satisfying the above boundary conditions is

$$\varphi = \sum_{n=1}^{\infty} d_n \sinh \frac{\alpha_n h}{a} J_0\!\left(\frac{\alpha_n \rho}{a}\right).$$

Now $\varphi = \varphi_0$ when $z = h$ and so

$$\varphi_0 = \sum_{n=1}^{\infty} d_n \sinh \frac{\alpha_n h}{a} J_0\!\left(\frac{\alpha_n \rho}{a}\right).$$

Hence

$$
\begin{aligned}
d_n \sinh\!\left(\frac{\alpha_n h}{a}\right) &= \frac{2}{[aJ_1(\alpha_n)]^2} \int_0^a \rho J_0\!\left(\frac{\alpha_n \rho}{a}\right) \varphi_0 \, d\rho \\
&= \frac{2\varphi_0}{[aJ_1(\alpha_n)]^2} \frac{a^2}{\alpha_n^2} \int_0^{\alpha_n} x J_0(x) \, dx \qquad \left(\text{where } x = \frac{\alpha_n \rho}{a}\right) \\
&= \frac{2\varphi_0}{[\alpha_n J_1(\alpha_n)]^2} \int_0^{\alpha_n} \frac{d}{dx}\,[x J_1(x)] \, dx \\
&= \frac{2\varphi_0}{[\alpha_n J_1(\alpha_n)]^2} \left[x J_1(x) \right]_0^{\alpha_n} \\
&= \frac{2\varphi_0}{\alpha_n J_1(\alpha_n)}.
\end{aligned}
$$

Therefore

$$\varphi = 2\varphi_0 \sum_{n=1}^{\infty} \frac{\sinh \dfrac{\alpha_n z}{a} J_0\!\left(\dfrac{\alpha_n \rho}{a}\right)}{\alpha_n J_1(\alpha_n) \sinh \dfrac{\alpha_n h}{a}}.$$

Exercises

1. Two coaxial and infinite cylindrical conductors of radii a and b ($a > b$) are kept at the potentials φ_a and φ_b respectively. Determine the potential φ at points in the region between the cylinders.

2. The function φ is harmonic inside an infinitely long circular cylinder of cross-sectional radius a. Using cylindrical polar coordinates (ρ,ψ,z) with z-axis along the axis of the cylinder, the function φ at points on the surface of the cylinder is given by

$$
\varphi = \begin{cases}
\varphi_0 & \text{for} \quad 0 \leqslant \psi < \pi/2, \\
-\varphi_0 & \text{for} \quad \pi/2 < \psi < 3\pi/2, \\
\varphi_0 & \text{for} \quad 3\pi/2 < \psi < 2\pi,
\end{cases}
$$

where φ_0 is a constant. Show that φ is given by

$$\varphi = \frac{4\varphi_0}{\pi} \sum_{n=1}^{\infty} \frac{(-1)^{n-1}}{2n-1} \left(\frac{\rho}{a}\right)^{2n-1} \cos(2n-1)\psi.$$

3. An infinite circular cylinder of radius a and dielectric constant κ is placed with its axis perpendicular to a uniform field E in free space. Find expressions for the potential inside and outside the cylinder.

4. Two infinitely long, coaxial, circular cylinders have radii a and b respectively with $b > a$. The material between the cylinders has permittivity ε and elsewhere there is free space. A uniform electric field of intensity E is applied with its direction perpendicular to the common axis of the cylinders. Find the resulting potential in the region outside the larger cylinder.

6.4 Solution using spherical polar coordinates (r,θ,ψ) (spherical harmonics)

Solutions of Laplace's equation in spherical polar coordinates are called *spherical harmonics*.

Laplace's equation is

$$\nabla^2\varphi = \frac{\partial^2\varphi}{\partial r^2} + \frac{2}{r}\frac{\partial\varphi}{\partial r} + \frac{1}{r^2}\frac{\partial^2\varphi}{\partial\theta^2} + \frac{1}{r^2}\cot\theta\frac{\partial\varphi}{\partial\theta} + \frac{1}{r^2\sin^2\theta}\frac{\partial^2\varphi}{\partial\psi^2} = 0.$$

(see Chapter 1).

Now in many practical problems there is symmetry about the axis $\theta = 0$ and so the solution φ is independent of the angle ψ. We shall only consider this important special case. The equation to be solved is then

$$\frac{\partial^2\varphi}{\partial r^2} + \frac{2}{r}\frac{\partial\varphi}{\partial r} + \frac{1}{r^2}\frac{\partial^2\varphi}{\partial\theta^2} + \frac{1}{r^2}\cot\theta\frac{\partial\varphi}{\partial\theta} = 0.$$

Put $\varphi = R(r)F(\theta)$.
Then

$$F\frac{d^2R}{dr^2} + \frac{2}{r}F\frac{dR}{dr} + \frac{1}{r^2}R\frac{d^2F}{d\theta^2} + \frac{1}{r^2}\cot\theta\,R\frac{dF}{d\theta} = 0.$$

Therefore

$$\frac{1}{R}\left(r^2\frac{d^2R}{dr^2} + 2r\frac{dR}{dr}\right) + \frac{1}{F}\left(\frac{d^2F}{d\theta^2} + \cot\theta\frac{dF}{d\theta}\right) = 0.$$

The first term is a function of r only and the second term is a function of θ only. Therefore each must be constant.
Put

$$\frac{1}{R}\left(r^2\frac{d^2R}{dr^2} + 2r\frac{dR}{dr}\right) = -k.$$

Therefore

$$r^2 \frac{d^2R}{dr^2} + 2r \frac{dR}{dr} + kR = 0.$$

Now put $r = e^t$ and this equation becomes

$$\frac{d^2R}{dt^2} + \frac{dR}{dt} + kR = 0.$$

The auxiliary equation is $p^2 + p + k = 0$ and so the sum of the roots is -1.

Therefore denote the roots by n and $-(n + 1)$. (Note that effectively this means writing k in the form $-n(n + 1)$.)

Hence we obtain

$$R = Ae^{nt} + Be^{-(n+1)t}$$

$$= Ar^n + \frac{B}{r^{n+1}}$$

for arbitrary constants A and B.

Also, the equation for the function $F(\theta)$ is

$$\frac{1}{F}\left(\frac{d^2F}{d\theta^2} + \cot \theta \frac{dF}{d\theta}\right) + n(n + 1) = 0$$

or

$$\frac{1}{\sin \theta}\frac{d}{d\theta}\left(\sin \theta \frac{dF}{d\theta}\right) + n(n + 1)F = 0.$$

If now we make the substitution $x = \cos \theta$ this equation reduces to

$$\frac{d}{dx}\left\{(1 - x^2)\frac{dF}{dx}\right\} + n(n + 1)F = 0$$

or

$$(1 - x^2)\frac{d^2F}{dx^2} - 2x\frac{dF}{dx} + n(n + 1)F = 0$$

which is Legendre's equation of degree n. In most practical problems n is integral* and so the solution of this equation is

$$F = CP_n(x) + DQ_n(x)$$

$$= CP_n(\cos \theta) + DQ_n(\cos \theta)$$

for arbitrary constants C and D.

---

* It can be shown that n must be integral for the solution of the equation to be continuous along with its first order derivative in the interval $-1 \leqslant x \leqslant 1$, that is in the interval $0 \leqslant \theta \leqslant \pi$ when $x = \cos \theta$.

Now $Q_n(\cos\theta)$ is only bounded for $|\cos\theta| < 1$, that is it is unbounded along the axis of symmetry for there $|\cos\theta| = 1$. Hence for a solution finite along the axis of symmetry $D = 0$ and

$$F = CP_n(\cos\theta).$$

Therefore a solution of Laplace's equation which is finite along the axis and also symmetric about this axis is

$$\varphi = \left(Ar^n + \frac{B}{r^{n+1}}\right)CP_n(\cos\theta)$$

$$= \left(ar^n + \frac{b}{r^{n+1}}\right)P_n(\cos\theta)$$

where a and b are arbitrary constants. A more general solution satisfying the above conditions is then

$$\varphi = \sum_{n=0}^{\infty}\left(a_n r^n + \frac{b_n}{r^{n+1}}\right)P_n(\cos\theta).$$

The arbitrary constants are then determined by the boundary conditions of particular problems. For example if $\varphi \to 0$ as $r \to \infty$ then the b_n's must be zero, or if φ remains finite as $r \to \infty$ then the a_n's are zero ($n > 0$).

Example An incompressible, inviscid fluid is flowing with uniform velocity U in a fixed direction. If a sphere of radius a is now held fixed in the fluid, determine the velocity potential for the resulting fluid flow.

The resulting flow is obviously symmetric about an axis in the direction of the undisturbed flow and passing through the centre of the sphere. Choose spherical polar coordinates (r,θ,ψ) with origin at the centre of the sphere and axis $\theta = 0$ as the axis of symmetry. Then if φ is the required velocity potential, we have that φ satisfies Laplace's equation, is independent of the spherical polar coordinate ψ and is finite along the axis $\theta = 0$.

Hence take φ in the form

$$\varphi = \sum_{n=0}^{\infty}\left(A_n r^n + \frac{B_n}{r^{n+1}}\right)P_n(\cos\theta)$$

for arbitrary constants A_n and B_n. Now at very large distances from the sphere, the effect of the sphere will be negligible, that is the flow will approximate to the undisturbed flow. Thus for very large r the flow approximates uniform flow with velocity U in the direction of $\theta = 0$ and the velocity potential for this flow is $-Ur\cos\theta$. (Then $\mathbf{q} = -\text{grad}\,(-Ur\cos\theta) = (U\cos\theta, -U\sin\theta, 0)$.) In other words, as $r \to \infty$, $\varphi \to -Ur\cos\theta$. Hence $A_1 = -U$ and $A_n = 0$ for $n \neq 1$.

Therefore

$$\varphi = -Ur\cos\theta + \sum_{n=0}^{\infty} \frac{B_n}{r^{n+1}} P_n(\cos\theta).$$

Now since no fluid flows out of (or into) the sphere, the radial component of the velocity of the fluid in contact with the surface of the sphere must be zero. Therefore

$$-\left(\frac{\partial\varphi}{\partial r}\right)_{r=a} = 0.$$

Hence

$$U\cos\theta + \sum_{n=0}^{\infty} \frac{(n+1)B_n}{a^{n+2}} P_n(\cos\theta) = 0.$$

Then, on comparing coefficients of $P_n(\cos\theta)$ for different values of n in this equation we see that

$$\frac{2B_1}{a^3} = -U \quad \text{and} \quad B_n = 0 \quad \text{for } n \neq 1.$$

Therefore

$$\varphi = -Ur\cos\theta - \frac{Ua^3}{2}\frac{\cos\theta}{r^2}$$

$$= -U\left(r + \frac{a^3}{2r^2}\right)\cos\theta.$$

Example   A spherical surface of radius a has a given potential distribution maintained on it. This potential distribution is symmetrical about a diameter of the sphere and is such that if the radius to any point P (figure 103) of the surface makes an angle θ with the axis of symmetry then the potential at P is $\varphi_0 \sin^2\theta$ where φ_0 is a constant. Determine the potential distribution for the regions inside and outside the surface.

Choose spherical polar coordinates with $\theta = 0$ as the axis of symmetry.
Let φ_1 and φ_2 be the potentials inside and outside the surface respectively.
Then φ_1 and φ_2 satisfy Laplace's equation, are independent of the spherical polar coordinate ψ, and are finite along the axis $\theta = 0$.

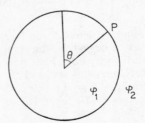

FIGURE 103

φ_1 is finite for $r = 0$ and so has the general form

$$\varphi_1 = \sum_{n=0}^{\infty} A_n r^n P_n(\cos\theta).$$

φ_2 tends to zero as r tends to infinity and so has the general form

$$\varphi_2 = \sum_{n=0}^{\infty} \frac{B_n}{r^{n+1}} P_n(\cos\theta).$$

In addition, φ is continuous on the spherical surface $r = a$ and so

$$\varphi_1 = \varphi_2 = \varphi_0 \sin^2\theta \text{ when } r = a.$$

Therefore

$$\sum_{n=0}^{\infty} A_n a^n P_n(\cos\theta) = \varphi_0 \sin^2\theta = \varphi_0(1 - \cos^2\theta).$$

The function $(1 - \cos^2\theta)$ must now be expanded as a series of Legendre polynomials in the form

$$(1 - \cos^2\theta) = \alpha P_0 + \beta P_1 + \gamma P_2$$
$$= \alpha + \beta\cos\theta + \gamma\tfrac{1}{2}(3\cos^2\theta - 1).$$

Hence $1 = \alpha - \tfrac{1}{2}\gamma$, $0 = \beta$ and $-1 = \tfrac{3}{2}\gamma$.

Therefore

$$(1 - \cos^2\theta) = \tfrac{2}{3}P_0 - \tfrac{2}{3}P_2.$$

Thus we now have

$$\sum_{n=0}^{\infty} A_n a^n P_n(\cos\theta) = \tfrac{2}{3}\varphi_0 P_0(\cos\theta) - \tfrac{2}{3}\varphi_0 P_2(\cos\theta)$$

from which we obtain

$$A_0 = \tfrac{2}{3}\varphi_0, \quad A_2 a^2 = -\tfrac{2}{3}\varphi_0 \quad \text{and} \quad A_n = 0 \qquad \text{(for } n \neq 0,2\text{)}.$$

Hence

$$\varphi_1 = \tfrac{2}{3}\varphi_0\left(P_0(\cos\theta) - \frac{r^2}{a^2}P_2(\cos\theta)\right)$$
$$= \tfrac{2}{3}\varphi_0\left\{1 - \frac{1}{2}\frac{r^2}{a^2}(3\cos^2\theta - 1)\right\}.$$

Also, since $\varphi_2 = \varphi_0 \sin^2\theta$ when $r = a$,

$$\sum_{n=0}^{\infty} \frac{B_n}{a^{n+1}} P_n(\cos\theta) = \tfrac{2}{3}\varphi_0 P_0(\cos\theta) - \tfrac{2}{3}\varphi_0 P_2(\cos\theta).$$

Therefore

$$\frac{B_0}{a} = \tfrac{2}{3}\varphi_0, \quad \frac{B_2}{a^3} = -\tfrac{2}{3}\varphi_0 \quad \text{and} \quad B_n = 0 \qquad \text{(for } n \neq 0,2\text{)}.$$

Hence

$$\varphi_2 = \tfrac{2}{3}\varphi_0\left(\frac{a}{r} P_0(\cos\theta) - \frac{a^3}{r^3} P_2(\cos\theta)\right)$$

$$= \tfrac{2}{3}\varphi_0\left\{\frac{a}{r} - \frac{1}{2}\frac{a^3}{r^3}(3\cos^2\theta - 1)\right\}.$$

Example   *A unit charge is placed in free space at* A *a distance c from the centre of a solid sphere of radius* $a < c$ *and dielectric constant* κ. *Obtain the induced potential field outside the sphere.*

Choose spherical polar coordinates (r,θ,ψ) with origin at O the centre of the sphere and OA as the axis $\theta = 0$ (figure 104).

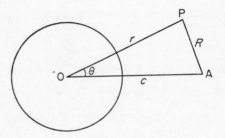

FIGURE 104

The problem is obviously symmetric about the axis $\theta = 0$ and so the potential φ must be independent of the coordinate ψ.

Let φ_i and φ_0 be the potentials inside and outside the sphere respectively, and write φ_0 in the form $1/(4\pi\varepsilon_0 R) + \varphi_1$ where R is radial distance from the unit charge at A and φ_1 is the potential due to the induced charge on the sphere.

Then both φ_i and φ_1 satisfy Laplace's equation, are independent of the coordinate ψ and are finite along the axis $\theta = 0$. Hence they may be written in the form

$$\varphi_i = \sum_{n=0}^{\infty}\left(A_n r^n + \frac{B_n}{r^{n+1}}\right)P_n(\cos\theta)$$

and

$$\varphi_1 = \sum_{n=0}^{\infty}\left(C_n r^n + \frac{D_n}{r^{n+1}}\right)P_n(\cos\theta).$$

But φ_i remains finite at $r = 0$ and so $B_n = 0$ for $n = 0, 1, 2, \ldots$. Also, $\varphi_1 \to 0$ as $r \to \infty$ and so $C_n = 0$ for $n = 0, 1, 2, \ldots$.

Therefore

$$\varphi_i = \sum_{n=0}^{\infty} A_n r^n P_n(\cos\theta)$$

and

$$\varphi_0 = \frac{1}{4\pi\varepsilon_0 R} + \sum_{n=0}^{\infty} \frac{D_n}{r^{n+1}} P_n(\cos\theta).$$

Now φ is continuous on the surface of the sphere and so

$$\sum_{n=0}^{\infty} A_n a^n P_n(\cos\theta) = \frac{1}{4\pi\varepsilon_0 (c^2 + a^2 - 2ac\cos\theta)^{\frac{1}{2}}} + \sum_{n=0}^{\infty} \frac{D_n}{a^{n+1}} P_n(\cos\theta)$$

since for a point on the surface of the sphere $r = a$ and $R^2 = c^2 + a^2 - 2ac\cos\theta$.
But

$$\frac{1}{(c^2 + a^2 - 2ac\cos\theta)^{\frac{1}{2}}} = \frac{1}{c} \sum_{n=0}^{\infty} \left(\frac{a}{c}\right)^n P_n(\cos\theta) \quad \text{for} \quad a > c.$$

(See Chapter 5 p. 278.)
Therefore

$$\sum_{n=0}^{\infty} A_n a^n P_n(\cos\theta) = \sum_{n=0}^{\infty} \left(\frac{1}{4\pi\varepsilon_0 c}\left(\frac{a}{c}\right)^n + \frac{D_n}{a^{n+1}}\right) P_n(\cos\theta).$$

Hence, equating coefficients of P_n in this equation, we obtain

$$A_n = \frac{1}{4\pi\varepsilon_0 c^{n+1}} + \frac{D_n}{a^{2n+1}}.$$

Also, $\kappa\, \partial\varphi_i/\partial r = \partial\varphi_0/\partial r$ on the surface of the sphere. Therefore

$$\kappa \sum_{n=1}^{\infty} n A_n a^{n-1} P_n(\cos\theta) = \frac{1}{4\pi\varepsilon_0} \left\{\frac{\partial}{\partial r}\left(\frac{1}{R}\right)\right\}_{r=a} - \sum_{n=0}^{\infty} \frac{(n+1)D_n}{a^{n+2}} P_n(\cos\theta).$$

But, for $r < c$,

$$\frac{1}{R} = \frac{1}{c} \sum_{n=0}^{\infty} \left(\frac{r}{c}\right)^n P_n(\cos\theta).$$

(See Chapter 5 p. 278.)
Therefore

$$\frac{\partial}{\partial r}\left(\frac{1}{R}\right) = \frac{1}{c} \sum_{n=1}^{\infty} \frac{n}{c}\left(\frac{r}{c}\right)^{n-1} P_n(\cos\theta).$$

Therefore

$$\left\{\frac{\partial}{\partial r}\left(\frac{1}{R}\right)\right\}_{r=a} = \frac{1}{c} \sum_{n=1}^{\infty} \frac{n}{c}\left(\frac{a}{c}\right)^{n-1} P_n(\cos\theta) \quad \text{for} \quad a < c.$$

Hence

$$\kappa \sum_{n=1}^{\infty} n A_n a^{n-1} P_n(\cos\theta)$$

$$= \sum_{n=1}^{\infty} \frac{1}{4\pi\varepsilon_0} \frac{n}{c^2}\left(\frac{a}{c}\right)^{n-1} P_n(\cos\theta) - \sum_{n=0}^{\infty} \frac{(n+1)D_n}{a^{n+2}} P_n(\cos\theta).$$

Then, equating coefficients of $P_n(\cos \theta)$ in this equation, we obtain

$$\kappa n A_n = \frac{n}{4\pi\varepsilon_0 c^{n+1}} - \frac{(n+1)D_n}{a^{2n+1}}.$$

(Note that this relation also holds for $n = 0$.)

Eliminating A_n leads to

$$\kappa n \left(\frac{1}{4\pi\varepsilon_0 c^{n+1}} + \frac{D_n}{a^{2n+1}} \right) = \frac{n}{4\pi\varepsilon_0 c^{n+1}} - \frac{(n+1)D_n}{a^{2n+1}}$$

and so

$$D_n = \frac{-n(\kappa - 1)}{4\pi\varepsilon_0(\kappa n + n + 1)} \left(\frac{a}{c} \right)^{n+1} a^n.$$

Therefore

$$\varphi_0 = \frac{1}{4\pi\varepsilon_0 R} - \sum_{n=1}^{\infty} \frac{n(\kappa - 1)a^n}{4\pi\varepsilon_0(\kappa n + n + 1)} \left(\frac{a}{cr} \right)^{n+1} P_n(\cos \theta).$$

Hence the induced potential is

$$- \sum_{n=1}^{\infty} \frac{n(\kappa - 1)a^n}{4\pi\varepsilon_0(\kappa n + n + 1)} \left(\frac{a}{cr} \right)^{n+1} P_n(\cos \theta).$$

Note that when $\kappa = 1$ (that is the dielectric sphere is removed), this induced potential reduces to zero as expected.

Example   A spherical shell of internal radius a and external radius b is made of uniform material of permeability μ. If this shell is placed in a uniform magnetic field $\mathbf{H_0}$ show that the resulting magnetic field inside the hollow enclosed by the shell is

$$\frac{9\mu \mathbf{H_0}}{(2 + \mu)(1 + 2\mu) - 2(1 - \mu)^2(a/b)^3}.$$

The problem is obviously symmetric about the axis in the direction of $\mathbf{H_0}$ through the centre of the shell. Choose spherical polar coordinates (r, θ, ψ) with $\theta = 0$ as the axis of symmetry and let Ω_1, Ω_2 and Ω_3 be the magnetic potentials in the hollow enclosed by the shell, in the material of the shell itself and in the region outside the shell respectively (figure 105).

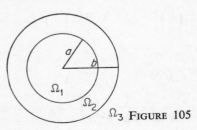

Ω_3 FIGURE 105

Then Ω_1, Ω_2 and Ω_3 satisfy Laplace's equation, are independent of ψ and are finite along the axis $\theta = 0$. Ω_1 remains finite for $r = 0$ and so has the general form

$$\Omega_1 = \sum_{n=0}^{\infty} A_n r^n P_n(\cos \theta).$$

Ω_2 has the general form

$$\Omega_2 = \sum_{n=0}^{\infty} \left(B_n r^n + \frac{C_n}{r^{n+1}} \right) P_n(\cos \theta).$$

Ω_3 can be written as $\Omega_0 + \Omega_i$ where Ω_0 is the potential of the uniform field H_0 (that is $H_0 = -\text{grad } \Omega_0$) and Ω_i is the potential induced due to the presence of the shell and so must tend to zero as r tends to infinity.

Hence Ω_3 has the general form

$$\Omega_3 = -H_0 r \cos \theta + \sum_{n=0}^{\infty} \frac{D_n}{r^{n+1}} P_n(\cos \theta).$$

The boundary conditions to be satisfied arise from the continuity of Ω and $\mu(\partial \Omega / \partial r)$ on the inner and outer surfaces of the shell and thus take the mathematical form:

1. $\Omega_1 = \Omega_2$ when $r = a$.
2. $\Omega_2 = \Omega_3$ when $r = b$.
3. $\partial \Omega_1 / \partial r = \mu(\partial \Omega_2 / \partial r)$ when $r = a$.
4. $\mu(\partial \Omega_2 / \partial r) = \partial \Omega_3 / \partial r$ when $r = b$.

From 1 we obtain

$$\sum_{n=0}^{\infty} A_n a^n P_n(\cos \theta) = \sum_{n=0}^{\infty} \left(B_n a^n + \frac{C_n}{a^{n+1}} \right) P_n(\cos \theta)$$

and so

$$A_n a^n = B_n a^n + \frac{C_n}{a^{n+1}}.$$

From 2 we obtain

$$\sum_{n=0}^{\infty} \left(B_n b^n + \frac{C_n}{b^{n+1}} \right) P_n(\cos \theta) = -H_0 b \cos \theta + \sum_{n=0}^{\infty} \frac{D_n}{b^{n+1}} P_n(\cos \theta)$$

and so

$$B_1 b + \frac{C_1}{b^2} = -H_0 b + \frac{D_1}{b^2}$$

(since $P_1(\cos \theta) = \cos \theta$) and

$$B_n b^n + \frac{C_n}{b^{n+1}} = \frac{D_n}{b^{n+1}}$$

(for $n \neq 1$.)

From 3 we obtain

$$\sum_{n=0}^{\infty} n A_n a^{n-1} P_n(\cos \theta) = \mu \sum_{n=0}^{\infty} \left\{ n B_n a^{n-1} - (n+1) \frac{C_n}{a^{n+2}} \right\} P_n(\cos \theta)$$

and so

$$nA_n a^{n-1} = \mu \left\{ nB_n a^{n-1} - (n+1) \frac{C_n}{a^{n+2}} \right\}.$$

From 4 we obtain

$$\mu \sum_{n=0}^{\infty} \left\{ nB_n b^{n-1} - (n+1) \frac{C_n}{b^{n+2}} \right\} P_n(\cos \theta)$$

$$= -H_0 \cos \theta - \sum_{n=0}^{\infty} (n+1) \frac{D_n}{b^{n+2}} P_n(\cos \theta)$$

and so

$$\mu B_1 - 2\mu \frac{C_1}{b^3} = -H_0 - \frac{2D_1}{b^3}$$

and

$$\mu n B_n b^{n-1} - \mu(n+1) \frac{C_n}{b^{n+2}} = -\frac{(n+1)D_n}{b^{n+2}}$$

for $n \neq 1$. Hence again we have the two systems of equations

$$\left.\begin{aligned} a^n A_n \quad - a^n B_n \quad\quad - \frac{C_n}{a^{n+1}} \quad\quad\quad\quad &= 0 \\ b^n B_n \quad\quad + \frac{C_n}{b^{n+1}} \quad - \frac{D_n}{b^{n+1}} &= 0 \\ na^{n-1}A_n - \mu n a^{n-1}B_n + \mu(n+1) \frac{C_n}{a^{n+2}} \quad\quad\quad\quad &= 0 \\ \mu n b^{n-1}B_n - \mu(n+1) \frac{C_n}{b^{n+2}} + (n+1) \frac{D_n}{b^{n+2}} &= 0 \end{aligned}\right\} \quad \text{for} \quad n \neq 1$$

and

$$\left.\begin{aligned} aA_1 - aB_1 - \frac{C_1}{a^2} \quad\quad\quad\quad &= 0 \\ bB_1 + \frac{C_1}{b^2} \quad - \frac{D_1}{b^2} &= -H_0 b \\ A_1 - \mu B_1 + \frac{2\mu C_1}{a^3} \quad\quad\quad\quad &= 0 \\ \mu B_1 - \frac{2\mu C_1}{b^3} + 2\frac{D_1}{b^3} &= -H_0 \end{aligned}\right\}.$$

The only solution of the first system is $A_n = B_n = C_n = D_n = 0$ (for $n \neq 1$). Solving the second system gives

$$A_1 = \frac{-9\mu H_0}{(2+\mu)(1+2\mu) - 2(\mu-1)^2(a/b)^3}.$$

Therefore

$$\Omega_1 = A_1 r P_1(\cos\theta)$$

$$= \frac{-9\mu H_0 r \cos\theta}{(2+\mu)(1+2\mu) - 2(1-\mu)^2(a/b)^3}.$$

Hence the magnetic field in the region enclosed by the inner surface of the shell is

$$\mathbf{H}_1 = -\operatorname{grad}\Omega_1 = \frac{9\mu\mathbf{H}_0}{(2+\mu)(1+2\mu) - 2(1-\mu)^2(a/b)^3}.$$

Example   Obtain the gravitational potential due to a uniform thin ring of radius a and total mass m.

The problem is symmetric about the axis of the ring. Choose spherical polar coordinates (r,θ,ψ) with the axis $\theta = 0$ coinciding with the axis of the ring.

Then the gravitational potential φ satisfies Laplace's equation (except on the ring), is independent of ψ and is finite along the axis $\theta = 0$.

Hence φ has the general form

$$\varphi = \sum_{n=0}^{\infty}\left(A_n r^n + \frac{B_n}{r^{n+1}}\right)P_n(\cos\theta).$$

Now it is easily seen (see Chapter 2) that the potential at a point distant r along the axis from the centre of the ring is

$$\frac{\gamma m}{\sqrt{r^2 + a^2}}.$$

Thus $\varphi = \gamma m/\sqrt{r^2 + a^2}$ when $\theta = 0$ and so, since $P_n(\cos\theta) = P_n(1) = 1$ when $\theta = 0$ we have

$$\sum_{n=0}^{\infty}\left(A_n r^n + \frac{B_n}{r^{n+1}}\right) = \frac{\gamma m}{\sqrt{r^2 + a^2}}.$$

But using the binomial theorem we obtain

$$\frac{\gamma m}{\sqrt{r^2 + a^2}} = \begin{cases} \dfrac{\gamma m}{a}\left(1 - \dfrac{1}{2}\dfrac{r^2}{a^2} + \dfrac{3}{8}\dfrac{r^4}{a^4} + \ldots\right) & \text{for } r < a, \\[3mm] \dfrac{\gamma m}{r}\left(1 - \dfrac{1}{2}\dfrac{a^2}{r^2} + \dfrac{3}{8}\dfrac{a^4}{r^4} + \ldots\right) & \text{for } r > a. \end{cases}$$

Hence for $r < a$,

$$B_n = 0,$$

$$A_0 = \frac{\gamma m}{a}, \qquad A_2 = -\frac{1}{2}\frac{\gamma m}{a^3}, \qquad A_4 = \frac{3}{8}\frac{\gamma m}{a^5}, \ldots$$

$$A_n = 0 \qquad \text{for odd integers } n.$$

Therefore

$$\varphi = \frac{\gamma m}{a}\left(1 - \frac{1}{2}\frac{r^2}{a^2}P_2(\cos\theta) + \frac{3}{8}\frac{r^4}{a^4}P_4(\cos\theta) + \cdots\right)$$

for $r < a$. For $r > a$,

$$A_n = 0$$
$$B_0 = \gamma m, \qquad B_2 = -\tfrac{1}{2}\gamma ma^2, \qquad B_4 = \tfrac{3}{8}\gamma ma^4, \cdots$$
$$B_n = 0 \qquad \text{for odd integers } n.$$

Therefore

$$\varphi = \gamma m\left(\frac{1}{r} - \frac{1}{2}\frac{a^2}{r^3}P_2(\cos\theta) + \frac{3}{8}\frac{a^4}{r^5}P_4(\cos\theta) + \cdots\right)$$

for $r > a$. Note that the obvious conditions $\varphi = \gamma m/a$ when $r = 0$ and $\varphi \to 0$ as $r \to \infty$ are satisfied.

The two series for φ are continuous on $r = a$ and either will suffice to give the potential on the surface, except when $\theta = \frac{1}{2}\pi$ (that is on the ring itself) where the potential is of course infinite and does not satisfy Laplace's equation.

Exercises

1. A spherical shell of material of dielectric constant κ has internal radius a and external radius b. The shell is placed, with a point charge e at its centre, in free space. Determine the form of the potential at points outside the shell and at points within the material of the shell itself.

2. A uniform solid sphere of radius a is surrounded by an infinite expanse of incompressible, inviscid fluid. If when the fluid is everywhere at rest the sphere suddenly starts to move with constant velocity U in a straight line, the problem of finding the initial velocity potential φ in the fluid reduces to solving Laplace's equation $\nabla^2\varphi = 0$ subject to the following boundary conditions

   (i) φ is independent of the spherical polar coordinate ψ
   (ii) $\varphi \to 0$ as $r \to \infty$ where r is distance measured from the initial position of the centre of the sphere.
   (iii) $(\partial\varphi/\partial r)_{r=a} = U\cos\theta$ where θ is the angle which the radius vector from the initial position of the centre of the sphere to the general point in the fluid makes with the direction of motion of the sphere.

Show that

$$\varphi = -\frac{1}{2}\frac{a^3}{r^2}U\cos\theta.$$

3. Find the potential due to a surface distribution of charge of density $k\cos^2\theta$ on the spherical surface $r = a$, where k is a constant and r and θ are spherical polar coordinates relative to the centre of the spherical surface.

4. A charge e is placed in free space at a distance c from the centre of a sphere of radius a ($< c$) and dielectric constant κ. Prove that the charge is attracted towards the sphere by a force

$$\frac{\kappa - 1}{4\pi\varepsilon_0} \cdot \frac{e^2}{c^2} \sum_{n=1}^{\infty} \frac{n(n + 1)}{1 + n(\kappa + 1)} \left(\frac{a}{c}\right)^{2n+1}.$$

5. An infinite, homogeneous material of permittivity ε_1 contains a spherical cavity of radius a filled with a homogeneous material of permittivity ε_2. A point charge e is placed in the spherical cavity at a distance b from its centre ($b < a$). Show that the potential in the region exterior to the cavity is

$$\frac{e}{4\pi r} \sum_{n=0}^{\infty} \frac{(2n + 1)P_n(\cos \theta)}{\{(n + 1)\varepsilon_1 + n\varepsilon_2\}} \left(\frac{b}{r}\right)^n,$$

where r is measured from the centre of the cavity and θ is the angle between the radius vector and the diameter through the point charge. Find the force acting on the point charge.

6. One half of a spherical surface of radius a is earthed while the other half is maintained at the constant potential φ_0. Show that, choosing a suitable system of spherical polar coordinates (r, θ, ψ), the potential φ inside the spherical surface is given by

$$\varphi = \varphi_0 \left(\frac{1}{2} + \frac{3}{4}\frac{r}{a} P_1(\cos \theta) - \frac{7}{16}\frac{r^3}{a^3} P_3(\cos \theta) + \frac{11}{32}\frac{r^5}{a^5} P_5(\cos \theta) + \ldots\right).$$

7. Obtain the gravitational potential due to a thin disc of radius a and total mass m.

8. A uniform sphere of radius a and dielectric constant κ is placed in a uniform electrostatic field. Find the potential at all points of the field and show that if the direction of the given uniform field is taken as the z-axis, then the lines of force form surfaces of revolution with equations of the form

$$\left(1 + \frac{2a^3(\mu - 1)}{r^3(\mu + 1)}\right)(x^2 + y^2) = \text{constant},$$

where $r^2 = x^2 + y^2 + z^2$.

9. A magnetic dipole of moment $\mathbf{m}$ is placed at the centre of a spherical hollow of radius a in an infinite homogeneous magnetic material of constant permeability μ. Obtain expressions for the magnetostatic potential at all points, assuming that it vanishes at infinity.

10. A magnetic dipole of moment $\mathbf{m}$ is in free space at the centre of a spherical shell of internal radius a and external radius b. If the material of the shell has

permeability μ and the region outside the shell is free space, determine the magnetic potential outside the shell.

11. An earthed conducting sphere of radius a is surrounded by a spherical shell of internal radius a and external radius b and dielectric constant κ. The whole system is placed in a uniform field of strength E. Obtain an expression for the electrostatic potential in the material of the shell and show that the charge induced on the sphere consists of equal positive and negative amounts

$$\frac{9\pi\kappa\varepsilon_0 Ea^2b^3}{[(\kappa + 2)b^3 + 2(\kappa - 1)a^3]}.$$

7

The heat conduction equation

7.1 Introduction

The temperature φ within a body forms a continuous scalar field. Through any point of the body passes a surface over which the temperature is constant. Such a surface is called an isothermal surface. At each point the temperature has a gradient grad φ in the direction normal to the isothermal surface at that point. This gradient defines a continuous vector field.

We shall assume that the direction of flow of heat at any point P is normal to the isothermal surface at that point (that is in the direction of grad φ) and that the rate of flow of heat per unit area across the isothermal surface at P is proportional to the temperature gradient at P.

Thus the vector $-k$ grad φ has the direction of the flow and magnitude equal to the quantity of heat per unit area per unit time crossing the isothermal surface at P. The scalar $k(> 0)$ measures the (thermal) conductivity of the body and may be a constant or a function of position. The negative sign is required since heat flows from points at higher to points at lower temperature.

Now consider an arbitrary region V of volume v completely enclosed by a surface S.

The total quantity of heat leaving the region V across the bounding surface S in unit time is equal to

$$\int_S (-k \text{ grad } \varphi) . \text{ds}.$$

Hence the rate at which heat is entering V is

$$\int_S k \text{ grad } \varphi . \text{ds} = \int_V \text{div} (k \text{ grad } \varphi) \, dv.$$

325

But the rate of absorption of heat by the region V is

$$\int_V c\rho \frac{\partial \varphi}{\partial t} \, dv$$

where c is the specific heat of the body and ρ its density. Therefore, assuming that there are no heat sources or sinks within the region V, we must have

$$\int_V \text{div} \, (k \, \text{grad} \, \varphi) \, dv = \int_V c\rho \frac{\partial \varphi}{\partial t} \, dv.$$

Now the region V was chosen arbitrarily and so the above relation must hold for every region in the body and hence the relation

$$\text{div} \, (k \, \text{grad} \, \varphi) = c\rho \frac{\partial \varphi}{\partial t}$$

must be satisfied at all points in the body.

If the body is homogeneous k is constant and this equation reduces to

$$k\nabla^2\varphi = c\rho \frac{\partial \varphi}{\partial t}$$

or

$$K\nabla^2\varphi = \frac{\partial \varphi}{\partial t}$$

where $K = k/c\rho$ is called the (thermal) diffusivity and, in terms of S.I. units, has the dimensions metres squared per second. This differential equation is called the *heat conduction equation*. Note that if a steady state has been reached, that is if

$$\frac{\partial \varphi}{\partial t} = 0,$$

then we obtain Laplace's equation

$$\nabla^2\varphi = 0.$$

Note also that, like Laplace's equation, this equation is linear so that if $\varphi_1, \varphi_2, \varphi_3, \ldots$ are all solutions then

$$\alpha_1\varphi_1 + \alpha_2\varphi_2 + \alpha_3\varphi_3 + \cdots$$

where $\alpha_1, \alpha_2, \alpha_3, \ldots$ are arbitrary constants is also a solution.

7.11 Other situations in which an equation of the above form is applicable

We have already seen (Chapter 2) that in soil mechanics the seepage equation takes the above form namely

$$c\nabla^2 p = \frac{\partial p}{\partial t}$$

where c is the coefficient of consolidation, p is the neutral pressure and t is time.

It can also be shown that the current density vector $\mathbf{j}$ satisfies the equation

$$\frac{1}{\lambda\mu}\nabla^2\mathbf{j} = \frac{\partial\mathbf{j}}{\partial t}$$

where μ is the permeability. This equation is known as the *'skin-effect' equation*.

The basic law of diffusion in a motionless medium is known as Fick's law. This states that the direction of diffusional flow at any point P is normal to the equi-concentration surface at that point, that is in the direction of grad C where C is the concentration, and that the rate of diffusional flow per unit area across the equi-concentration surface at P is proportional to the temperature gradient at P.

Thus the vector, $-K$ grad C, has the direction of the flow and magnitude equal to the quantity of diffusing substance per unit area per unit time crossing the equi-concentration surface at P. The scalar $K\,(>0)$ measures the diffusivity of the medium. The negative sign is required since the substance will diffuse from points at higher to points at lower concentration.

Clearly, continuing as in section 7.1 for heat conduction, we obtain the equation

$$K\nabla^2 C = \frac{\partial C}{\partial t}$$

which in this context is known as the *diffusion equation*.

Certain problems in the flow of incompressible, viscous fluids can also be shown to reduce to the solution, for the velocity u, of the equation

$$\nu\nabla^2 u = \frac{\partial u}{\partial t},$$

in which ν is the kinematic viscosity of the fluid.

7.2 Solutions of the heat conduction equation

As has already been observed, when a steady state has been reached the equation reduces to Laplace's equation and some time has already been devoted to solving this equation. We shall concentrate here mainly on obtaining non-steady state solutions, that is solutions φ for which $\partial \varphi / \partial t \neq 0$.

The method of separation of variables, used to obtain solutions of Laplace's equation, is again applicable for suitably shaped boundaries and suitable boundary conditions.

7.21 Cartesian coordinates (x,y,z)

The heat conduction equation is

$$\frac{\partial^2 \varphi}{\partial x^2} + \frac{\partial^2 \varphi}{\partial y^2} + \frac{\partial^2 \varphi}{\partial z^2} = \frac{1}{K} \frac{\partial \varphi}{\partial t}.$$

In particular, for a thin rod lying along the x-axis this reduces to

$$\frac{\partial^2 \varphi}{\partial x^2} = \frac{1}{K} \frac{\partial \varphi}{\partial t}.$$

Now put $\varphi(x,t) = X(x)T(t)$ where $X(x)$ is a function of x only and $T(t)$ is a function of t only.

Therefore

$$T \frac{\mathrm{d}^2 X}{\mathrm{d}x^2} = \frac{1}{K} X \frac{\mathrm{d}T}{\mathrm{d}t}$$

or

$$\frac{1}{X} \cdot \frac{\mathrm{d}^2 X}{\mathrm{d}x^2} = \frac{1}{KT} \frac{\mathrm{d}T}{\mathrm{d}t}.$$

Now the left hand side is a function of x only and the right hand side is a function of t only so each must be a constant. Put this constant equal to $-p^2$. (The choice of the minus sign here is arbitrary and even if chosen wrongly it can be rectified by allowing p to be imaginary. However, as will be seen, in most problems p as chosen above will be real.)

Therefore

$$\frac{1}{X} \frac{\mathrm{d}^2 X}{\mathrm{d}x^2} = -p^2$$

and so

$$\frac{\mathrm{d}^2 X}{\mathrm{d}x^2} + p^2 X = 0.$$

Hence

$$X = \begin{cases} A \cos px + B \sin px & p \neq 0, \\ A_0 x + B_0 & p = 0 \end{cases}$$

for arbitrary constants A, A_0, B and B_0. Also,

$$\frac{1}{T}\frac{dT}{dt} = -p^2 K$$

and this equation has the solution $T = Ce^{-p^2 Kt}$ where C is an arbitrary constant. When $p = 0$, this solution reduces to $T = C$.

Hence the solution for φ may now be written

$$\varphi = X(x)T(t) = \begin{cases} (a \cos px + b \sin px)e^{-p^2 Kt} & p \neq 0, \\ a_0 + b_0 x & p = 0, \end{cases}$$

for arbitrary constants a, a_0, b, b_0.

The boundary conditions of a particular problem then determine values for p and the corresponding arbitrary constants a, b, a_0 and b_0. For example if it is given that the ends of the rod at $x = 0$ and $x = l$ are maintained at zero temperature, then we have

$$\varphi = 0 \quad \text{when} \quad x = 0 \quad \text{and when} \quad x = l \quad \text{for all values of } t.$$

Therefore

$$0 = ae^{-p^2 Kt} \quad \text{for all } t \text{ if } p \neq 0$$

and

$$0 = a_0 \quad \text{for all } t \text{ if } p = 0.$$

Hence

$$a = 0 \quad \text{and} \quad a_0 = 0.$$

Therefore

$$\varphi = \begin{cases} b \sin px\, e^{-p^2 Kt} & p \neq 0, \\ b_0 x & p = 0. \end{cases}$$

Then

$$0 = b \sin pl\, e^{-p^2 Kt} \quad \text{for all } t \text{ if } p \neq 0$$

and

$$0 = b_0 l \quad \text{for all } t \text{ if } p = 0.$$

Therefore

$$b_0 = 0 \quad \text{and} \quad b \sin pl = 0.$$

But if $b = 0$ we obtain merely the trivial solution $\varphi = 0$ for all points of the rod at all times. Therefore, for a non-trivial solution, $\sin pl = 0$.

Hence $pl = n\pi$ for non-zero integer values of n.

Therefore

$$\varphi = b \sin \frac{n\pi x}{l} e^{-(n^2\pi^2 Kt)/l^2}$$

is a solution of the differential equation satisfying the given boundary conditions for all (non-zero) integer values of n. Then, since the differential equation is linear, a general solution of this type (satisfying the boundary conditions used so far) is

$$\varphi = \sum_{n=1}^{\infty} b_n \sin \frac{n\pi x}{l} e^{-(n^2\pi^2 Kt)/l^2}.$$

(There is no need to consider negative integer values for n since

$$b_n' \sin \frac{n\pi x}{l} e^{-(n^2\pi^2 Kt)/l^2} + b_{-n}' \sin\left(-\frac{n\pi x}{l}\right) e^{-(n^2\pi^2 Kt)/l^2}$$

$$= b_n \sin\frac{n\pi x}{l} e^{-(n^2\pi^2 Kt)/l^2}$$

where $b_n = b_n' - b_{-n}'$.)

If now it is also given that initially the temperature along the rod is given by the function $f(x)$, then

$$\varphi = f(x) \quad \text{when} \quad t = 0 \quad \text{and} \quad 0 \leqslant x \leqslant l.$$

Therefore

$$f(x) = \sum_{n=1}^{\infty} b_n \sin \frac{n\pi x}{l} \quad \text{for} \quad 0 \leqslant x \leqslant l.$$

For functions $f(x)$ satisfying the Dirichlet conditions the coefficients b_n can then be determined using Fourier's method. This gives

$$b_n = \frac{2}{l} \int_0^l f(x) \sin \frac{n\pi x}{l} \, dx.$$

The problem is now completely solved as we have an expression for the temperature at any point of the rod at any particular time.

In order to calculate this temperature in a specific practical situation however we are, in general, faced with the problem of evaluating the sum of an infinite series. This means that we are usually only able to obtain an approximation to the result by summing only a finite number of terms. The presence of the factor $-n^2$ in the argument of the exponential function however ensures that the magnitudes of the terms in the series, for most initial distributions $f(x)$, decrease rapidly with increasing n. Thus, unless very many significant figures are required in the result or K is very small or l is very large, it will not normally be required to sum too many terms in the series.

Example A thin uniform rod of length *l* is initially at the constant temperature φ_0. Its ends are then brought to and subsequently maintained at zero temperature. Determine the temperature distribution in the rod at any later time *t*.

As above, we obtain for the temperature $\varphi(x,t)$

$$\varphi = \sum_{n=1}^{\infty} b_n \sin \frac{n\pi x}{l} e^{-(n^2\pi^2 Kt)/l^2}.$$

Now when $t = 0$, $\varphi = \varphi_0$ and so

$$\varphi_0 = \sum_{n=1}^{\infty} b_n \sin \frac{n\pi x}{l} \quad \text{for} \quad 0 \leqslant x \leqslant l.$$

Therefore

$$b_n = \frac{2}{l} \int_0^l \varphi_0 \sin \frac{n\pi x}{l} \, \mathrm{d}x = \frac{2\varphi_0}{l} \left[-\frac{l}{n\pi} \cos \frac{n\pi x}{l} \right]_0^l$$

$$= \frac{2\varphi_0}{n\pi} (1 - \cos n\pi) = \begin{cases} \dfrac{4\varphi_0}{n\pi} & \text{for } n \text{ odd,} \\ 0 & \text{for } n \text{ even.} \end{cases}$$

Therefore

$$\varphi(x,t) = \frac{4\varphi_0}{\pi} \sin \frac{\pi x}{l} e^{-(\pi^2 Kt)/l^2} + \frac{4\varphi_0}{3\pi} \sin \frac{3\pi x}{l} e^{-(9\pi^2 Kt)/l^2} + \ldots$$

$$= \frac{4\varphi_0}{\pi} \sum_{m=1}^{\infty} \frac{1}{2m-1} \sin \frac{(2m-1)\pi x}{l} e^{-(2m-1)^2\pi^2 Kt/l^2}.$$

Note that as $t \to \infty$, $e^{-[(2m-1)^2\pi^2 Kt/l^2]} \to 0$ and $\varphi \to 0$, that is the whole rod has zero temperature. This of course agrees with what would be expected from physical considerations and is the solution of the steady state equation $\partial^2 \varphi / \partial x^2 = 0$ satisfying the given boundary conditions $\varphi = 0$ at $x = 0$ and $x = l$.

Example A thin uniform rod of length *l* is initially at the constant temperature φ_0. One of its ends is subsequently brought to and maintained at zero temperature while its other end is maintained at the temperature φ_0. Determine the temperature distribution in the rod at any later time *t*.

The temperature $\varphi(x,t)$ satisfies the equation

$$\frac{\partial^2 \varphi}{\partial x^2} = \frac{1}{K} \frac{\partial \varphi}{\partial t}.$$

Separating the variables in this equation in the usual way we obtain a solution of the form

$$\varphi = \begin{cases} (a \cos px + b \sin px)e^{-p^2 Kt} & \text{for } p \neq 0, \\ a_0 + b_0 x & \text{for } p = 0. \end{cases}$$

$\varphi = 0$ when $x = 0$ for all *t* and so $a = 0$ and $a_0 = 0$.

Therefore

$$\varphi = \begin{cases} b \sin px \, e^{-p^2 K t} & p \neq 0, \\ b_0 x & p = 0. \end{cases}$$

Now $\varphi = \varphi_0$ when $x = l$ for all t. This condition cannot be satisfied by the solution for $p \neq 0$ and constant b. When $p = 0$ however we have $\varphi_0 = b_0 l$ and so $b_0 = \varphi_0/l$, but the solution $\varphi = (\varphi_0/l)x$ itself cannot be made to satisfy the final condition that $\varphi = \varphi_0$ when $t = 0$ for all x in $0 \leqslant x \leqslant l$. However

$$\varphi = \frac{\varphi_0}{l} x + b \sin px \, e^{-p^2 K t}$$

is a solution of the equation and satisfies the condition $\varphi = \varphi_0$ when $x = l$ for all t if $b \sin pl \, e^{-p^2 K t} = 0$ for all t, that is if $b \sin pl = 0$.

Therefore $\sin pl = 0$ ($b = 0$ as we have seen is not a suitable solution).

Hence $pl = n\pi$ for integer values of n.

Then

$$\varphi = \frac{\varphi_0 x}{l} + b \sin \frac{n\pi x}{l} e^{-(n^2 \pi^2 K t/l^2)}$$

is a solution satisfying the given boundary conditions for all integer values of n.

Thus the general solution of this form is

$$\varphi = \frac{\varphi_0 x}{l} + \sum_{n=1}^{\infty} b_n \sin \frac{n\pi x}{l} e^{-(n^2 \pi^2 K t/l^2)}.$$

(As before it is not necessary to consider negative integer values of n.) Now $\varphi = \varphi_0$ when $t = 0$ and x is in the interval $0 \leqslant x \leqslant l$.

Therefore

$$\varphi_0 = \frac{\varphi_0}{l} x + \sum_{n=1}^{\infty} b_n \sin \frac{n\pi x}{l} \quad \text{for} \quad 0 \leqslant x \leqslant l.$$

Hence

$$\frac{\varphi_0}{l} (l - x) = \sum_{n=1}^{\infty} b_n \sin \frac{n\pi x}{l} \quad \text{for} \quad 0 \leqslant x \leqslant l$$

and so

$$\begin{aligned} b_n &= \frac{2}{l} \int_0^l \frac{\varphi_0}{l} (l - x) \sin \frac{n\pi x}{l} \, dx \\ &= \frac{2\varphi_0}{l^2} \left\{ \left[-\frac{l}{n\pi} \cos \frac{n\pi x}{l} (l - x) \right]_0^l - \frac{l}{n\pi} \int_0^l \cos \frac{n\pi x}{l} \, dx \right\} \\ &= \frac{2\varphi_0}{n\pi l} \left\{ l - \frac{l}{n\pi} \left[\sin \frac{n\pi x}{l} \right]_0^l \right\} = \frac{2\varphi_0}{n\pi}. \end{aligned}$$

Therefore

$$\varphi(x,t) = \frac{\varphi_0}{l} x + \frac{2\varphi_0}{\pi} \sum_{n=1}^{\infty} \frac{1}{n} \sin \frac{n\pi x}{l} e^{-n^2 \pi^2 K t/l^2}.$$

Note that as $t \to \infty$, $e^{-n^2\pi^2 Kt/l^2} \to 0$ and so the solution $\varphi \to (\varphi_0/l)x$. That is temperature varies linearly from 0 at one end to φ_0 at the other. This solution is of course what would be expected physically since the rod is uniform and the ends are maintained at these respective temperatures, and is the solution of the steady state equation $\partial^2\varphi/\partial x^2 = 0$ satisfying the given boundary conditions.

In particular, if the length of the rod is 1 metre, its diffusivity is 0·05 metres squared per second and its initial temperature is 100°C, then its subsequent temperature $\varphi(x,t)$ in degrees centigrade is

$$\varphi(x,t) = \frac{100}{1}x + \frac{200}{\pi}\sum_{n=1}^{\infty}\frac{1}{n}\sin n\pi x\, e^{-0\cdot05\pi^2 n^2 t}.$$

Hence at the midpoint of the rod the temperature $\varphi(\tfrac{1}{2},t)$ as a function of t is given by putting $x = \tfrac{1}{2}$ in this latter equation giving

$$\varphi(\tfrac{1}{2},t) = 50 + \frac{200}{\pi}\sum_{n=1}^{\infty}\frac{1}{n}\sin\frac{n\pi}{2}\,e^{-0\cdot05\pi^2 n^2 t}$$

$$= 50 + \frac{200}{\pi}\{e^{-0\cdot05\pi^2 t} - \tfrac{1}{3}e^{-0\cdot45\pi^2 t} + \tfrac{1}{5}e^{-1\cdot25\pi^2 t} + \ldots\}.$$

Now, using tables for the exponential function, we obtain the following values for the temperature in degrees centigrade correct to two decimal places

$$\varphi(\tfrac{1}{2},1) = 88\cdot62, \qquad \varphi(\tfrac{1}{2},2) = 73\cdot72, \qquad \varphi(\tfrac{1}{2},3) = 64\cdot49,$$
$$\varphi(\tfrac{1}{2},4) = 58\cdot84, \qquad \varphi(\tfrac{1}{2},5) = 55\cdot40, \qquad \varphi(\tfrac{1}{2},10) = 50\cdot46,$$
$$\varphi(\tfrac{1}{2},15) = 50\cdot04, \qquad \varphi(\tfrac{1}{2},20) = 50\cdot00.$$

In calculating these values to the given accuracy it was only necessary to include the second term, $-\tfrac{1}{3}e^{-0\cdot45\pi^2 t}$, in the infinite series in the calculation of the first two values $\varphi(\tfrac{1}{2},1)$ and $\varphi(\tfrac{1}{2},2)$. For the other values even this term was negligible.

The surface of a body is said to be *insulated* if it is impervious to heat. Thus no heat can flow across an insulated surface and so the component of the vector $-k$ grad φ normal to such a surface at any point on it must be zero. (That is the rate of flow of heat across an insulated surface is zero.) In particular, if a straight rod lying along the x-axis has the end $x = a$ insulated, then $\partial\varphi/\partial x$ must be equal to zero at $x = a$ for all time.

Example   A thin uniform rod of length l is initially at the constant temperature φ_0. One of its ends is subsequently brought to and maintained at zero temperature while the other end is kept insulated. Determine the temperature in the rod at any later time t.

As in the previous examples, starting with the solution

$$\varphi = \begin{cases} (a\cos px + b\sin px)e^{-p^2 Kt} & \text{for } p \neq 0, \\ a_0 + b_0 x & \text{for } p = 0, \end{cases}$$

and putting $\varphi = 0$ when $x = 0$ for all t we obtain

$$\varphi = \begin{cases} b \sin px\, e^{-p^2 Kt} & \text{for} \quad p \neq 0, \\ b_0 x & \text{for} \quad p = 0. \end{cases}$$

Now $\partial \varphi / \partial x = 0$ when $x = l$ for all t and so

$$bp \cos pl\, e^{-p^2 Kt} = 0 \quad \text{for all } t \text{ when} \quad p \neq 0$$

and

$$b_0 = 0 \quad \text{when} \quad p = 0.$$

Therefore $\cos pl = 0$ ($b = 0$ again gives the trivial solution $\varphi = 0$ at all points in the rod for all t).
Therefore

$$pl = \frac{(2n - 1)}{2}\pi$$

for integer values of n.
Therefore

$$\varphi = b \sin \frac{(2n - 1)\pi x}{2l}\, e^{-(2n-1)^2 \pi^2 Kt/4l^2}$$

is a solution for integer values of n. Therefore the general solution of this type (satisfying the boundary conditions) is

$$\varphi = \sum_{n=1}^{\infty} b_n \sin \frac{(2n - 1)\pi x}{2l}\, e^{-(2n-1)^2 \pi^2 Kt/4l^2}.$$

But when $t = 0$, $\varphi = \varphi_0$ for $0 \leqslant x \leqslant l$.
Therefore

$$\varphi_0 = \sum_{n=1}^{\infty} b_n \sin \frac{(2n - 1)\pi x}{2l} \quad \text{for} \quad 0 \leqslant x \leqslant l.$$

Therefore

$$\begin{aligned} b_n &= \frac{2}{l} \int_0^l \varphi_0 \sin \frac{(2n - 1)\pi x}{2l}\, dx \\ &= \frac{2\varphi_0}{l} \left[\frac{-2l}{(2n - 1)\pi} \cos \frac{(2n - 1)\pi x}{2l} \right]_0^l \\ &= \frac{4\varphi_0}{(2n - 1)\pi}. \end{aligned}$$

Therefore

$$\varphi(x,t) = \frac{4\varphi_0}{\pi} \sum_{n=1}^{\infty} \frac{1}{2n - 1} \sin \frac{(2n - 1)\pi x}{2l}\, e^{-(2n-1)^2 \pi^2 Kt/l^2}.$$

Once again it is observed that as $t \to \infty$, $\varphi(x,t) \to 0$ as expected.

Example *A tube of length l has both its ends sealed. A substance is diffusing along the tube (from which it cannot escape) with diffusivity K. If the initial concentration C of the diffusing substance varies linearly from zero at one end of the tube to C_0 at the other, find the concentration throughout the tube at any subsequent time t.*

The equation to be solved is

$$\frac{\partial C}{\partial t} = K \frac{\partial^2 C}{\partial x^2}.$$

The conditions to be satisfied by the solution C are

(i) $\dfrac{\partial C}{\partial x} = 0$ when $x = 0$,

(ii) $\dfrac{\partial C}{\partial x} = 0$ when $x = l$ and

(iii) $C = \dfrac{C_0}{l} x$ when $t = 0$.

As in the previous examples we obtain as a solution of the differential equation

$$C = \begin{cases} (a \cos px + b \sin px)e^{-p^2 Kt} & p \neq 0, \\ a_0 + b_0 x & p = 0. \end{cases}$$

Therefore

$$\frac{\partial C}{\partial x} = \begin{cases} p(-a \sin px + b \cos px)e^{-p^2 Kt} & p \neq 0, \\ b_0 & p = 0. \end{cases}$$

But $\partial C/\partial x = 0$ when $x = 0$ for all $t > 0$ and so $b = 0$ and $b_0 = 0$. Then, since $\partial C/\partial x = 0$ when $x = l$ for all $t > 0$, $\sin pl = 0$.

Therefore $pl = n\pi$ for integer values of n.

Hence the general solution of separated form (satisfying the boundary conditions) is

$$C = a_0 + \sum_{n=1}^{\infty} a_n \cos \frac{n\pi x}{l} e^{-n^2 \pi^2 Kt/l^2}.$$

But when $t = 0$, $C = (C_0/l)x$ for $0 \leqslant x \leqslant l$.

Therefore

$$\frac{C_0}{l} x = a_0 + \sum_{n=1}^{\infty} a_n \cos \frac{n\pi x}{l}.$$

Hence

$$a_0 = \frac{1}{l} \int_0^l \frac{C_0}{l} x \, dx = \frac{C_0}{2},$$

and

$$a_n = \frac{2}{l} \int_0^l \frac{C_0}{l} x \cos \frac{n\pi x}{l} \, dx = \begin{cases} -\dfrac{4C_0}{n^2\pi^2} & n \text{ odd and } \geqslant 1 \\ 0 & n \text{ even.} \end{cases}$$

Therefore

$$C(x,t) = \frac{C_0}{2} - \frac{4C_0}{\pi^2} \sum_{n=1}^{\infty} \frac{1}{(2n-1)^2} \cos \frac{(2n-1)\pi x}{l} e^{-(2n-1)^2 \pi^2 Kt/l^2}.$$

Note that as $t \to \infty$, $C(x,t) \to C_0/2$ as expected from simple physical considerations.

*Example   The region $x \geqslant 0$ is completely occupied by a material having thermal
diffusivity K. The temperature φ on the surface $x = 0$ of the material is a function
of time t only given by $\varphi_0 \sin \omega t$ in which φ_0 and ω are constants. Determine the
temperature distribution in the region $x \geqslant 0$ assuming that it is independent of
the rectangular coordinates y and z.*

Once again the equation to be solved is $\partial \varphi / \partial t = K(\partial^2 \varphi / \partial x^2)$.
The conditions to be satisfied by the solution φ are

(i) $\varphi = \varphi_0 \sin \omega t$ when $x = 0$, and
(ii) φ remains finite as $x \to \infty$.

To separate the variables put $\varphi = X(x)T(t)$.
Then

$$\frac{1}{T} \frac{dT}{dt} = \frac{K}{X} \frac{d^2 X}{dx^2}.$$

Now, since we are here concerned with a solution valid as $x \to \infty$, we look
for an exponential form of solution in x rather than a trigonometric one.
    Put

$$\frac{K}{X} \frac{d^2 X}{dx^2} = p^2.$$

Then

$$X = \begin{cases} Ae^{px/\sqrt{K}} + Be^{-px/\sqrt{K}} & \text{for } p \neq 0, \\ A_0 + B_0 x & \text{for } p = 0. \end{cases}$$

Also

$$\frac{1}{T} \frac{dT}{dt} = p^2 \quad \text{and so} \quad T = Ce^{p^2 t}.$$

Hence

$$\varphi = \begin{cases} (Ae^{px/\sqrt{K}} + Be^{-px/\sqrt{K}})Ce^{p^2 t} & \text{for } p \neq 0, \\ (A_0 + B_0 x)C & \text{for } p = 0. \end{cases}$$

$$= \begin{cases} (ae^{px/\sqrt{K}} + be^{-px/\sqrt{K}})e^{p^2 t} & \text{for } p \neq 0 \\ a_0 + b_0 x & \text{for } p = 0 \end{cases}$$

for arbitrary constants a, b, a_0 and b_0.

Now when $x = 0$,

$$\varphi = \begin{cases} (a + b)e^{p^2 t} & \text{for } p \neq 0, \\ a_0 & \text{for } p = 0. \end{cases}$$

Thus condition (i) cannot be satisfied when $p = 0$ and can only be satisfied for $p \neq 0$ if $p^2 = i\omega$, $(a + b) = \varphi_0$ and we consider only the imaginary part of $(a + b)e^{-p^2 t}$. Thus we take as our solution, up to this stage,

$$\varphi = \text{Im}\{(ae^{px/\sqrt{K}} + be^{-px/\sqrt{K}})e^{p^2 t}\}$$

with $(a + b) = \varphi_0$ and $p^2 = i\omega$.

But $p^2 = i\omega$ implies that

$$p = \pm\sqrt{\frac{\omega}{2}}(1 + i)$$

and so we have

$$\varphi = \text{Im}\{(ae^{\sqrt{\omega}(1+i)x/\sqrt{2K}} + be^{-\sqrt{\omega}(1+i)x/\sqrt{2K}}e^{i\omega t}\}.$$

Now condition (ii) gives $a = 0$ (and hence $b = \varphi_0$). Therefore

$$\varphi = \text{Im}\{\varphi_0 e^{-\sqrt{\omega}(1+i)x/\sqrt{2K}}e^{i\omega t}$$
$$= \text{Im}\{\varphi_0 e^{-\sqrt{\omega}x/\sqrt{2K}}e^{i(\omega t - \sqrt{\omega}x/\sqrt{2K})}\}$$
$$= \varphi_0 e^{-\sqrt{\omega}x/2\sqrt{K}}\cos\left(\omega t - \sqrt{\frac{\omega}{2K}}x\right).$$

This expression satisfies the differential equation and the given conditions and so is the required temperature distribution.

Note that as $x \to \infty$, $\varphi(x,t) \to 0$ as expected from physical considerations.

For two-dimensional heat flow the equation to be solved (in terms of cartesian coordinates) is

$$\frac{\partial^2 \varphi}{\partial x^2} + \frac{\partial^2 \varphi}{\partial y^2} = \frac{1}{K}\frac{\partial \varphi}{\partial t}.$$

The method of solution is separation of variables, similar to the above, and will be illustrated by a specific example.

Example   A thin rectangular plate has sides of length a and b. Using a rectangular cartesian coordinate system its edges lie along the lines $x = 0$, $x = a$, $y = 0$ and $y = b$ (figure 106). The temperature φ is maintained at zero round the edges. Assuming that no heat is lost across the rectangular faces, determine the temperature distribution in the plate at any time t if at time $t = 0$ it is given by

$$\varphi = \varphi_0(x - y).$$

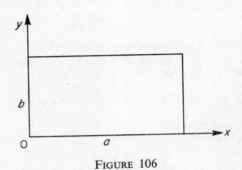

FIGURE 106

Put $\varphi(x,y,t) = X(x)\,Y(y)\,T(t)$ in the equation

$$\frac{\partial^2 \varphi}{\partial x^2} + \frac{\partial^2 \varphi}{\partial y^2} = \frac{1}{K}\frac{\partial \varphi}{\partial t}.$$

Therefore

$$\frac{1}{X}\frac{d^2 X}{dx^2} + \frac{1}{Y}\frac{d^2 Y}{dy^2} = \frac{1}{KT}\frac{dT}{dt}.$$

Thus a function of x only plus a function of y only is equal to a function of t only and so each function must be equal to a constant.
    Put

$$\frac{1}{X}\frac{d^2 X}{dx^2} = -p^2.$$

Therefore

$$\frac{d^2 X}{dx^2} + p^2 X = 0$$

and so

$$X = \begin{cases} A\cos px + B\sin px & p \neq 0, \\ A_0 + B_0 x & p = 0 \end{cases}$$

for arbitrary constants A, A_0, B and B_0. Similarly, put

$$\frac{1}{Y}\frac{d^2Y}{dy^2} = -q^2.$$

Then

$$Y = \begin{cases} C\cos qy + D\sin qy & q \neq 0, \\ C_0 + D_0y & q = 0 \end{cases}$$

for arbitrary constants C, C_0, D and D_0. Then

$$\frac{1}{KT}\cdot\frac{dT}{dt} = -(p^2 + q^2)$$

and so $T = Ee^{-K(p^2+q^2)t}$ for arbitrary constant E.

Now since $\varphi = 0$ when $x = 0$ for $0 \leqslant y \leqslant b$ and all $t > 0$,

$$X(x) = 0 \quad\text{when}\quad x = 0.$$

Therefore

$$A = 0 \quad\text{and}\quad A_0 = 0.$$

Similarly,

$$X(x) = 0 \quad\text{when}\quad x = a$$

and so

$$B\sin pa = 0 \quad\text{and}\quad B_0 = 0.$$

Now if we are to obtain a non-trivial solution, $B \neq 0$ and so $\sin pa = 0$. Therefore $pa = n\pi$ for integer values of n.

Hence

$$X = B\sin\frac{n\pi x}{a} \quad\text{for integer values of } n.$$

Similarly, since $\varphi = 0$ when $y = 0, b$ for $0 \leqslant x \leqslant a$ and all $t > 0$, we obtain

$$Y = D\sin\frac{m\pi y}{b} \quad\text{for integer values of } m.$$

Thus

$$\varphi = B\sin\frac{n\pi x}{a} D\sin\frac{m\pi y}{b}Ee^{-K\pi^2(n^2/a^2+m^2/b^2)t}$$

is a solution for integer values of n and m.

This can be written as

$$\varphi = c\sin\frac{n\pi x}{a}\sin\frac{m\pi y}{b}e^{-K\pi^2(n^2/a^2+m^2/b^2)t}$$

for constant c and integer values of n and m. Therefore the general solution of this type (satisfying the boundary conditions) is

$$\varphi = \sum_{n=1}^{\infty}\sum_{m=1}^{\infty}c_{m,n}\sin\frac{n\pi x}{a}\sin\frac{m\pi y}{b}e^{-K\pi^2(n^2/a^2+m^2/b^2)t}.$$

Now when $t = 0$, $\varphi = \varphi_0(x - y)$ for $0 \leqslant x \leqslant a$ and $0 \leqslant y \leqslant b$. Therefore

$$\varphi_0(x - y) = \sum_{n=1}^{\infty} \sum_{m=1}^{\infty} c_{m,n} \sin \frac{n\pi x}{a} \sin \frac{m\pi y}{b}$$

for $0 \leqslant x \leqslant a$ and $0 \leqslant y \leqslant b$. Therefore

$$\int_0^b \int_0^a \varphi_0(x - y) \sin \frac{n'\pi x}{a} \sin \frac{m'\pi y}{b} \, dx \, dy$$

$$= \sum_{n=1}^{\infty} \sum_{m=1}^{\infty} c_{m,n} \int_0^b \int_0^a \sin \frac{n\pi x}{a} \sin \frac{m\pi y}{a} \sin \frac{n'\pi x}{a} \sin \frac{m'\pi y}{b} \, dx \, dy$$

assuming that the order of the summations can be interchanged with the integrals. But

$$\int_0^b \int_0^a \sin \frac{n\pi x}{a} \sin \frac{m\pi y}{b} \sin \frac{n'\pi x}{a} \sin \frac{m'\pi y}{b} \, dx \, dy$$

$$= \int_0^b \sin \frac{m\pi y}{b} \sin \frac{m'\pi y}{b} \, dy \int_0^a \sin \frac{n\pi x}{a} \sin \frac{n'\pi x}{a} \, dx$$

$$= \begin{cases} \frac{1}{4}ab & \text{if} \quad m = m' \quad \text{and} \quad n = n', \\ 0 & \text{otherwise,} \end{cases}$$

and

$$\int_0^a x \sin \frac{n'\pi x}{a} \, dx = \frac{-a^2}{n'\pi} \cos n'\pi$$

(on integrating by parts). Therefore

$$\int_0^b \int_0^a x \sin \frac{n'\pi x}{a} \sin \frac{m'\pi y}{b} \, dx \, dy = \frac{-a^2}{n'\pi} \cos n'\pi \int_0^b \sin \frac{m'\pi y}{b} \, dy$$

$$= \frac{ba^2}{n'm'\pi^2} \cos n'\pi (\cos m'\pi - 1).$$

Similarly

$$-\int_0^b \int_0^a y \sin \frac{n'\pi x}{a} \sin \frac{m'\pi y}{b} \, dx \, dy = \frac{-ab^2}{n'm'\pi^2} \cos m'\pi (\cos n'\pi - 1).$$

Hence

$$\frac{\varphi_0 ab}{mn\pi^2} \{a \cos n\pi (\cos m\pi - 1) - b \cos m\pi (\cos n\pi - 1)\} = \frac{ab}{4} c_{m,n}.$$

Therefore

$$c_{m,n} = \frac{4\varphi_0}{\pi^2} \frac{1}{mn} \{a(-1)^n((-1)^m - 1) - b(-1)^m((-1)^n - 1)\}.$$

Thus $\varphi(x,y,t)$ is now completely determined for the plate.

Example A *semi-infinite rectangular plate of width h has its edges lying along the lines* $x = 0$, $y = 0$, $y = h$ *(figure 107). The edge* $x = 0$ *is maintained at temperature* φ_0 *while the long edges* $y = 0$ *and* $y = h$ *are held at zero temperature. Show that when the steady state has been reached, the temperature* φ *at any point* (x,y) *of the plate is given by*

$$\varphi = \frac{4\varphi_0}{\pi} \sum_{m=0}^{\infty} \frac{1}{2m-1} \, e^{-(2m-1)\pi x/h} \sin \frac{(2m-1)\pi y}{h}.$$

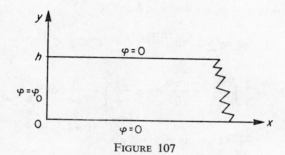

FIGURE 107

When the steady state has been reached the equation to be solved is Laplace's equation

$$\frac{\partial^2 \varphi}{\partial x^2} + \frac{\partial^2 \varphi}{\partial y^2} = 0.$$

Using separation of variables, we obtain the form of solution

$$\varphi = \begin{cases} (A \cos py + B \sin py)(Ce^{px} + De^{-px}) & p \neq 0, \\ (A_0 + B_0 y)(C_0 + D_0 x) & p = 0 \end{cases}$$

for arbitrary constants A, B, C, D, A_0, B_0, C_0 and D_0.

Now φ remains finite as $x \to \infty$ and so $C = 0$ and $D_0 = 0$. Therefore the solution has the form

$$\varphi = \begin{cases} (A \cos py + B \sin py)De^{-px} & p \neq 0, \\ (A_0 + B_0 y)C_0 & p = 0, \end{cases}$$

or

$$\varphi = \begin{cases} (a \cos py + b \sin py)e^{-px} & p \neq 0, \\ a_0 + b_0 y & p = 0 \end{cases}$$

for arbitrary constants a, b, a_0 and b_0.

Now $\varphi = 0$ when $y = 0$ and $x > 0$. Hence $a = 0$ and $a_0 = 0$. Also $\varphi = 0$ when $y = h$ and $x > 0$. Hence $b_0 = 0$ and $b \sin ph = 0$. Therefore $b = 0$

(trivial solution) or $\sin ph = 0$ and so $ph = n\pi$ for integers n. Hence for a non-trivial solution we have

$$\varphi = b \sin \frac{n\pi y}{h} e^{-n\pi x/h}.$$

Therefore the general solution of this form is

$$\varphi = \sum_{n=1}^{\infty} b_n \sin \frac{n\pi y}{h} e^{-n\pi x/h}.$$

Now when $x = 0$, $\varphi = \varphi_0$ for $0 < y < h$ and so

$$\varphi_0 = \sum_{n=1}^{\infty} b_n \sin \frac{n\pi y}{h} \quad \text{for} \quad 0 < y < h.$$

Hence

$$b_n = \frac{2}{h} \int_0^h \varphi_0 \sin \frac{n\pi y}{h} \, dy = \frac{2\varphi_0}{h} \left[-\frac{h}{n\pi} \cos \frac{n\pi y}{h} \right]_0^h$$

$$= \frac{2\varphi_0}{n\pi} (1 - \cos n\pi) = \begin{cases} \dfrac{4\varphi_0}{n\pi} & \text{if } n \text{ is odd} \\ 0 & \text{if } n \text{ is even.} \end{cases}$$

Therefore

$$\varphi(x,y) = \frac{4\varphi_0}{\pi} \sum_{m=0}^{\infty} \frac{1}{2m+1} e^{-(2m+1)\pi x/h} \sin \frac{(2m+1)\pi y}{h}.$$

7.22 Cylindrical polar coordinates (ρ, ψ, z)

The heat conduction equation is

$$\frac{\partial^2 \varphi}{\partial \rho^2} + \frac{1}{\rho} \frac{\partial \varphi}{\partial \rho} + \frac{1}{\rho^2} \frac{\partial^2 \varphi}{\partial \psi^2} + \frac{\partial^2 \varphi}{\partial z^2} = \frac{1}{K} \frac{\partial \varphi}{\partial t}.$$

The method of solution is again separation of variables and will be illustrated by an example.

Example   An infinitely long thermally conducting rod has a circular cross-section of radius a and its temperature φ is given by $\varphi = f(\rho)$ where ρ is distance from the axis of the rod. Determine the temperature distribution within the rod at any time t after its surface has been reduced to and maintained at zero temperature.

The temperature φ is independent of the cylindrical polar coordinates ψ and z and so satisfies the equation

$$\frac{\partial^2 \varphi}{\partial \rho^2} + \frac{1}{\rho} \frac{\partial \varphi}{\partial \rho} = \frac{1}{K} \frac{\partial \varphi}{\partial t}.$$

Put $\varphi = R(\rho)T(t)$.

Therefore

$$\frac{1}{R}\frac{d^2R}{d\rho^2} + \frac{1}{\rho R}\frac{dR}{d\rho} = \frac{1}{KT}\frac{dT}{dt}.$$

Then since the left hand side is a function of ρ only and the right hand side is a function of t only, each must be constant.

Therefore, put

$$\frac{1}{KT}\frac{dT}{dt} = -p^2 \qquad (p \neq 0).$$

Then

$$T = Ae^{-p^2Kt}$$

for arbitrary constant A. Also

$$\rho^2\frac{d^2R}{d\rho^2} + \rho\frac{dR}{d\rho} + p^2\rho^2R = 0.$$

This is Bessel's equation of order zero and has solution

$$R = BJ_0(p\rho) + CY_0(p\rho)$$

for arbitrary constants B and C. But since φ remains finite when $\rho = 0$, C must be zero.

Therefore

$$\varphi = bJ_0(p\rho)e^{-p^2Kt}$$

for arbitrary constant b. Now the surface of the rod is maintained at zero temperature and so $\varphi = 0$ when $\rho = a$ for all $t \geqslant 0$.

Therefore

$$J_0(pa) = 0 \qquad (b = 0 \text{ leads to } \varphi = 0)$$

and so $p = \alpha_n/a$ $(n = 1,2,3,\ldots)$ where the α_n are the roots of $J_0(x) = 0$. Therefore a more general solution satisfying the above boundary conditions is

$$\varphi = \sum_{n=1}^{\infty} b_nJ_0\left(\frac{\alpha_n\rho}{a}\right)e^{-\alpha^2 nKt/a^2}.$$

(It is easily shown that the solution corresponding to $p = 0$ and satisfying the above boundary conditions is $\varphi = 0$.)

When $t = 0$, $\varphi = f(\rho)$ and so

$$f(\rho) = \sum_{n=1}^{\infty} b_nJ_0\left(\frac{\alpha_n\rho}{a}\right).$$

Therefore

$$b_n = \frac{2}{[aJ_1(\alpha_n)]^2}\int_0^a \rho f(\rho)J_0\left(\frac{\alpha_n\rho}{a}\right) d\rho.$$

In particular, if the rod has radius 1 cm and initially its temperature at all points is 10°C,

$$a = 1, \qquad f(\rho) = 10$$

and

$$b_n = \frac{2}{[J_1(\alpha_n)]^2} \int_0^1 10\rho J_0(\alpha_n\rho)\,d\rho$$

$$= \frac{20}{[J_1(\alpha_n)]^2}\frac{1}{\alpha_n^2} \int_0^{\alpha_n} xJ_0(x)\,dx \quad \text{where} \quad x = \alpha_n\rho$$

$$= \frac{20}{[\alpha_n J_1(\alpha_n)]^2} \Big[xJ_1(x)\Big]_0^{\alpha_n} = \frac{20}{\alpha_n J_1(\alpha_n)}.$$

Hence the temperature φ in degrees centigrade is

$$\varphi = 20 \sum_{n=1}^{\infty} \frac{J_0(\alpha_n\rho)}{\alpha_n J_1(\alpha_n)} e^{-\alpha_n^2 Kt}.$$

K is a known constant for the material of the rod and the values of α_n and $J_1(\alpha_n)$ are obtained from tables.†

7.23 Spherical polar coordinates (r, θ, ψ).

The heat conduction equation is

$$\frac{\partial^2\varphi}{\partial r^2} + \frac{2}{r}\frac{\partial\varphi}{\partial r} + \frac{1}{r^2}\frac{\partial^2\varphi}{\partial\theta^2} + \frac{1}{r^2}\cot\theta\frac{\partial\varphi}{\partial\theta} + \frac{1}{r^2\sin^2\theta}\frac{\partial^2\varphi}{\partial\psi^2} = \frac{1}{K}\frac{\partial\varphi}{\partial t}.$$

The method of solution is again separation of variables and will be illustrated by examples.

Example   A uniform sphere of radius a is at the constant temperature φ_0. Determine the temperature distribution within the sphere at any time t after its surface has been reduced to and maintained at zero temperature.

The temperature φ is independent of the spherical polar coordinates θ and ψ and so satisfies the equation

$$\frac{\partial^2\varphi}{\partial r^2} + \frac{2}{r}\frac{\partial\varphi}{\partial r} = \frac{1}{K}\frac{\partial\varphi}{\partial t}.$$

The subsequent working is somewhat simplified if the change of variable $u = r\varphi$ is made at this stage. The above equation then becomes

$$\frac{\partial^2 u}{\partial r^2} = \frac{1}{K}\frac{\partial u}{\partial t}.$$

† For example British Association for the Advancement of Science Mathematical Tables, Volume VI, Cambridge University Press.

(That is the equation for heat conduction in a thin straight rod.) Separating the variables by putting $u = R(r)T(t)$ we obtain

$$\frac{1}{KT}\frac{dT}{dt} = -p^2$$

for some constant $-p^2$ and

$$\frac{1}{R}\cdot\frac{d^2R}{dr^2} = -p^2.$$

Therefore

$$T = Ce^{-p^2Kt}$$

and

$$R = A\cos pr + B\sin pr$$

for arbitrary constants A, B and C and $p \neq 0$.

Therefore

$$\varphi = \frac{1}{r}u = \frac{1}{r}R(r)T(t)$$

$$= \frac{1}{r}(A\cos pr + B\sin pr)Ce^{-p^2Kt}$$

for $p \neq 0$ is a solution of the heat conduction equation. When $p = 0$, the solution is

$$\varphi = \frac{A_0}{r} + B_0$$

where A_0 and B_0 are arbitrary constants.

But φ must remain finite when $r = 0$ and so A, $A_0 = 0$. Therefore

$$\varphi = \frac{BC}{r}\sin pr\, e^{-p^2Kt} = \frac{b}{r}\sin pr\, e^{-p^2Kt}\dagger$$

for $p \neq 0$ and arbitrary constant b, and $\varphi = B_0$ when $p = 0$. Now $\varphi = 0$ when $r = a$ and so $B_0 = 0$, and $\sin pa = 0$ ($b = 0$ gives $\varphi = 0$ everywhere within the sphere).

Therefore $p = n\pi/a$ for non-zero integers n, and

$$\varphi = \frac{b}{r}\sin\frac{n\pi r}{a}\, e^{-n^2\pi^2Kt/a^2}.$$

A more general solution, satisfying the above boundary conditions, is then

$$\varphi = \sum_{n=1}^{\infty}\frac{b_n}{r}\sin\frac{n\pi r}{a}\, e^{-n^2\pi^2Kt/a^2}.$$

Now $\varphi = \varphi_0$ when $t = 0$ and so

$$\varphi_0 = \frac{1}{r}\sum_{n=1}^{\infty}b_n\sin\frac{n\pi r}{a}.$$

$\dagger$ Note that $\dfrac{\sin \rho r}{r} \to p$ as $r \to 0$ and so φ remains finite at $r = 0$.

Therefore

$$b_n = \frac{2}{a} \int_0^a r\varphi_0 \sin \frac{n\pi r}{a} \, dr$$

$$= \frac{2\varphi_0 a}{n\pi} (-1)^{n+1}.$$

Hence

$$\varphi = \frac{2\varphi_0 a}{r\pi} \sum_{n=1}^{\infty} \frac{(-1)^{n+1}}{n} \sin \frac{n\pi r}{a} \, e^{-n^2\pi^2 Kt/a^2}.$$

Example   A conducting sphere of radius a has zero temperature throughout. Its surface is now maintained at the constant temperature φ_0 for a time t_0 and then reduced to and maintained at zero temperature. Find the temperature distribution within the sphere at any time t after the surface has been brought to zero temperature.

Consider firstly the time during which the surface of the sphere is maintained at the temperature φ_0. The temperature φ is independent of the spherical polar coordinates θ and ψ and so satisfies the equation

$$\frac{\partial^2 \varphi}{\partial r^2} + \frac{2}{r} \frac{\partial \varphi}{\partial r} = \frac{1}{K} \frac{\partial \varphi}{\partial t} .$$

The other conditions to be satisfied by φ are

(i) $\varphi = \varphi_0$ when $r = a$ for $0 < t < t_0$,
(ii) $\varphi = 0$ when $t = 0$ for $0 \leqslant r \leqslant a$.

The physical condition that φ remains finite for $r = 0$ for all t must of course also be satisfied.

Putting $u = r\varphi$ and then separating the variables by putting $u = R(r)T(t)$, we obtain as in the previous example that

$$\varphi = \begin{cases} \dfrac{1}{r}(A \cos pr + B \sin pr)Ce^{-p^2 Kt} & \text{for } p \neq 0, \\[2mm] \dfrac{A_0}{r} + B_0 & \text{for } p = 0 \end{cases}$$

where A, A_0, B, B_0, C and p are constants.

Then since φ remains finite when $r = 0$, A and $A_0 = 0$ and so

$$\varphi = \begin{cases} \dfrac{b}{r} \sin pr \, e^{-p^2 Kt} & p \neq 0, \\[2mm] B_0 & p = 0 \end{cases}$$

for constant b.

Now condition (i) cannot be satisfied by the solution for $p \neq 0$, but when $p = 0$ we have $\varphi_0 = B_0$. However, the resulting solution $\varphi = \varphi_0$ itself cannot be made to satisfy condition (ii) that $\varphi = 0$ when $t = 0$ and $0 \leqslant r \leqslant a$. But

$$\varphi = \varphi_0 + \frac{b}{r} \sin pr \, e^{-p^2 K t}$$

is a solution of the differential equation and satisfies condition (i) if

$$\frac{b}{a} \sin pa \, e^{-p^2 K t} = 0 \quad \text{for} \quad 0 < t < t_0.$$

That is if

$$b \sin pa = 0.$$

Therefore $\sin pa = 0$ ($b = 0$ gives again the solution $\varphi = \varphi_0$ and condition (ii) cannot be satisfied).

Hence $p = n\pi/a$ for non-zero integers n, and so

$$\varphi = \varphi_0 + \frac{b}{r} \sin \frac{n\pi r}{a} \, e^{-n^2 \pi^2 K t/a^2}$$

is a solution satisfying the condition (i) for all non-zero integer values of n.

A more general solution of this type is

$$\varphi = \varphi_0 + \sum_{n=1}^{\infty} \frac{b_n}{r} \sin \frac{n\pi r}{a} \, e^{-n^2 \pi^2 K t/a^2}.$$

Now from condition (ii) we obtain

$$0 = \varphi_0 + \sum_{n=1}^{\infty} \frac{b_n}{r} \sin \frac{n\pi r}{a}.$$

Hence, as in the previous example,

$$b_n = -\frac{2}{a} \int_0^a r \varphi_0 \sin \frac{n\pi r}{a} \, dr = \frac{2\varphi_0 a}{n\pi} \, (-1)^n.$$

Therefore

$$\varphi = \varphi_0 + \frac{2\varphi_0 a}{r\pi} \sum_{n=1}^{\infty} \frac{(-1)^n}{n} \sin \frac{n\pi r}{a} \, e^{-n^2 \pi^2 K t/a^2}.$$

Therefore, when $t = t_0$,

$$\varphi = \varphi_0 + \frac{2\varphi_0 a}{r\pi} \sum_{n=1}^{\infty} \frac{(-1)^n}{n} \sin \frac{n\pi r}{a} \, e^{-n^2 \pi^2 K t/oa^2}.$$

Now consider the problem after the surface of the sphere has again been reduced to zero temperature. Taking the origin of time at the instant when the surface temperature is reduced to zero, we must again solve

$$\frac{\partial^2 \varphi}{\partial r^2} + \frac{2}{r} \frac{\partial \varphi}{\partial r} = \frac{1}{K} \frac{\partial \varphi}{\partial t}$$

subject to the conditions

(iii) $\varphi = 0$ when $r = a$ for all $t > 0$,

(iv) $\varphi = \varphi_0 + \dfrac{2\varphi_0 a}{r\pi} \displaystyle\sum_{n=1}^{\infty} \dfrac{(-1)^n}{n} \sin \dfrac{n\pi r}{a} e^{-n^2\pi^2 K t_0/a^2}$

$$\text{when} \quad t = 0 \quad \text{for} \quad 0 \leqslant r \leqslant a.$$

Also φ must remain finite for $r = 0$. Proceeding as in the first part, we obtain

$$\varphi = \begin{cases} \dfrac{c}{r} \sin qr \, e^{-q^2 K t} & q \neq 0, \\ C_0 & q = 0 \end{cases}$$

for arbitrary constants c, C_0, q. Now condition (iii) gives $C_0 = 0$ and

$$\frac{c}{a} \sin qa = 0.$$

Therefore $\sin qa = 0$ ($c = 0$ gives merely $\varphi = 0$). Hence $q = n\pi/a$ for non-zero integer values of n. Therefore

$$\varphi = \frac{c}{r} \sin \frac{n\pi r}{a} e^{-n^2\pi^2 K t/a^2}.$$

A more general solution is then

$$\varphi = \sum_{n=1}^{\infty} \frac{c_n}{r} \sin \frac{n\pi r}{a} e^{-n^2\pi^2 K t/a^2}.$$

Now condition (iv) gives

$$\varphi_0 + \frac{2\varphi_0 a}{r\pi} \sum_{n=1}^{\infty} \frac{(-1)^n}{n} \sin \frac{n\pi r}{a} e^{-n^2\pi^2 K t_0/a^2} = \sum_{n=1}^{\infty} \frac{c_n}{r} \sin \frac{n\pi r}{a}.$$

But in the previous example it was shown that

$$\varphi_0 = \frac{2\varphi_0 a}{r\pi} \sum_{n=1}^{\infty} \frac{(-1)^{n+1}}{n} \sin \frac{n\pi r}{a}.$$

Thus we have

$$\frac{2\varphi_0 a}{r\pi} \sum_{n=1}^{\infty} \frac{(-1)^{n+1}}{n} \sin \frac{n\pi r}{a} + \frac{2\varphi_0 a}{r\pi} \sum_{n=1}^{\infty} \frac{(-1)^n}{n} \sin \frac{n\pi r}{a} e^{-n^2\pi^2 K t_0/a^2} = \sum_{n=1}^{\infty} \frac{c_n}{r} \sin \frac{n\pi r}{a}.$$

Hence, equating coefficients of $(1/r) \sin (n\pi r/a)$ on both sides, we obtain

$$\frac{2\varphi_0 a}{n\pi} (-1)^n \left\{ e^{-n^2\pi^2 K t_0/a^2} - 1 \right\} = c_n$$

for positive integers n.

Thus the temperature φ at points inside the sphere at any time t after the surface temperature has been reduced to zero is given by

$$\varphi = \frac{2\varphi_0 a}{r\pi} \sum_{n=1}^{\infty} \frac{(-1)^n}{n} (e^{-n^2\pi^2 K t_0/a^2} - 1) \sin \frac{n\pi r}{a} e^{-n^2\pi^2 K t/a^2}.$$

Exercises

1. A thin uniform rod of length l lies along the x-axis with its ends at $x = 0$ and $x = l$. Initially the temperature φ in the rod is given by

$$\varphi = \begin{cases} x & \text{for} \quad 0 \leqslant x \leqslant l/2, \\ l - x & \text{for} \quad l/2 \leqslant x \leqslant l. \end{cases}$$

Find the temperature distribution in the rod at any time t if

(i) the ends are maintained at zero temperature and
(ii) the ends are insulated.

Show that in both cases the solution as $t \to \infty$ agrees with what would be expected from physical considerations.

2. The function φ is a function of x and t only and satisfies the differential equation

$$\frac{\partial^2 \varphi}{\partial x^2} = \frac{\partial \varphi}{\partial t}.$$

Use the method of separation of variables to show that if $\partial \varphi / \partial t \neq 0$, then φ may be written in the form

$$\varphi = A e^{px + p^2 t} + B e^{-px + p^2 t}$$

for constants A, B and p. Hence obtain the solution of the given equation, valid in the range $-a \leqslant x \leqslant a$ and satisfying the conditions

(i) $\varphi \to 0$ as $t \to \infty$,
(ii) $\varphi = 0$ when $x = \pm a \ (t > 0)$,
(iii) $\varphi = x$ when $t = 0$ and $-a < x < a$.

3. A thin uniform rod of length l is initially at zero temperature. One end is kept at zero temperature whilst the other end is suddenly raised to and maintained at the constant temperature φ_0. Determine the temperature distribution in the rod at any subsequent time t. If the rod has length 1 metre, diffusivity 0·05 metres squared per second and if $\varphi_0 = 100°C$, determine correct to 2 decimal places the temperature at the midpoint of the rod at intervals of 2 seconds up to 10 seconds after the temperature of the end is raised.

4. A thin rectangular plate defined by $0 \leqslant x \leqslant a$, $0 \leqslant y \leqslant b$ has its edges $x = 0$, $x = a$, $y = 0$ maintained at zero temperature and the edge $y = b$ maintained at temperature φ_0. Determine the temperature distribution in the plate when a steady state has been reached.

5. A square slab of conducting material with sides of length a has its faces insulated. Taking cartesian coordinates x and y with origin at one corner of the slab and x and y axes along a pair of edges of the slab the temperature distribution φ in the slab is given by $\varphi = xy$ for $0 \leqslant x \leqslant a$, $0 \leqslant y \leqslant a$. At time $t = 0$, one side of the slab is insulated and the others are reduced to and maintained at zero temperature. Determine the subsequent temperature distribution throughout the slab.

6. A uniform solid circular cylinder of height h and radius a has constant thermal conductivity throughout. One of its plane ends is insulated and the other is maintained at the constant temperature φ_0. If the curved surface is maintained at zero temperature, determine the steady state temperature distribution within the cylinder.

7. A spherical shell of internal radius a and external radius b has constant thermal conductivity. The inner spherical surface is maintained at zero temperature and the outer spherical surface is maintained at temperature φ_0. Show that in the steady state the temperature at a point at distance r from the centre of the shell ($a < r < b$) is

$$\frac{(r - a)b\varphi_0}{r(b - a)}.$$

The above spherical shell is initially at temperature zero throughout and has its inner surface maintained at zero temperature. At time $t = 0$ the shell is plunged into a boiling liquid so that its outer surface is now maintained at temperature φ_0. If the thermal conductivity and the specific heat per unit volume are k and σ respectively, show that at time $t > 0$ the temperature at a point within the shell at distance r from its centre is

$$\frac{2b\varphi_0}{\pi r} \left(\tfrac{1}{2}u - e^{-\alpha} \sin u + \tfrac{1}{2}e^{-4\alpha} \sin 2u - \tfrac{1}{3}e^{-9\alpha} \sin 3u + \cdots \right)$$

where

$$u = \frac{\pi(r - a)}{b - a} \quad \text{and} \quad \alpha = \frac{\pi^2 kt}{\sigma(b - a)^2}.$$

8. Find the steady state temperature distribution inside a uniform solid sphere of radius a if the temperature at any point P on its surface is $\varphi_0 \sin^4 \theta$ where φ_0 is a constant and θ is the angle between the radius of the sphere through P and a fixed diameter of the sphere.

9. The voltage V in volts and current I in amperes in a telephone wire or submarine cable of negligible inductance lying along the x-axis are related to the

resistance R per unit length in ohms per metre and the capacitance C per unit length in farads per metre by the equation

$$\frac{\partial^2 V}{\partial x^2} = RC \frac{\partial V}{\partial t}.$$

Such a cable AB of length l metres, resistance R ohms/metre and capacitance C farads/metre has one end A maintained at a potential 10 volts and the other B at 5 volts and the voltage varies linearly along its length. Suddenly the end B is earthed (that is its voltage is reduced to zero). Find the subsequent voltage distribution along the cable.

10. A cable AB has length $1\cdot5 \times 10^6$ metres, resistance $1\cdot25 \times 10^{-3}$ ohms per metre and capacitance 2 farads per metre. In the steady state the voltages at A and B are 1200 and 1100 volts respectively. Determine the voltage in the cable t seconds after the end B is suddenly earthed.

11. The region $x \geqslant 0$ is completely occupied by a substance having electrical conductivity λ and magnetic permeability μ. The current density vector $\mathbf{j}$ on the surface $x = 0$ of the material is $(0, a \sin \omega t, 0)$ where a and ω are constants and t is the time. Determine the distribution of current density within the material. [The equation to be satisfied is $(1/\lambda\mu)\nabla^2 \mathbf{j} = \partial \mathbf{j}/\partial t$, and since in this problem $\mathbf{j} = (0, j(x,t), 0)$ we obtain

$$\frac{\partial j}{\partial t} = \frac{1}{\lambda\mu} \nabla^2 j = \frac{1}{\lambda\mu} \frac{\partial^2 j}{\partial x^2}.$$

The conditions to be satisfied by the solution j are (i) j remains finite as $x \to \infty$, (ii) $j = a \sin \omega t$ for all y when $x = 0$.]

8

The wave equation

Everyone is familiar with some form of wave motion. When a stone is dropped into a pond water waves can be observed moving radially outwards; when a violin string is bowed it vibrates and sends out sound waves; electromagnetic waves are sent out from radio and T.V. transmitters etc. All these forms of wave motion have the following properties in common:

1. they transmit energy with finite velocity from the source and
2. they do not give the medium through which they travel any permanent displacement.

8.1 Waves on strings (the one-dimensional wave equation)

We shall consider waves for which the displacement of each particle of the string is normal to the direction of transmission of the wave itself. Such waves are called *transverse* waves (figure 108).

It will be assumed that the force of gravity acting on the string is negligible and that the string is maintained at a constant tension T throughout. The former assumption is valid for motions on a smooth horizontal plane, while the latter can be shown to be reasonable for 'smooth' waves of small amplitude.

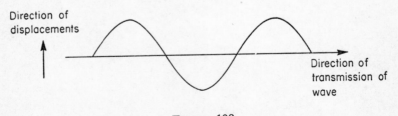

Direction of displacements

Direction of transmission of wave

FIGURE 108

352

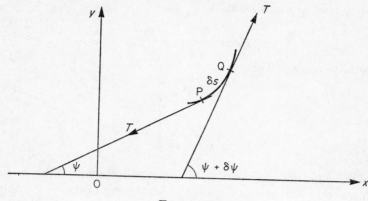

FIGURE 109

Consider the small element PQ of length δs of the string (figure 109). In the undisturbed position this element will have length δx and its mass will be $\rho\delta x$ if ρ is the mass per unit length of the string.

Using the notation shown in figure 109 and resolving forces parallel to the y-axis, we obtain

$$T \sin (\psi + \delta\psi) - T \sin \psi = \rho\delta x \frac{\partial^2 y}{\partial t^2}$$

where $\partial^2 y/\partial t^2$ is the acceleration in the direction of the y-axis of the centre of mass of the element of string PQ. (δs is assumed to be so small that PQ satisfies Newton's law of particle motion.)

Therefore

$$T\{\sin \psi \cos \delta\psi + \cos \psi \sin \delta\psi\} - T \sin \psi = \rho\delta x \frac{\partial^2 y}{\partial t^2}.$$

But

$$\cos \delta\psi = 1 - \frac{(\delta\psi)^2}{2!} + \cdots$$

and

$$\sin \delta\psi = \delta\psi - \frac{(\delta\psi)^3}{3!} + \cdots$$

Hence

$$T \cos \psi \cdot \delta\psi + 0\{(\delta\psi)^2\} = \rho\delta x \frac{\partial^2 y}{\partial t^2}$$

where $0\{(\delta\psi)^2\}$ means the sum of terms each of which has $(\delta\psi)^2$ or a higher power of the small quantity $\delta\psi$ as a factor. Then proceeding to the limit as

δx and $\delta \psi$ tend to zero we obtain

$$T \cos \psi \frac{\partial \psi}{\partial x} = \rho \frac{\partial^2 y}{\partial t^2}.$$

But $\partial y / \partial x = \tan \psi$ and so

$$\frac{\partial^2 y}{\partial x^2} = \sec^2 \psi \frac{\partial \psi}{\partial x}.$$

Therefore

$$T \cos^3 \psi \frac{\partial^2 y}{\partial x^2} = \rho \frac{\partial^2 y}{\partial t^2}.$$

If now we restrict consideration to small displacements of the string for which $\cos^3 \psi \simeq 1$ (that is to cases for which the angle between the tangent to the string at any point and the x-axis is very small) we obtain the one-dimensional wave equation

$$\frac{\partial^2 y}{\partial x^2} = \frac{1}{c^2} \frac{\partial^2 y}{\partial t^2}$$

where $c^2 = T/\rho$. To solve this equation put $u = x - ct$ and $v = x + ct$. Then

$$\frac{\partial y}{\partial x} = \frac{\partial y}{\partial u}\frac{\partial u}{\partial x} + \frac{\partial y}{\partial v}\frac{\partial v}{\partial x} = \frac{\partial y}{\partial u} + \frac{\partial y}{\partial v}$$

and so

$$\begin{aligned}
\frac{\partial^2 y}{\partial x^2} &= \frac{\partial}{\partial x}\left(\frac{\partial y}{\partial x}\right) \\
&= \frac{\partial}{\partial u}\left(\frac{\partial y}{\partial x}\right) + \frac{\partial}{\partial v}\left(\frac{\partial y}{\partial x}\right) \\
&= \frac{\partial}{\partial u}\left(\frac{\partial y}{\partial u} + \frac{\partial y}{\partial v}\right) + \frac{\partial}{\partial v}\left(\frac{\partial y}{\partial u} + \frac{\partial y}{\partial v}\right) \\
&= \frac{\partial^2 y}{\partial u^2} + 2\frac{\partial^2 y}{\partial u\, \partial v} + \frac{\partial^2 y}{\partial v^2}.
\end{aligned}$$

Similarly,

$$\begin{aligned}
\frac{\partial^2 y}{\partial t^2} &= \frac{\partial}{\partial t}\left(\frac{\partial y}{\partial u}(-c) + \frac{\partial y}{\partial v}c\right) \\
&= -c\frac{\partial}{\partial u}\left\{-c\frac{\partial y}{\partial u} + c\frac{\partial y}{\partial v}\right\} + c\frac{\partial}{\partial v}\left\{-c\frac{\partial y}{\partial u} + c\frac{\partial y}{\partial v}\right\} \\
&= c^2\left\{\frac{\partial^2 y}{\partial u^2} - 2\frac{\partial^2 y}{\partial u\, \partial v} + \frac{\partial^2 y}{\partial v^2}\right\}.
\end{aligned}$$

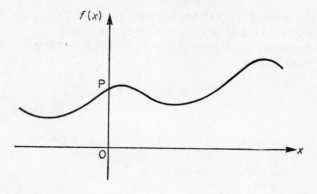

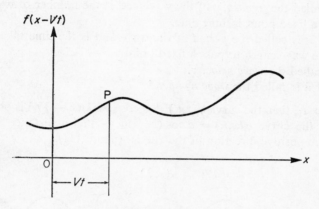

FIGURE 110

Thus the differential equation becomes

$$\frac{\partial^2 y}{\partial u\, \partial v} = 0.$$

Therefore

$$y = f(u) + g(v) = f(x - ct) + g(x + ct)$$

where f and g are arbitrary functions.

Hence the function $\varphi = f(x - Vt)$ where f is *any* continuous function and V is a constant represents mathematically a one-dimensional wave. Consideration of figure 110 illustrates that this wave moves with the constant velocity V in the positive direction of the x-axis without change of shape. The curve $\varphi = f(x - Vt)$ is obtained by moving the curve $\varphi = f(x)$ a distance Vt along the positive direction of the x-axis.

Similarly, $\varphi = g(x + Vt)$ represents a one-dimensional wave moving with velocity V in the negative direction of the x-axis.

In particular we shall be concerned mainly with harmonic waves for which φ has the form

$$a \cos \frac{2\pi}{\lambda}(x - Vt)$$

or

$$a \cos 2\pi \left(\frac{x}{\lambda} - nt\right) \quad \text{where} \quad V = n\lambda.$$

a is called the *amplitude* of the wave.

λ is called the *wavelength*.

n is called the *frequency* of the wave and is the number of wavelengths passing a fixed point in unit time.

$\tau = 1/n$ is called the *period* of the wave and is the time taken for one complete wavelength to pass a fixed point.

V is called the *wave velocity*.

$k = 1/\lambda$ is called the *wave number*.*

If $t_2 > t_1$, then the curve $\varphi(x,t_2) = a \cos (2\pi/\lambda)(x - Vt_2)$ is obtained by moving the curve $\varphi(x,t_1) = a \cos (2\pi/\lambda)(x - Vt_1)$ a distance $V(t_2 - t_1)$ along the positive direction of the x-axis, (figure 111).

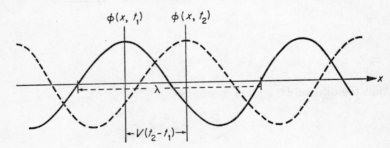

FIGURE 111

The function $\varphi(x,t) = a \cos (2\pi/\lambda)(x - Vt)$ represents the movement in the positive direction of the x-axis with velocity V of the cosine curve of amplitude a and wavelength λ.

If

$$\varphi_1 = a \cos \frac{2\pi}{\lambda}(x - Vt)$$

* Some texts define the wave number as $2\pi/\lambda$.

and

$$\varphi_2 = a \cos \frac{2\pi}{\lambda}\left(x - Vt + \frac{\lambda\varepsilon}{2\pi}\right) = a \cos \left\{\frac{2\pi}{\lambda}\left(x - Vt\right) + \varepsilon\right\},$$

we see that φ_1 and φ_2 each represent waves with amplitude a, wavelength λ and velocity V. In fact the wave represented by φ_2 is the same as the wave represented by φ_1 displaced a distance $\lambda\varepsilon/2\pi$. ε is called the *phase difference* between the waves represented by φ_1 and φ_2. If $\varepsilon = 0$, 2π, 4π, ... φ_1 and φ_2 are said to be in phase and their maxima coincide. If $\varepsilon = \pi$, 3π, 5π, ... φ_1 and φ_2 are said to be exactly out of phase and the maxima of φ_1 coincide with the minima of φ_2.

8.2 Plane waves and spherical waves

We have seen that the one-dimensional wave equation has the form

$$\frac{\partial^2 \varphi}{\partial x^2} = \frac{1}{V^2} \frac{\partial^2 \varphi}{\partial t^2}.$$

For one, two or three dimensions this equation may be written as

$$\nabla^2 \varphi = \frac{1}{V^2} \frac{\partial^2 \varphi}{\partial t^2}$$

where ∇^2 is the Laplacian operator of the appropriate dimension, and its solutions take different forms depending upon the coordinate system used.

Like Laplace's equation and the heat conduction equation this equation is linear so that if φ_1, φ_2, φ_3, ... are all solutions then

$$\alpha_1\varphi_1 + \alpha_2\varphi_2 + \alpha_3\varphi_3 + \ldots$$

where α_1, α_2, α_3, ... are arbitrary constants is also a solution.

8.21 Plane Waves

Plane waves are given by solutions of the wave equation using rectangular cartesian coordinates. The equation to be solved in three dimensions is

$$\frac{\partial^2 \varphi}{\partial x^2} + \frac{\partial^2 \varphi}{\partial y^2} + \frac{\partial^2 \varphi}{\partial z^2} = \frac{1}{V^2} \frac{\partial^2 \varphi}{\partial t^2}.$$

Put $u = lx + my + nz - Vt$ and $v = lx + my + nz + Vt$ where l, m, n are constants. Then

$$\frac{\partial^2 \varphi}{\partial x^2} = \frac{\partial}{\partial x}\left(\frac{\partial \varphi}{\partial u}\frac{\partial u}{\partial x} + \frac{\partial \varphi}{\partial v}\frac{\partial v}{\partial x}\right)$$

$$= \frac{\partial}{\partial x}\left(l\frac{\partial \varphi}{\partial u} + l\frac{\partial \varphi}{\partial v}\right)$$

$$= l^2 \frac{\partial}{\partial u}\left(\frac{\partial \varphi}{\partial u} + \frac{\partial \varphi}{\partial v}\right) + l^2 \frac{\partial}{\partial v}\left(\frac{\partial \varphi}{\partial u} + \frac{\partial \varphi}{\partial v}\right)$$

$$= l^2\left(\frac{\partial^2 \varphi}{\partial u^2} + 2\frac{\partial^2 \varphi}{\partial u\, \partial v} + \frac{\partial^2 \varphi}{\partial v^2}\right).$$

Similarly,

$$\frac{\partial^2 \varphi}{\partial y^2} = m^2\left(\frac{\partial^2 \varphi}{\partial u^2} + 2\frac{\partial^2 \varphi}{\partial u\, \partial v} + \frac{\partial^2 \varphi}{\partial v^2}\right),$$

$$\frac{\partial^2 \varphi}{\partial z^2} = n^2\left(\frac{\partial^2 \varphi}{\partial u^2} + 2\frac{\partial^2 \varphi}{\partial u\, \partial v} + \frac{\partial^2 \varphi}{\partial v^2}\right)$$

and

$$\frac{\partial^2 \varphi}{\partial t^2} = V^2\left(\frac{\partial^2 \varphi}{\partial u^2} - 2\frac{\partial^2 \varphi}{\partial u\, \partial v} + \frac{\partial^2 \varphi}{\partial v^2}\right).$$

Then the equation becomes

$$(l^2 + m^2 + n^2)\left(\frac{\partial^2 \varphi}{\partial u^2} + 2\frac{\partial^2 \varphi}{\partial u\, \partial v} + \frac{\partial^2 \varphi}{\partial v^2}\right) = \left(\frac{\partial^2 \varphi}{\partial u^2} - 2\frac{\partial^2 \varphi}{\partial u\, \partial v} + \frac{\partial^2 \varphi}{\partial v^2}\right).$$

This reduces to

$$\frac{\partial^2 \varphi}{\partial u\, \partial v} = 0$$

when $l^2 + m^2 + n^2 = 1$. The solution is then

$$\varphi = f(u) + g(v) = f(lx + my + nz - Vt) + g(lx + my + nz + Vt)$$

where f and g are arbitrary functions. From the form of this solution it is clear that, at any given time, the solution will be constant for all points on the surface $lx + my + nz = $ constant. This surface is called a *wave front*. Thus the wave fronts here are parallel planes and the constants l, m and n are the direction cosines of their normal.

In two dimensions the form of this solution of the wave equation in cartesian coordinates is

$$\varphi = f(lx + my - Vt) + g(lx + my - Vt)$$

where $l^2 + m^2 = 1$.

The equations of the wave fronts are $lx + my = $ constant for different values of the constant. Thus the wave fronts are parallel straight lines.

8.22 Spherical waves

Spherical waves are given by solutions of the wave equation using spherical polar coordinates independent of θ and ψ.

The equation to be solved is

$$\frac{\partial^2 \varphi}{\partial r^2} + \frac{2}{r}\frac{\partial \varphi}{\partial r} = \frac{1}{V^2}\frac{\partial^2 \varphi}{\partial t^2}$$

or

$$\frac{\partial^2}{\partial r^2}(r\varphi) = \frac{1}{V^2}\frac{\partial^2}{\partial t^2}(r\varphi).$$

By analogy with the one-dimensional cartesian equation, this has the solution

$$r\varphi = f(r - Vt) + g(r + Vt)$$

for arbitrary functions f and g.

Now it is clear that, at any given time, φ will have the same value at all points on the surface given by the equation $r = $ constant. Thus the wave fronts all have equations of the form $r = $ constant for different values of the constant and so are concentric spherical surfaces.

8.3 Stationary waves

The above solutions of the equation of wave motion are all what are called progressive solutions representing *progressive waves*, that is waves whose form 'progresses' (forwards or backwards) in space as time increases. We now consider the sum of two such harmonic waves having the same amplitude and moving in opposite directions with the same speed.

$$\varphi = a \cos 2\pi(kx - nt) + a \cos 2\pi(kx + nt)$$
$$= (2a \cos 2\pi nt) \cos 2\pi kx.$$

This is known as a *stationary wave* since its form does not move forward (or backward) in space as t increases.

The bracketed factor may be looked upon as the time-dependent amplitude of the wave. The wave form is sketched in figure 112 for four different values of the time t.

The points where $\varphi = 0$ for all t, that is where $\cos 2\pi kx$ or $\cos (2\pi x/\lambda)$ is zero and so $x = \pm \dfrac{\lambda}{4}$, $\pm \dfrac{3\lambda}{4}$, $\pm \dfrac{5\lambda}{4}$, ... are called *nodes* and the points where φ has its maximum numerical values with respect to x, that is

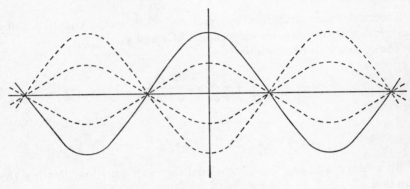

FIGURE 112

where $\cos 2\pi kx$ or $\cos (2\pi x/\lambda)$ is equal to ± 1 and so $x = 0$, $\pm \dfrac{\lambda}{2}$, $\pm \lambda$, $\pm \dfrac{3\lambda}{2}$, ... are called *antinodes*.

It is also worth noting here that the progressive wave represented by

$$\varphi = a \cos 2\pi(kx - nt)$$

may be written as

$$\varphi = (a \cos 2\pi nt) \cos 2\pi kx + (a \sin 2\pi nt) \sin 2\pi kx$$

that is as the sum of two stationary waves.

Solutions of the wave equation in terms of stationary waves may be obtained using the method of separation of variables as for Laplace's equation and the heat conduction equation. As an illustration of the general method consider the following example.

A string of length l is fixed at both ends. If the initial displacement at each point of the string is given by $f(x)$ for $0 \leqslant x \leqslant l$ and the initial normal velocity of each point of the string is given by $g(x)$ for $0 \leqslant x \leqslant l$, find the displacement y of all points of the string at any subsequent time t.

The problem is essentially to obtain a solution of the equation

$$\frac{\partial^2 y}{\partial x^2} = \frac{1}{c^2} \frac{\partial^2 y}{\partial t^2}$$

satisfying the boundary conditions $y = 0$ at $x = 0$, l for all $t \geqslant 0$ and the initial conditions $y = f(x)$, $\partial y/\partial t = g(x)$ when $t = 0$ for $0 \leqslant x \leqslant l$.

Using the method of separation of variables, we put $y = X(x)T(t)$. Therefore

$$\frac{1}{X} \frac{d^2 X}{dx^2} = \frac{1}{c^2 T} \frac{d^2 T}{dt^2}.$$

Separate the variables by putting

$$\frac{1}{X} \frac{d^2 X}{dx^2} = -p^2.$$

Hence

$$X = A \cos px + B \sin px$$

for arbitrary constants A and B and $p \neq 0$. Also

$$\frac{1}{c^2 T} \frac{d^2 T}{dt^2} = -p^2$$

and so

$$T = C \cos pct + D \sin pct$$

for arbitrary constants C and D and $p \neq 0$. Therefore

$$y = (A \cos px + B \sin px)(C \cos pct + D \sin pct) \quad \text{for} \quad p \neq 0.\dagger$$

When $x = 0$, $0 = A(C \cos pct + D \sin pct)$ for $t \geqslant 0$.

Therefore $A = 0$, and so

$$y = \sin px(E \cos pct + F \sin pct) \qquad (E = BC \quad \text{and} \quad F = BD).$$

When $x = l$, $0 = \sin pl(E \cos pct + F \sin pct)$ for $t \geqslant 0$.

Therefore $\sin pl = 0$ and so

$$p = \frac{n\pi}{l}$$

for integer values of n. Hence

$$y = \sin \frac{n\pi x}{l} \left(E_n \cos \frac{n\pi ct}{l} + F_n \sin \frac{n\pi ct}{l} \right)$$

† The solution corresponding to $p = 0$ is $y = (A_0 x + B_0)(C_0 t + D_0)$. All the arbitrary constants must be zero however since y does not increase indefinitely with time and is zero at both ends of the string.

is a solution satisfying the boundary conditions of the problem. Then, since the differential equation is linear, the general solution of the above form satisfying the above boundary conditions is

$$y = \sum_{n=1}^{\infty} \sin \frac{n\pi x}{l} \left(E_n \cos \frac{n\pi ct}{l} + F_n \sin \frac{n\pi ct}{l} \right)$$

(As before it is not necessary to consider negative integer values of n). Now when $t = 0$, we have

$$f(x) = \sum_{n=1}^{\infty} E_n \sin \frac{n\pi x}{l}$$

and so for suitable functions $f(x)$ the E_n's are found by Fourier's method as

$$E_n = \frac{2}{l} \int_0^l f(x) \sin \frac{n\pi x}{l} \, dx.$$

Also,

$$g(x) = \left(\frac{\partial y}{\partial t} \right)_{t=0} = \sum_{n=1}^{\infty} \frac{n\pi c}{l} F_n \sin \frac{n\pi x}{l}.$$

Then, again using Fourier's method, we obtain

$$F_n = \frac{2}{n\pi c} \int_0^l g(x) \sin \frac{n\pi x}{l} \, dx$$

and so the given problem has now been solved. The solution

$$y = \sin \frac{n\pi x}{l} \left(E_n \cos \frac{n\pi ct}{l} + F_n \sin \frac{n\pi ct}{l} \right)$$

for any positive integer n is called a *normal mode* of vibration.
The particular solution corresponding to $n = 1$, that is

$$y = \sin \frac{\pi x}{l} \left(E_1 \cos \frac{\pi ct}{l} + F_1 \sin \frac{\pi ct}{l} \right)$$

is called the *fundamental mode* of vibration. This may be written as

$$y = a \sin \frac{\pi x}{l} \sin \left(\frac{\pi ct}{l} + \varepsilon \right)$$

for constants a and ε and so we see that it corresponds to a stationary wave of period

$$\frac{2l}{c} = 2l \sqrt{\frac{\rho}{T}}.$$

The only nodes occur at the ends of the string (figure 113). Also we see that the frequency of the fundamental mode of a string of fixed density can be increased by decreasing the length l or increasing its tension T. Both these effects are used in musical instruments. For example a violin, a guitar and a piano are all tuned by adjusting the tensions in their strings. Also, a guitarist sometimes effectively uses a shorter length of string by

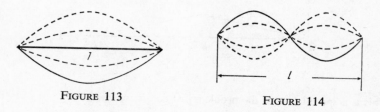

FIGURE 113   FIGURE 114

clamping it tightly at some point of its length. Decreasing the density of a string of fixed length will increase the period of the fundamental note and this effect is also used in musical instruments. For example the different strings of a violin or guitar have different densities. The *second normal mode* is the particular solution corresponding to $n = 2$ and may be written

$$y = a_1 \sin \frac{2\pi x}{l} \sin \left(\frac{2\pi ct}{l} + \varepsilon_1 \right)$$

for constants a_1 and ε_1. (In music this normal mode is referred to as the *first harmonic* or *first overtone*.) In this case there is also a node at the midpoint of the string (figure 114). The period of vibration is l/c, that is half that of the fundamental. Similarly higher order modes (or harmonics) are defined. The mth normal mode or $(m - 1)$th harmonic is the solution corresponding to $n = m$ and may be written

$$y = a_m \sin \frac{m\pi x}{l} \sin \left(\frac{m\pi ct}{l} + \varepsilon_m \right)$$

and has nodes at the points

$$x = 0, \frac{l}{m}, \frac{2l}{m}, \ldots, l.$$

Example   A string of length l is fixed at both ends. Its midpoint is displaced a small distance h normal to the straight line between the end points and then released from rest in this position (figure 115). Determine the displacement of all points of the string at any subsequent time t.

Determine also the resulting displacement if initially, instead of displacing the midpoint, a point of trisection of the string is displaced a distance h (figure 116).

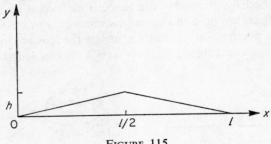

FIGURE 115

This is a special case of the problem just considered with

$$f(x) = \begin{cases} 2hx/l & 0 \leqslant x \leqslant l/2, \\ (2h/l)(l-x) & l/2 < x \leqslant l \end{cases}$$

and

$$g(x) = 0 \qquad 0 \leqslant x \leqslant 1.$$

Hence, as above, we obtain a solution of the form

$$y = \sum_{n=1}^{\infty} E_n \sin \frac{n\pi x}{l} \cos \frac{n\pi ct}{l}$$

where

$$\begin{aligned} E_n &= \frac{2}{l} \int_0^l f(x) \sin \frac{n\pi x}{l} \, dx \\ &= \frac{2}{l} \left\{ \int_0^{l/2} \frac{2hx}{l} \sin \frac{n\pi x}{l} \, dx + \int_{l/2}^l \frac{2h}{l} (l-x) \sin \frac{n\pi x}{l} \, dx \right\} \\ &= \begin{cases} \dfrac{8h(-1)^{(n-1)/2}}{n^2\pi^2} & n \text{ odd}, \\ 0 & n \text{ even}. \end{cases} \end{aligned}$$

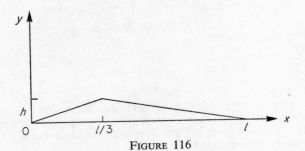

FIGURE 116

That is

$$y = \frac{8h}{\pi^2} \sum_{m=1}^{\infty} \frac{(-1)^{m-1}}{(2m-1)^2} \sin \frac{(2m-1)\pi x}{l} \cos \frac{(2m-1)\pi ct}{l}.$$

Note that the 2nd, 4th, 6th, . . . normal modes are all absent from this solution. When the point of trisection $x = \frac{1}{3}l$ is displaced initially,

$$f(x) = \begin{cases} (3h/l)x & 0 \leqslant x \leqslant l/3, \\ (3h/2l)(l-x) & l/3 < x \leqslant l \end{cases}$$

and

$$g(x) = 0 \qquad 0 \leqslant x \leqslant 1.$$

Hence

$$y = \sum_{n=1}^{\infty} E'_n \sin \frac{n\pi x}{l} \cos \frac{n\pi ct}{l}$$

where

$$E'_n = \frac{2}{l} \int_0^l f(x) \sin \frac{n\pi x}{l} \, dx$$

$$= \frac{2}{l} \left\{ \int_0^{l/3} \frac{3hx}{l} \sin \frac{n\pi x}{l} \, dx + \int_{l/3}^l \frac{3h}{2l} (l-x) \sin \frac{n\pi x}{l} \, dx \right\}$$

$$= \frac{9h}{n^2\pi^2} \sin \frac{n\pi}{3} = \begin{cases} \frac{9\sqrt{3}h}{2n^2\pi^2} (-1)^{[n/3]} & n \neq 3m, \\ 0 & n = 3m \end{cases}$$

for positive integers m, where $[n/3]$ denotes the integer part of $n/3$.

Note that the 3rd, 6th, 9th, . . . normal modes are all absent from this solution.

Note also that while, in the first case the 5th normal mode is in phase with the fundamental it is completely out of phase with the fundamental in the second case.

The absence of different normal modes from the two solutions together with the above phase differences means that musically a different note is produced when a string is plucked at its midpoint from what is produced when it is plucked at a point of trisection.

In general, due to the different normal modes produced, the quality of a note from a stretched string will depend upon where it is plucked (although the fundamental frequency will remain the same no matter where it is plucked).

8.31 A string with a mass attached at some point of its length

Let a mass m be attached to the string at a distance d from one end. The wave equation will be satisfied by the displacements of the portions of the string on each side of the mass and so the general form of the solution in each of these regions can be determined using the method of separation of variables. It remains however to determine the boundary conditions linking these two solutions together at the point of attachment of the mass.

Let y_1 and y_2 be the displacements of the string to the left and to the right of the mass m respectively. Since the string is continuous, we must have $y_1 = y_2$ at $x = d$ for all t.

Then, with the notation of figure 117, we see that the equation of

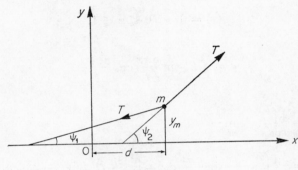

FIGURE 117

motion of the mass m is

$$T \sin \psi_2 - T \sin \psi_1 = m\ddot{y}_m$$

where y_m is its displacement.

But for small displacements

$$\sin \psi_1 \simeq \tan \psi_1 = \frac{\partial y_1}{\partial x}$$

and

$$\sin \psi_2 \simeq \tan \psi_2 = \frac{\partial y_2}{\partial x}.$$

Therefore to this approximation we have

$$T\left(\frac{\partial y_2}{\partial x} - \frac{\partial y_1}{\partial x}\right) \simeq m\ddot{y}_m \quad \text{at} \quad x = d.$$

Note that since $y_1 = y_2$ at $x = d$ for all t, $\ddot{y}_1 = \ddot{y}_2$ at $x = d$ and each is equal to $\ddot{y}_m$.

Example   *Obtain the normal modes of vibration of a string of length l and total mass M fixed at both ends and having a mass m attached at a point of trisection (figure 118).*

The displacement y_1 satisfies the wave equation

$$\frac{\partial^2 y_1}{\partial x^2} = \frac{1}{c^2} \frac{\partial^2 y_1}{\partial t^2}.$$

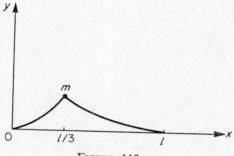

FIGURE 118

Using the method of separation of variables we obtain the solution

$$y_1 = (A_1 \cos px + B_1 \sin px)(C_1 \cos pct + D_1 \sin pct)$$

for arbitrary constants A_1, B_1, C_1 and D_1. This may be rewritten in the alternative form

$$y_1 = (a_1 \cos px + b_1 \sin px) \cos (pct + \varepsilon).$$

Now $y_1 = 0$ at $x = 0$ for all t, and so $a_1 = 0$.
    Therefore

$$y_1 = b_1 \sin px \cos (pct + \varepsilon).$$

Since y_2 also satisfies the wave equation and is zero at $x = l$, it is readily seen that y_2 may be written in the form

$$y_2 = b_2 \sin p(l - x) \cos (pct + \varepsilon).$$

Note that the same values of p and ε are used for both y_1 and y_2. This is essential, for if the two sections of the string are not to break apart, then the vibrations in both sections must be in phase and have the same frequency.
    Now $y_1 = y_2$ at $x = l/3$ for all t, and so

$$b_1 \sin \frac{pl}{3} = b_2 \sin \frac{2pl}{3}.$$

Also

$$T\left(\frac{\partial y_2}{\partial x} - \frac{\partial y_1}{\partial x}\right) = m\ddot{y}_m \quad \text{at} \quad x = \frac{l}{3}.$$

Therefore

$$T\left\{-pb_2 \cos \frac{2pl}{3} - pb_1 \cos \frac{pl}{3}\right\} \cos (pct + \varepsilon)$$

$$= -mp^2c^2b_1 \sin \frac{pl}{3} \cos (pct + \varepsilon)$$

for all t. (Here we have used $\ddot{y}_m = \ddot{y}_1$ at $x = l/3$, but the same final result would be obtained by using $\ddot{y}_m = \ddot{y}_2$ at $x = l/3$.)
Therefore

$$\frac{Mc^2}{l} b_1\left\{\frac{\sin pl/3}{\sin 2pl/3} \cos \frac{2pl}{3} + \cos \frac{pl}{3}\right\} = mpc^2b_1 \sin \frac{pl}{3}$$

since $T = c^2\rho = c^2\dfrac{M}{l}$.

Therefore

$$\sin \frac{pl}{3} \cos \frac{2pl}{3} + \cos \frac{pl}{3} \sin \frac{2pl}{3} = \frac{mpl}{M} \sin \frac{pl}{3} \sin \frac{2pl}{3}.$$

Hence

$$\sin pl = \frac{mpl}{M} \sin \frac{pl}{3} \sin \frac{2pl}{3}$$

or, on putting $\theta = pl/3$ and $m/M = k$,

$$\sin 3\theta = 3k\theta \sin \theta \sin 2\theta.$$

Solutions of this equation may be determined graphically or numerically. One set of solutions is obviously given by $\theta = n\pi$ and so $p = 3n\pi/l$ for integers n.

The periods of the normal modes are of course given by $2\pi/cp_i$ where the values of p_i are the solutions of the above equation for p.

Note that the particular displacement of any point of the string at any subsequent time t cannot be determined until initial conditions are specified enabling values for the constants b_1, b_2 and ε to be obtained.

8.32 *Two strings with unequal densities tied together*

Let the two strings, having densities ρ_1 and ρ_2, be joined together at $x = d$ (figure 119). The displacement of each string will again satisfy the

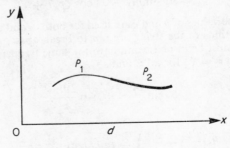

FIGURE 119

wave equation, so that the general form of the solution can be obtained for each using the method of separation of variables. It then remains to determine the boundary conditions linking these two solutions together at the join of the strings.

Let y_1 and y_2 be the displacements of the strings of densities ρ_1 and ρ_2 respectively.

Since the two strings remain joined together, we must have $y_1 = y_2$ at $x = d$ for all t.

Also, since there is no discrete mass attached at the join of the strings, we see (on putting $m = 0$ in the corresponding condition of the previous section) that

$$\frac{\partial y_1}{\partial x} = \frac{\partial y_2}{\partial x}$$

at $x = d$ for all t.

Example   Obtain the normal modes of vibration of a string of length l formed by joining together a string of length l/3 and density ρ_1 and a string of length 2l/3 and density ρ_2 when the other ends of the two strings are fixed (figure 120).

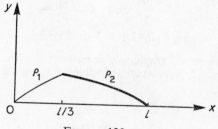

FIGURE 120

Let y_1 and y_2 be the displacements of the strings of densities ρ_1 and ρ_2 respectively.

The displacement y_1 satisfies the wave equation and so, using the method of separation of variables and inserting the condition that $y_1 = 0$ at $x = 0$ for all t, we obtain (as in the last example) that

$$y_1 = b_1 \sin px \cos (pc_1 t + \varepsilon)$$

where $c_1 = \sqrt{T/\rho_1}$ and T is the tension in the string.

Since y_2 also satisfies the wave equation and is zero at $x = l$, we have that

$$y_2 = b_2 \sin p'(l - x) \cos (p'c_2 t + \varepsilon)$$

where $c_2 = \sqrt{T/\rho_2}$. Now, since the two strings do not break apart, the vibrations in both must have the same frequency and so

$$pc_1 = p'c_2.$$

Therefore

$$y_2 = b_2 \sin \frac{pc_1}{c_2} (l - x) \cos (pc_1 t + \varepsilon).$$

Now $y_1 = y_2$ at $x = l/3$ for all t and so

$$b_1 \sin \frac{pl}{3} = b_2 \sin \frac{2plc_1}{3c_2}.$$

Also

$$\frac{\partial y_1}{\partial x} = \frac{\partial y_2}{\partial x} \quad \text{at} \quad x = \frac{l}{3}$$

for all t and so

$$pb_1 \cos \frac{pl}{3} = -pb_2 \frac{c_1}{c_2} \cos \frac{2plc_1}{3c_2}.$$

Therefore

$$\cos \frac{pl}{3} \sin \left(\frac{2pl}{3} \frac{c_1}{c_2}\right) + \frac{c_1}{c_2} \sin \frac{pl}{3} \cos \left(\frac{2pl}{3} \frac{c_1}{c_2}\right) = 0$$

or, on putting

$$\theta = \frac{pl}{3} \quad \text{and} \quad k = \frac{c_1}{c_2} = \sqrt{\frac{\rho_2}{\rho_1}},$$

$$\cos \theta \sin^2 k\theta + k \sin \theta \cos^2 k\theta = 0.$$

This can again be solved graphically or numerically. The periods of the normal modes are given by $2\pi/c_1 p_i$ where the values of p_i are the solutions of the above equation for p.

Exercises

1. A string of length l has both its ends fixed on a smooth horizontal table. If the points at distance $\frac{1}{4}l$ from either end of the string are pulled aside each a small distance h on opposite sides of the equilibrium position and then released from rest, find the subsequent configuration of the string.

2. A string of total mass M and length $4l$ is fixed at both ends and has a particle of mass m attached at a distance l from one end. Show that for small transverse oscillations the wave lengths of the normal modes are given by $2\pi/p$ where p satisfies the equation

$$M(\cot 3pl + \cot pl) = 4mpl.$$

3. A string of length $(1 + k)l$ consists of two parts, one of length l and density ρ and the other of length kl and density ρ'. Show that the periods $2\pi/p$ of normal oscillations are given by

$$\alpha \cot \alpha l + \alpha' \cot \alpha' kl = 0$$

where $\alpha^2 = p^2\rho/T$, $\alpha' = p^2\rho'/T$ and T is the tension in the string. Show also that normal oscillations in which the join of the two parts of the string remains at rest are possible only if $\sqrt{(\rho/\rho')}/k$ is a rational fraction.

4. At $t = 0$ a uniform stretched string is released from rest in the shape $y = e^{-x^2}$. Show that at any subsequent time t its shape is given by

$$y = \tfrac{1}{2}\{e^{-(x-ct)^2} + e^{-(x+ct)^2}\}$$

in which $c^2 = T/\rho$ and T is the tension in the string and ρ is the mass per unit length of the string.

8.4 Waves in membranes

In the present context a membrane is a thin piece of perfectly elastic material.

A membrane is said to be under constant tension T if when a cut of length d is made anywhere in the membrane the total force required to hold either edge in its original position is Td.

We consider the transverse vibrations of a two-dimensional membrane whose weight is negligible compared to its tension and assume that its tension T remains constant throughout the motion. As in the case of waves on a string, this assumption is reasonable for 'smooth' waves of small amplitude.

Choose cartesian coordinate axes $Oxyu$ such that Oxy is the plane of the undisplaced membrane and all displacements are parallel to the axis Ou.

Consider a small rectangular element ABCD of the undisturbed membrane having edges of lengths δx and δy (figure 121).

In the displaced position, the component of force parallel to Ou due to the tensions $T\,\delta y$ on the edges BC and AD is (with the notation of figure 121)

$$T\,\delta y \sin(\psi + \delta\psi) - T\,\delta y \sin\psi = T\,\delta y(\cos\psi\,\delta\psi + O(\delta\psi)^2)$$

Similarly, the component of force parallel to Ou due to the tensions $T\,\delta x$ on the edges AB and DC is

$$T\,\delta x \sin(\chi + \delta\chi) - T\,\delta x \sin\chi = T\,\delta x(\cos\chi\,\delta\chi + O(\delta\chi)^2).$$

Then from Newton's law of motion, we obtain

$$T\,\delta y(\cos\psi\,\delta\psi + O(\delta\psi)^2) + T\,\delta x(\cos\chi\,\delta\chi + O(\delta\chi)^2) \simeq \rho\,\delta x\,\delta y\frac{\partial^2 u}{\partial t^2}$$

where ρ is the mass per unit area of the membrane.

13

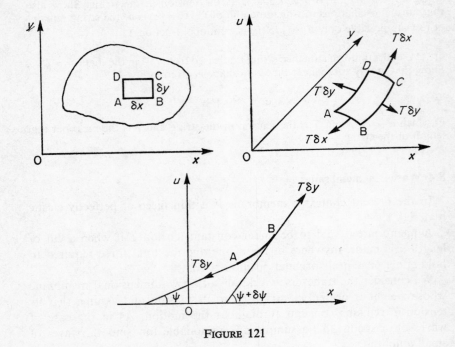

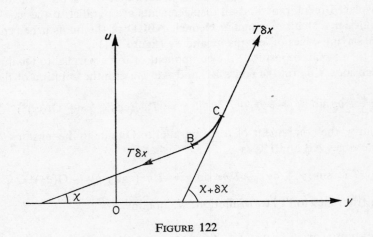

FIGURE 121

FIGURE 122

Hence

$$T \cos \psi \, \frac{\partial \psi}{\partial x} + T \cos \chi \, \frac{\partial \chi}{\partial y} = \rho \, \frac{\partial^2 u}{\partial t^2}.$$

But

$$\tan \psi = \frac{\partial u}{\partial x} \quad \text{and} \quad \tan \chi = \frac{\partial u}{\partial y}.$$

Therefore

$$\sec^2 \psi \, \frac{\partial \psi}{\partial x} = \frac{\partial^2 u}{\partial x^2} \quad \text{and} \quad \sec^2 \chi \, \frac{\partial \chi}{\partial y} = \frac{\partial^2 u}{\partial y^2}.$$

Hence

$$T \left(\frac{\partial^2 u}{\partial x^2} \cos^3 \psi + \frac{\partial^2 u}{\partial y^2} \cos^3 \chi \right) = \rho \, \frac{\partial^2 u}{\partial t^2}.$$

Now, restricting consideration to small displacements of the membrane for which $\cos^3 \psi \simeq 1$ and $\cos^3 \chi \simeq 1$, we obtain the two-dimensional wave equation

$$\frac{\partial^2 u}{\partial x^2} + \frac{\partial^2 u}{\partial y^2} = \frac{1}{c^2} \frac{\partial^2 u}{\partial t^2}$$

where once again $c^2 = T/\rho$.

As an illustration of the method of solution using separation of variables consider the following example.

A rectangular membrane fixed round its edge has sides of length a and b. Taking rectangular cartesian coordinate axes Ox and Oy along an adjacent pair of perpendicular edges (figure 123) the displacement of any point of the membrane is given by $f(x,y)$ for $0 \leqslant x \leqslant a$, $0 \leqslant y \leqslant b$, while the normal velocity of every point of the membrane is zero. Obtain an expression giving the displacement u at any point (x,y) of the membrane at any subsequent time t.

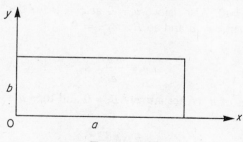

FIGURE 123

The displacement u satisfies the equation

$$\frac{\partial^2 u}{\partial x^2} + \frac{\partial^2 u}{\partial y^2} = \frac{1}{c^2}\frac{\partial^2 u}{\partial t^2},$$

the boundary conditions

$$u = 0 \quad \text{for} \quad x = 0, a \quad \text{and} \quad 0 \leqslant y \leqslant b,$$
$$u = 0 \quad \text{for} \quad y = 0, b \quad \text{and} \quad 0 \leqslant x \leqslant a,$$

and the initial conditions $u = f(x,y)$ for $0 \leqslant x \leqslant a$ and $0 \leqslant y \leqslant b$ when $t = 0$ and $\partial u/\partial t = 0$ for $0 \leqslant x \leqslant a$ and $0 \leqslant y \leqslant b$ when $t = 0$. Putting $u = X(x)Y(y)T(t)$ in the differential equation we obtain

$$\frac{1}{X}\frac{\mathrm{d}^2 X}{\mathrm{d}x^2} + \frac{1}{Y}\frac{\mathrm{d}^2 Y}{\mathrm{d}y^2} = \frac{1}{c^2 T}\frac{\mathrm{d}^2 T}{\mathrm{d}t^2}.$$

Put

$$\frac{1}{X}\frac{\mathrm{d}^2 X}{\mathrm{d}x^2} = -p^2.$$

Therefore

$$X = A\cos px + B\sin px.$$

Put

$$\frac{1}{Y}\frac{\mathrm{d}^2 Y}{\mathrm{d}y^2} = -q^2.$$

Therefore

$$Y = C\cos qy + D\sin qy.$$

Then

$$\frac{1}{T}\frac{\mathrm{d}^2 T}{\mathrm{d}t^2} = -c^2(p^2 + q^2) = -c^2 r^2$$

where $r^2 = p^2 + q^2$.

Hence

$$T = E\cos rct + F\sin rct.$$

Now since $u = 0$ at $x = 0$ for $0 \leqslant y \leqslant b$ and all $t \geqslant 0$, $X = 0$ when $x = 0$ and so $A = 0$. Also, since $u = 0$ at $x = a$ for $0 \leqslant y \leqslant b$ and all $t \geqslant 0$, $X = 0$ when $x = a$ and so $B\sin pa = 0$.

Hence

$$p = \frac{n\pi}{a}$$

for integer values of n. (Alternatively $B = 0$ and then $u \equiv 0$.)

Therefore

$$X = B\sin\frac{n\pi x}{a}.$$

Similarly, since $u = 0$ at $y = 0, b$ for $0 \leqslant x \leqslant a$ and all $t \geqslant 0$,

$$Y = D \sin \frac{m\pi y}{b}$$

for integer values of m. Hence the solution for u may now be written

$$u = \sin \frac{n\pi x}{a} \sin \frac{m\pi y}{b} (E' \cos rct + F' \sin rct)$$

where $E' = BDE$ and $F' = BDF$ are arbitrary constants and

$$r^2 = \pi^2(n^2/a^2 + m^2/b^2)$$

Therefore the general solution of this form is

$$u = \sum_{n,m} \sin \frac{n\pi x}{a} \sin \frac{m\pi y}{b} (E_{n,m} \cos rct + F_{n,m} \sin rct).$$

The coefficients $E_{n,m}$ and $F_{n,m}$ can now be found from the initial conditions using a double Fourier series as follows.

$$\frac{\partial u}{\partial t} = \sum_{n,m} \left(\sin \frac{n\pi x}{a} \sin \frac{m\pi y}{b} \right)(-E_{n,m} \sin rct + F_{n,m} \cos rct)rc.$$

Hence, since $\partial u/\partial t = 0$ when $t = 0$ and for all x, y such that $0 \leqslant x \leqslant a$ and $0 \leqslant y \leqslant b$, $F_{n,m} = 0$ for all integers n and m.

Therefore

$$u = \sum_{n,m} E_{n,m} \sin \frac{n\pi x}{a} \sin \frac{m\pi y}{b} \cos rct.$$

But $u = f(x,y)$ when $t = 0$ and so

$$f(x,y) = \sum_{n,m} E_{n,m} \sin \frac{n\pi x}{a} \sin \frac{m\pi y}{b}$$

for $0 \leqslant x \leqslant a$ and $0 \leqslant y \leqslant b$. Hence, as in Chapter 4, we obtain

$$E_{n,m} = \frac{4}{ab} \int_0^b \int_0^a f(x,y) \sin \frac{n\pi x}{a} \sin \frac{m\pi y}{b} \, dx \, dy$$

and so the problem is solved.

For particular values of n, m the expression for u is the displacement corresponding to the (n,m) mode. Its frequency is

$$\frac{rc}{2\pi} = \sqrt{\frac{T}{4\rho} \left(\frac{n^2}{a^2} + \frac{m^2}{b^2} \right)}.$$

For the (4,3) mode

$$u = E_{4,3} \sin \frac{4\pi x}{a} \sin \frac{3\pi y}{b} \sin rct$$

where

$$r^2 = \pi^2 \left(\frac{16}{a^2} + \frac{9}{b^2} \right).$$

Hence the lines $x = 0, \frac{1}{4}a, \frac{1}{2}a, \frac{3}{4}a, a$ and the lines $y = 0, \frac{1}{3}b, \frac{2}{3}b, b$ remain permanently at rest and are called nodal lines.

The displacements on opposite sides of a nodal line have opposite senses. This normal mode is depicted in figure 124 where the shaded parts

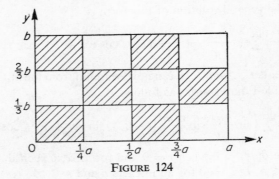

FIGURE 124

indicate regions for which the displacement is in the opposite sense to that in the unshaded parts.

It can happen that two different modes have the same frequency (for example in the case of a square the (1,3) and the (3,1) modes will have the same frequency). This is called a *degenerate* case.

Example   A circular drum skin of radius a is fixed round its rim. In terms of plane polar coordinates (r,θ) with origin at the centre of the drum skin, the displacement of any point at $t = 0$ is given by $f(r,\theta)$. If the drum skin is also at rest at this time, obtain an expression giving the displacement u at any point (r,θ) of the drum skin at any subsequent time.

The displacement u satisfies the equation

$$\nabla^2 u = \frac{\partial^2 u}{\partial r^2} + \frac{1}{r} \frac{\partial u}{\partial r} + \frac{1}{r^2} \frac{\partial^2 u}{\partial \theta^2} = \frac{1}{c^2} \frac{\partial^2 u}{\partial t^2}.$$

The boundary condition is $u = 0$ for $r = a$, all θ and all $t \geqslant 0$. The initial conditions are

$$u = f(r,\theta) \quad \text{for} \quad 0 \leqslant r \leqslant a \quad \text{and all } \theta \quad \text{when} \quad t = 0$$

and

$$\partial u / \partial t = 0 \quad \text{for} \quad 0 \leqslant r \leqslant a \quad \text{and all } \theta \text{ when } t = 0.$$

Putting $u = R(r)F(\theta)T(t)$ in the differential equation we obtain

$$\frac{1}{R}\frac{d^2R}{dr^2} + \frac{1}{rR}\frac{dR}{dr} + \frac{1}{r^2F}\frac{d^2F}{d\theta^2} = \frac{1}{c^2T}\frac{d^2T}{dt^2}.$$

Then separating the variables by putting

$$\frac{1}{c^2T}\frac{d^2T}{dt^2} = -p^2 \quad \text{and} \quad \frac{1}{F}\frac{d^2F}{d\theta^2} = -m^2$$

we obtain

$$T = A \cos pct + B \sin pct$$

and

$$F = C \cos m\theta + D \sin m\theta$$

where m must be an integer in order that F should be single valued in θ. Also

$$\frac{1}{R}\frac{d^2R}{dr^2} + \frac{1}{rR}\frac{dR}{dr} - \frac{m^2}{r^2} = -p^2$$

and therefore

$$r^2\frac{d^2R}{dr^2} + r\frac{dR}{dr} + (p^2r^2 - m^2)R = 0.$$

This is Bessel's equation of order m and independent variable pr. The general solution finite at the origin is therefore

$$R = EJ_m(pr).$$

Therefore $u = J_m(pr)(C' \cos m\theta + D' \sin m\theta)(A \cos pct + B \sin pct)$ where $C' = EC$ and $D' = ED$ are arbitrary constants.

Now by proper choice of the axis $\theta = 0$ we can ensure that $D' = 0$ and so obtain

$$u = J_m(pr) \cos m\theta (A' \cos pct + B' \sin pct)$$

where $A' = AC$ and $B' = BC'$.

Since $u = 0$ at $r = a$ for all θ and all $t \geqslant 0$,

$$J_m(pa) = 0.$$

The roots of this equation can be found from tables of Bessel functions. Let the corresponding values of p be $p_{m,1}, p_{m,2}, \ldots$. Then the general solution of the above form satisfying the boundary conditions is

$$u = \sum_{m,k} J_m(p_{m,k}r) \cos m\theta (A_{m,k} \cos p_{m,k}ct + B_{m,k} \sin p_{m,k}ct).$$

Now since $\partial u / \partial t = 0$ for $0 \leqslant r \leqslant a$ and all θ when $t = 0$, $B_{m,k} = 0$ for all positive integers m and k.

Therefore

$$u = \sum_m \sum_k A_{m,k} J_m(p_{m,k} r) \cos m\theta \cos p_{m,k} ct.$$

But $u = f(r,\theta)$ when $t = 0$ and so

$$f(r,\theta) = \sum_{m,k} [A_{m,k} J_m(p_{m,k} r)] \cos m\theta.$$

Hence, using the result for Fourier series (Chapter 4), we obtain

$$\int_0^\pi f(r,\theta) \cos m\theta \, d\theta = \frac{\pi}{2} \sum_k A_{m,k} J_m(p_{m,k} r).$$

Then, using the corresponding result for expansion of a function as a series of Bessel functions (Chapter 5), we obtain

$$A_{m,k} = \frac{4}{\pi a^2 [J_{m+1}(p_{m,k} a)]^2} \int_0^a \int_0^\pi r f(r,\theta) \cos m\theta J_m(p_{m,k} r) \, d\theta \, dr$$

and so the problem is solved.

For particular values of m and k the expression for u is the displacement corresponding to the (m,k) mode.

Its frequency is

$$\frac{p_{m,k} c}{2\pi}.$$

For the (4,3) mode

$$u = A_{4,3} J_4(p_{4,3} r) \cos 4\theta \cos p_{4,3} ct$$

where $ap_{4,3}$ is the third zero of J_4 and so from tables is approximately equal to $14 \cdot 37$. Since $7 \cdot 59$ and $11 \cdot 06$ are also zeros of J_4, it follows that $r = (7 \cdot 59/14 \cdot 37)a$, $(11 \cdot 06/14 \cdot 37)a$ are nodal lines for this mode. It is also clearly seen that the lines $\theta = (\frac{1}{8}\pi + i\frac{1}{4}\pi)$ for $i = 0, 1, 2, \ldots, 7$ are also nodal lines.

This normal mode is depicted in figure 125.

Normal modes are also called *eigenfunctions* and it can be shown that for most reasonable vibrations the displacement can be represented by a sum of normal modes.

The concept of eigenfunctions is a very important one and deserves further comment. When considering stationary waves on strings or in membranes separation of variables, in the cases we considered, invariably led to solving an equation of the form

$$u'' + \lambda u = 0$$

with boundary conditions $u = 0$ or $u' = 0$ when $x = 0$ and $x = l$, and it was found that a non-trivial solution existed only for certain values of the parameter λ. The same problem arose when solving Laplace's equation and the heat-conduction equation using separation of variables and is an example of the class of problems called *eigenvalue problems*. More

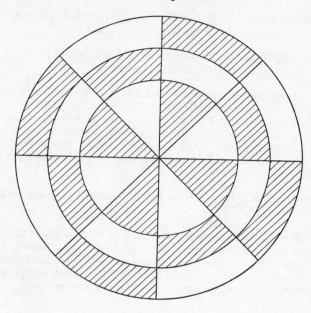

FIGURE 125

generally, if a second-order differential equation for $u(x)$ can be written in the form

$$p(x)u''(x) + p'(x)u'(x) + (q(x) + \lambda h(x))u(x) = 0, \qquad \alpha < x < \beta$$

and if the solution must satisfy boundary conditions of the form

$$a_1u(\alpha) + b_1u'(\alpha) = 0,$$
$$a_2u(\beta) + b_2u'(\beta) = 0$$

for constants a_1, a_2, b_1, b_2 then a non-trivial solution will only exist for certain values of the parameter λ. These special values of λ are called *eigenvalues* and the corresponding solutions $u(x)$ are called *eigenfunctions* of the problem.

Exercises

1. A membrane is stretched to tension T over a rectangular frame whose edges have lengths $3a$ and $4a$. Show that the period of the slowest vibration is

$$\frac{24a}{5} \sqrt{\frac{\rho}{T}}$$

where the total mass of the membrane is $12a^2\rho$.

2. Illustrate graphically (as in the text) the (3,2) and the (2,3) normal modes of oscillation of a rectangular membrane of sides a and b.

3. Find the nodal lines for a square membrane of side a when the (3,1) and the (1,3) normal modes are combined in the ratio $1:-1$.

4. A thin square membrane when stretched to tension T has sides of unit length and density ρ per unit area where $T/\rho = c^2$. Taking origin at one corner of the membrane and x and y axis along a pair of adjacent sides, the initial displacement of any point (x,y) of the membrane is given by $kxy(1 - x)(1 - y)$ where k is a small constant. If the membrane is now released from rest but its edges are kept fixed, determine the subsequent displacement of the point (x,y) of the membrane.

5. Illustrate graphically the (3,2) and the (2,3) normal modes of oscillation of a circular membrane of radius a.

6. A thin circular membrane when stretched to tension T has unit radius and density ρ per unit area where $T/\rho = c^2$. The initial displacement of all points of the membrane distant $r(<a)$ from its centre is $k(1 - r^2)$ for constant k. The membrane is now released from rest, but its circumference is kept fixed. Determine the subsequent displacements of points of the membrane.

8.5 Longitudinal waves

Longitudinal waves are waves in which the displacements of the particles of the transmitting medium are parallel to the direction of wave propagation. Such waves are usually associated with the elasticity of the medium.

We consider the passage of a longitudinal wave in a uniform bar of mass ρ per unit length in which the tension is initially zero. We shall assume at this stage that the force of gravity can be neglected or alternatively that the bar lies on a smooth horizontal table. In the example to follow however the force of gravity is not neglected.

Let the element PQ of the bar be of length δx and suppose that, during the passage of the wave, P and Q are displaced to P_1 and Q_1 respectively (figure 126) where P_1Q_1 has length $\delta x + \delta \xi$.

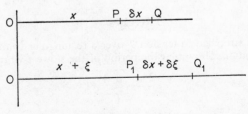

FIGURE 126

Now Hooke's law states that for a uniform extension of a bar the tension is proportional to the extension of the bar divided by its unstretched length. The constant of proportionality, which is of course dimensionless, is called the *modulus of elasticity* of the bar. Thus, considering the element PQ, if T is its tension and λ its modulus then

$$T = \lambda \frac{\delta \xi}{\delta x}.$$

This result implies that the tension T is constant throughout the element PQ and will apply if the extension is uniform throughout its length. For extensions which are not necessarily uniform, we have that the tension at the point P_1 (denoted by T_{P_1}) is given by

$$T_{P_1} = \lim_{\delta x \to 0} \lambda \frac{\delta \xi}{\delta x} = \lambda \frac{\partial \xi}{\partial x}.$$

Then, using Taylor's series, we see that

$$T_{Q_1} = T_{P_1} + \left(\frac{\partial T}{\partial x}\right)_{P_1} \delta x + \mathrm{O}[(\delta x)^2] = T_{P_1} + \lambda \left(\frac{\partial^2 \xi}{\partial x^2}\right)_{P_1} \delta x + \mathrm{O}[(\delta x)^2].$$

Now the mass of the element $P_1 Q_1$ is equal to the mass of PQ and so is $\rho \, \delta x$ and the resultant force acting on this element due to the tension is

$$T_{Q_1} - T_{P_1} = \lambda \left(\frac{\partial^2 \xi}{\partial x^2}\right)_{P_1} \delta x + \mathrm{O}[(\delta x)^2].$$

Hence, assuming that Newton's law of particle motion can be applied to the element, we obtain

$$\rho \, \delta x \frac{\partial^2 \xi}{\partial t^2} = \lambda \left(\frac{\partial^2 \xi}{\partial x^2}\right)_{P_1} \delta x + \mathrm{O}[(\delta x)^2]$$

where $\partial^2 \xi / \partial t^2$ is to be evaluated at the centre of mass of the element.[*] Therefore

$$\rho \frac{\partial^2 \xi}{\partial t^2} = \lambda \left(\frac{\partial^2 \xi}{\partial x^2}\right)_{P_1} + \mathrm{O}(\delta x).$$

Then, taking the limit as $\delta x \to 0$, we see that the equation

$$\frac{\partial^2 \xi}{\partial x^2} = \frac{1}{c^2} \frac{\partial^2 \xi}{\partial t^2}$$

---

[*] Indeed $\partial^2 \xi / \partial t^2$ can be evaluated at any point R of the element and in particular at P_1 since, using Taylor's theorem, $(\partial^2 \xi / \partial t^2)_R = (\partial^2 \xi / \partial t^2)_{P_1} + \mathrm{O}(\delta x)$.

where $c^2 = \lambda/\rho$, must be satisfied at all points of the bar. That is the one-dimensional wave equation.

If the bar has a free end then the tension there must be zero and so

$$\frac{\partial \xi}{\partial x} = 0$$

at a free end.

If the bar has a fixed end then the displacement there must be zero and so

$$\xi = 0$$

at a fixed end.

Hence for a bar of length l fixed at both ends, the solution of this equation obtained by separation of variables as in section 8.3 may be written in the form

$$\xi = A \sin \frac{n\pi x}{l} \cos \left(\frac{n\pi ct}{l} + \varepsilon \right)$$

for positive integers n and constants A and ε. Thus the fundamental period is

$$\frac{2l}{c} = 2l \sqrt{\frac{\rho}{\lambda}} \ .$$

Now suppose that before the passage of the wave the bar is uniformly stretched until the tension has the constant value T_0 at all points of its length.

Let the natural length of the bar be l and its mass per unit length be ρ. After stretching until the tension becomes T_0, let the length be l_0. Suppose that the element PQ of the unstretched bar has length δx and that after stretching it becomes the element P_0Q_0 (figure 127) of length $\delta x + \delta X$.

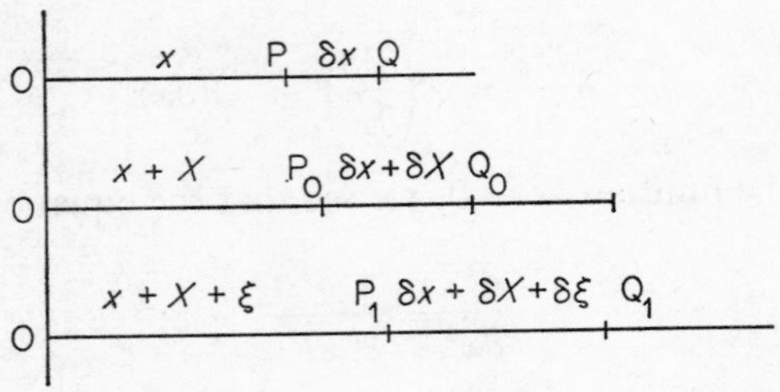

FIGURE 127

Since the stretching is uniform,

$$T_0 = \lambda\left(\frac{l_0 - l}{l}\right)$$

$$= \lim_{\delta x \to 0} \lambda\left(\frac{\delta X}{\delta x}\right)$$

$$= \lambda\frac{\partial X}{\partial x}.$$

Now suppose that, during the passage of the wave, P and Q are again displaced to P_1 and Q_1 respectively (figure 127) where P_1Q_1 has length $(\delta x + \delta X + \delta\xi)$. Then

$$T_{P_1} = \lim_{\delta x \to 0} \lambda\left(\frac{\delta X + \delta\xi}{\delta x}\right)$$

$$= \lim_{\delta x \to 0} \lambda\left(\frac{\delta X}{\delta x} + \frac{\delta\xi}{\delta x}\right)$$

$$= \lambda\frac{\partial X}{\partial x} + \lambda\frac{\partial\xi}{\partial x}$$

$$= T_0 + \lambda\frac{\partial\xi}{\partial x}.$$

Again, using Taylor's series, we have

$$T_{Q_1} = T_{P_1} + \left(\frac{\partial T}{\partial x}\right)_{P_1}\delta x + O[(\delta x)^2]$$

$$= T_{P_1} + \lambda\left(\frac{\partial^2\xi}{\partial x^2}\right)_{P_1}\delta x + O[(\delta x)^2].$$

Now the mass of the element P_1Q_1 is equal to the mass of PQ and so is $\rho\,\delta x$ and the resultant force acting on this element due to the tension is

$$T_{Q_1} - T_{P_1} = \lambda\left(\frac{\partial^2\xi}{\partial x^2}\right)_{P_1}\delta x + O[(\delta x)^2].$$

Then, again assuming that Newton's law of particle motion can be applied to the element, we obtain

$$\rho\,\delta x\frac{\partial^2\xi}{\partial t^2} = \lambda\left(\frac{\partial^2\xi}{\partial x^2}\right)_{P_1}\delta x + O[(\delta x)^2]$$

where $\partial^2\xi/\partial t^2$ is to be evaluated at the centre of mass of the element.
  Therefore

$$\rho\frac{\partial^2\xi}{\partial t^2} = \lambda\left(\frac{\partial^2\xi}{\partial x^2}\right)_{P_1} + O(\delta x).$$

Then, taking the limit as $\delta x \to 0$, we see that the equation

$$\frac{\partial^2 \xi}{\partial x^2} = \frac{1}{c^2} \frac{\partial^2 \xi}{\partial t^2}$$

where $c^2 = \lambda/\rho$ must be satisfied at all points of the bar. That is the same equation as before.

Hence for a bar of natural length l with ends fixed a distance l_0 apart, the solutions obtained by separation of variables again have the form

$$\xi = A \sin \frac{n\pi x}{l} \cos \left(\frac{n\pi c t}{l} + \varepsilon \right)$$

for positive integers n and constants A and ε, and the fundamental period

is $2l \sqrt{\dfrac{\rho}{\lambda}}$ which is independent of the tension T_0 and length l_0.

Now, since the initial stretching of the bar is uniform,

$$\frac{l_0}{l} = \frac{\delta x + \delta X}{\delta x}$$

and so putting $y = x + X$, this becomes

$$\frac{l_0}{l} = \frac{\delta y}{\delta x} \, .$$

Hence, using the variable y instead of x, the above wave equation becomes

$$\frac{\partial^2 \xi}{\partial y^2} = \frac{1}{c_1^2} \frac{\partial^2 \xi}{\partial t^2}$$

where $c_1^2 = \lambda l_0^2 / \rho l^2$. This result can of course be obtained directly in the same manner as above.

The solutions, obtained by separation of variables, for the above bar now have the form

$$\xi = A \sin \frac{n\pi y}{l_0} \cos \left(\frac{n\pi c_1 t}{l_0} + \varepsilon \right)$$

for positive integers n and constants A and ε. The fundamental period is

$$\frac{2l_0}{c_1} = 2l_0 \left(\frac{l}{l_0} \sqrt{\frac{\rho}{\lambda}} \right) = 2l \sqrt{\frac{\rho}{\lambda}}$$

which is, of course, the same as that obtained above.

Putting $y = (l_0/l)x$ and $c_1 = (l_0/l)c$ in this latter form of solution we obtain the previous form

$$\xi = A \sin \frac{n\pi x}{l} \cos \left(\frac{n\pi ct}{l} + \varepsilon \right).$$

Most problems can be solved using either the coordinate x or the coordinate y provided of course that the boundary conditions are applied correctly.

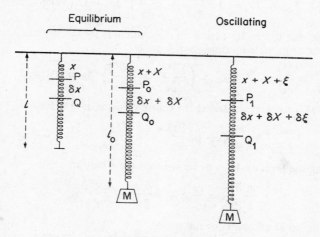

FIGURE 128

Example   Consider the vertical oscillations of a spring of natural length l, mass m and modulus λ with a mass M attached to its end.

The notation is similar to that used above.

The element PQ of length δx of the spring becomes the element $P_0 Q_0$ of length $(\delta x + \delta X)$ after the mass M is hung on the end. Then, during the oscillations, the element $P_0 Q_0$ becomes the element $P_1 Q_1$ of length $(\delta x + \delta X + \delta \xi)$ (figure 128).

As above,

$$T_{P_1} = \lambda \frac{\partial X}{\partial x} + \lambda \frac{\partial \xi}{\partial x}.$$

In this case however, due to the effects of gravity, the term $\lambda (\partial X/\partial x)$ representing the equilibrium tension is not constant throughout the length of the spring. It is perhaps intuitively obvious that this equilibrium tension must be greater near the top of the spring as a greater weight of spring has to be supported there.

From Taylor's series, we see that

$$T_{Q_1} = T_{P_1} + \left(\frac{\partial T}{\partial x}\right)_{P_1} \delta x + O[(\delta x)^2]$$

$$= T_{P_1} + \left(\lambda \frac{\partial^2 X}{\partial x^2} + \lambda \frac{\partial^2 \xi}{\partial x^2}\right)_{P_1} \delta x + O[(\delta x)^2].$$

Hence the resultant force acting on the element due to the tension is

$$T_{Q_1} - T_{P_1} = \left(\lambda \frac{\partial^2 X}{\partial x^2} + \lambda \frac{\partial^2 \xi}{\partial x^2}\right)_{P_1} \delta x + O[(\delta x)^2].$$

But the force due to gravity acting on the element is $(m/l)g \, \delta x$ and so, applying Newton's law of particle motion,

$$\frac{m}{l} \delta x \frac{\partial^2 \xi}{\partial t^2} = \frac{m}{l} g \, \delta x + \left(\lambda \frac{\partial^2 X}{\partial x^2} + \lambda \frac{\partial^2 \xi}{\partial x^2}\right)_{P_1} \delta x + O[(\delta x)^2]$$

where $\partial^2 \xi / \partial t^2$ is to be evaluated at the centre of mass of the element. Therefore

$$\frac{m}{l} \frac{\partial^2 \xi}{\partial t^2} = \frac{m}{l} g + \left(\lambda \frac{\partial^2 X}{\partial x^2} + \lambda \frac{\partial^2 \xi}{\partial x^2}\right)_{P_1} + O(\delta x).$$

Then, taking the limit as $\delta x \to 0$, we see that the equation

$$\frac{m}{l} \frac{\partial^2 \xi}{\partial t^2} = \frac{m}{l} g + \lambda \frac{\partial^2 X}{\partial x^2} + \lambda \frac{\partial^2 \xi}{\partial x^2}$$

must be satisfied at all points of the spring.

But in the equilibrium position $\xi = 0$ at all points of the spring and so

$$0 = \frac{m}{l} g + \lambda \frac{\partial^2 X}{\partial x^2} .$$

Subtracting this equation from that above then gives

$$\frac{m}{l} \frac{\partial^2 \xi}{\partial t^2} = \lambda \frac{\partial^2 \xi}{\partial x^2}$$

or

$$\frac{\partial^2 \xi}{\partial x^2} = \frac{1}{c^2} \frac{\partial^2 \xi}{\partial t^2}$$

where $c^2 = \lambda l/m$. That is, once again, the one-dimensional wave equation.

Now, as in section 8.3, we can write the solution of this equation, obtained by separation of variables, in the form

$$\xi = \begin{cases} (A \sin px + B \cos px) \cos (pct + \varepsilon) & \text{for } p \neq 0, \\ (A_0 x + B_0)(C_0 t + D_0) & \text{for } p = 0, \end{cases}$$

in which A, B, ε, A_0, B_0, C_0 and D_0 are all constants.

But since physically ξ does not increase indefinitely with t, $C_0 = 0$. Now since the top end of the spring is held fixed, $\xi = 0$ at $x = 0$ for all t and so B and B_0 are both zero.

Therefore

$$\xi = \begin{cases} A \sin px \cos (pct + \varepsilon) & \text{for } p \neq 0, \\ (A_0 D_0)x & \text{for } p = 0. \end{cases}$$

Since Newton's law must hold for the mass M on the end of the spring, we must have,

$$M\left(\frac{\partial^2 \xi}{\partial t^2}\right)_{x=l} = Mg - (T)_{x=l}.$$

Therefore

$$\left(\frac{\partial^2 \xi}{\partial t^2}\right)_{x=l} = g - \frac{\lambda}{M}\left(\frac{\partial X}{\partial x} + \frac{\partial \xi}{\partial x}\right)_{x=l}.$$

But in the equilibrium position $\xi = 0$ and so

$$0 = g - \frac{\lambda}{M}\left(\frac{\partial X}{\partial x}\right)_{x=l}.$$

Hence

$$\left(\frac{\partial^2 \xi}{\partial t^2}\right)_{x=l} = -\frac{\lambda}{M}\left(\frac{\partial \xi}{\partial x}\right)_{x=l}.$$

When $p = 0$ this implies that $(A_0 D_0) = 0$ and when $p \neq 0$ we obtain

$$-p^2 c^2 A \sin pl \cos (pct + \varepsilon) = -\frac{\lambda}{M} Ap \cos pl \cos (pct + \varepsilon).$$

Therefore

$$pc^2 \sin pl = \frac{\lambda}{M} \cos pl$$

since $A \cos (pct + \varepsilon)$ cannot be zero or the solution reduces to $\xi = 0$ everywhere along the spring.

Hence

$$p\frac{\lambda l}{m} \sin pl = \frac{\lambda}{M} \cos pl$$

or

$$pl \tan pl = \frac{m}{M}$$

and so

$$\tan pl = \frac{m}{M}\frac{1}{pl}.$$

The natural frequencies of the system are given by $pc/2\pi$ for each root p of this equation. It is clear that as p becomes very large these roots tend to those of $\tan pl = 0$ that is to $n\pi/l$ for large positive integers n.

The roots may be estimated graphically by determining the points of inter-section of the curves $y = \tan pl$ and $y = (m/M)(1/pl)$.

This is done in figure 129 in which (pl) is taken as the independent variable.

Now if m is very small compared with M, then the least root of the above equation will itself be small and so putting $pl = \theta$, and expanding $\tan \theta$ in

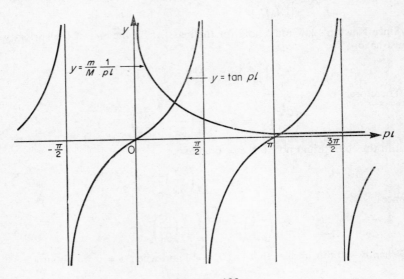

FIGURE 129

ascending powers of θ we have

$$\theta\left(\theta + \frac{\theta^3}{3} + \ldots\right) = \frac{m}{M}.$$

Thus, if θ^2 is small compared with unity, a first approximation to θ at the least root is given by

$$\theta^2 \simeq \frac{m}{M}.$$

Thus

$$\theta \simeq \sqrt{\frac{m}{M}} \quad \text{and} \quad p \simeq \frac{1}{l}\sqrt{\frac{m}{M}}.$$

A better approximation can be obtained as follows:

$$\theta\left(\theta + \frac{\theta^3}{3}\right) \simeq \frac{m}{M}$$

(neglecting terms of order θ^6).

Therefore

$$\theta^2 \simeq \frac{m}{M\left(1 + \dfrac{\theta^2}{3}\right)} \simeq \frac{m}{M\left(1 + \dfrac{1}{3}\dfrac{m}{M}\right)}$$

(using the first approximation to θ^2).†
Hence

$$p \simeq \frac{1}{l}\left(\frac{m}{M + \tfrac{1}{3}m}\right)^{\frac{1}{2}}.$$

The period of this oscillation is then approximately equal to

$$\frac{2\pi}{\dfrac{c}{l}\left(\dfrac{m}{M + \tfrac{1}{3}m}\right)^{\frac{1}{2}}} = 2\pi\left\{\frac{l}{\lambda}\left(M + \tfrac{1}{3}m\right)\right\}^{\frac{1}{2}} \qquad \left(\text{since } c^2 = \frac{\lambda l}{m}\right).$$

The periods of the oscillations corresponding to the other roots for p are very small compared with this period.

Note that if the mass of the spring is completely neglected then the period of the oscillations is $2\pi\sqrt{lM/\lambda}$. Thus, in general, for a spring whose mass is small compared with the mass M hung on the end, the dominant period of oscillation is the same as the period of oscillation of a similar spring of negligible mass with a mass ($M + \tfrac{1}{3}$ the mass of the spring) hung on the end.

Exercises

1. A uniform elastic string of natural length l and total mass m is stretched between two fixed points a distance l_0 apart on a smooth horizontal table. Show that the fundamental frequency for small longitudinal waves in the string is

$$\frac{1}{2}\left(\frac{T_0}{m(l_0 - l)}\right)^{\frac{1}{2}}$$

where T_0 is the tension in the string when it is in the equilibrium position.

2. An elastic string of modulus λ, natural length l and mass ρ per unit length is stretched on a smooth horizontal table. If, after stretching, one end of the string is fixed while the other end is given, for subsequent time t, a displacement along the length of the string of $A \sin qct$ for constants A and q and $c = \sqrt{\lambda/l}$, show that the displacement of the midpoint of the string at time t is

$$\tfrac{1}{2}A \sec \frac{ql}{2} \sin qct.$$

† This method of obtaining an improved approximation from a power one is known as iteration and has wide application in numerical analysis.

3. A uniform bar of length a, modulus λ_1 and mass ρ_1 per unit length is joined to another uniform bar of the same length, modulus λ_2 and mass ρ_2 per unit length to form a bar of length $2a$. The ends of this composite bar are now firmly clamped a distance $2a$ apart on a smooth horizontal table. Show that the periods of the normal modes of longitudinal vibrations are given by $2\pi/n$ for values of n satisfying the equation

$$\frac{1}{\sqrt{\rho_1\lambda_1}}\tan\left(na\sqrt{\frac{\rho_1}{\lambda_1}}\right) + \frac{1}{\sqrt{\rho_2\lambda_2}}\tan\left(na\sqrt{\frac{\rho_2}{\lambda_2}}\right) = 0.$$

(Hint: at the join of the two bars the displacement and the tension are continuous.)

4. A uniform bar of natural length l, modulus λ and mass ρ per unit length hangs vertically from a fixed point. Determine the frequencies of the normal longitudinal vibrations of the bar.

5. A spring of natural length $2l$, mass $2m$ and modulus λ is suspended vertically from a fixed point. A mass M is attached to its midpoint and another mass M is attached to the free end. Show that the periods of the normal longitudinal vibrations of the spring are given by $2\pi/nc$ for $c = \sqrt{\lambda l/m}$ and values of n which satisfy the equation

$$k^2 n^2 - 3kn \cot nl + \cot^2 nl = 1$$

in which $k = Ml/m$.

8.6 Sound waves

The passage of sound waves through a fluid is associated with the compressibility of the fluid. Since, on the passage of a sound wave, the particles of the fluid experience an oscillatory motion in the direction of wave propagation, these waves are thus another example of longitudinal waves.

In Chapter 2 we saw that the equation of motion for an inviscid fluid for which all external forces are negligible compared with that due to the pressure may be written

$$\frac{D\mathbf{q}}{Dt} = -\frac{1}{\rho}\operatorname{grad} p$$

and that the continuity equation takes the form

$$\operatorname{div}\mathbf{q} = -\frac{1}{\rho}\frac{D\rho}{Dt}.$$

Now considering only 'small' displacements for which second order terms (that is products of two small quantities) may be neglected the equation of motion becomes

$$\frac{\partial \mathbf{q}}{\partial t} = -\frac{1}{\rho} \operatorname{grad} p.$$

If also the motion is irrotational, curl $\mathbf{q} = 0$ and we can define a scalar potential φ such that $\mathbf{q} = \operatorname{grad} \varphi$. Hence

$$\frac{\partial}{\partial t}(\operatorname{grad} \varphi) = -\frac{1}{\rho} \operatorname{grad} p$$

or

$$\operatorname{grad}\left(\frac{\partial \varphi}{\partial t}\right) = -\frac{1}{\rho} \operatorname{grad} p.$$

Now integrate this equation along an element of a trajectory.
Therefore

$$\int \operatorname{grad}\left(\frac{\partial \varphi}{\partial t}\right) \cdot \mathbf{dl} = -\int \frac{1}{\rho} \operatorname{grad} p \cdot \mathbf{dl}.$$

Hence

$$\int d\left(\frac{\partial \varphi}{\partial t}\right) = -\int \frac{dp}{\rho}.$$

Therefore

$$\frac{\partial \varphi}{\partial t} = -\int_{p_0}^{p} \frac{dp}{\rho}$$

where $\partial \varphi/\partial t = 0$ when $p = p_0$ and p_0, ρ_0 are the undisturbed pressure and density respectively. But

$$\int_{p_0}^{p} \frac{dp}{\rho} = \int_{p_0}^{p} \left(\frac{dp}{d\rho}\right)\frac{d\rho}{\rho}$$

and if the density variations are small, $dp/d\rho$ will be approximately constant. Put this constant $= c^2$. Then

$$\int_{p_0}^{p} \frac{dp}{\rho} \simeq c^2 \int_{\rho_0}^{\rho} \frac{d\rho}{\rho} = c^2 \ln \frac{\rho}{\rho_0} = c^2 \ln (1 + s)$$

where $\rho = \rho_0(1 + s)$ and s is called the *condensation*. Hence if s is small,

$$\int_{p_0}^{p} \frac{dp}{\rho} \simeq c^2 s$$

Therefore, to this approximation we now have

$$\frac{\partial \varphi}{\partial t} + c^2 s = 0.$$

But neglecting 2nd order terms in the continuity equation we obtain

$$\nabla^2 \varphi = -\frac{1}{\rho}\frac{\partial \rho}{\partial t} = -\frac{\rho_0}{\rho}\frac{\partial s}{\partial t} \simeq -\frac{\partial s}{\partial t}.$$

Then, eliminating s from the last two equations, we obtain

$$\nabla^2 \varphi = \frac{1}{c^2}\frac{\partial^2 \varphi}{\partial t^2}$$

which is the wave equation with velocity of wave propagation c.

8.61 One-dimensional case

The one-dimensional wave equation has the form

$$\frac{\partial^2 \varphi}{\partial x^2} = \frac{1}{c^2}\frac{\partial^2 \varphi}{\partial t^2}.$$

Therefore

$$\frac{\partial^3 \varphi}{\partial x^3} = \frac{1}{c^2}\frac{\partial^2}{\partial t^2}\left(\frac{\partial \varphi}{\partial x}\right)$$

and so

$$\frac{\partial^2 u}{\partial x^2} = \frac{1}{c^2}\frac{\partial^2 u}{\partial t^2} \qquad \left(\text{since } u = \frac{\partial \varphi}{\partial x}\right).$$

Now consider the motion of a sound wave through the fluid in a uniform straight tube parallel to the x-axis. We assume that the motion of the fluid is entirely in the x-direction and that the velocity and displacement are the same for all points of the same cross-section.

Before displacement cross-sections through P and Q are a distance δx apart. After displacement the cross-sections through P and Q are displaced to the cross-sections through P' and Q' respectively, a distance $(\delta x + \delta \xi)$ apart (figure 130). The mass of fluid between the cross-sections through P

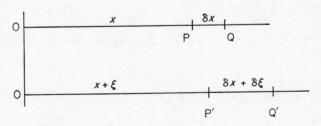

FIGURE 130

and Q and between the cross-sections through P' and Q' is the same and so

$$\rho_0 \, \delta x A = \rho(\delta x + \delta \xi)A$$

where A is the area of cross-section of the tube.

Hence

$$\rho_0 = \rho\left(1 + \frac{\partial \xi}{\partial x}\right)$$

and so

$$\rho = \rho_0\left(1 - \frac{\partial \xi}{\partial x}\right)$$

to the first order in $\partial \xi / \partial x$.

Therefore

$$s = -\frac{\partial \xi}{\partial x}$$

to this approximation.

Hence

$$\frac{\partial \varphi}{\partial t} = -c^2 s = c^2 \frac{\partial \xi}{\partial x}.$$

Therefore

$$c^2 \frac{\partial^2 \xi}{\partial x^2} = \frac{\partial^2 \varphi}{\partial x \, \partial t} = \frac{\partial}{\partial t}\left(\frac{\partial \varphi}{\partial x}\right) = \frac{\partial}{\partial t}(u) = \frac{\partial}{\partial t}\left(\frac{\partial \xi}{\partial t}\right).$$

Hence

$$\frac{\partial^2 \xi}{\partial x^2} = \frac{1}{c^2}\frac{\partial^2 \xi}{\partial t^2}.$$

Thus we can solve any particular problem in terms of the displacement ξ or the velocity potential φ and these solutions are related by the equation

$$\frac{\partial \xi}{\partial x} = \frac{1}{c^2}\frac{\partial \varphi}{\partial t}.$$

In particular, for a simple harmonic wave,

$$\frac{\partial \xi}{\partial t} = u = \frac{\partial \varphi}{\partial x} = -\frac{2\pi a}{\lambda}\sin\frac{2\pi}{\lambda}(x - ct)$$

and so

$$\xi = -\frac{a}{c}\cos\frac{2\pi}{\lambda}(x - ct) = -\frac{\varphi}{c}.$$

Boundary conditions for a straight tube

At a closed end the fluid displacement and velocity must both be zero so that

$$\xi = 0 \quad \text{and} \quad u = \partial \varphi / \partial x = 0.$$

At an open end the density and pressure must have approximately the undisturbed values ρ_0 and p_0 respectively so that $s = 0$ and hence

$$\frac{\partial \xi}{\partial x} = 0 \quad \text{and} \quad \frac{\partial \varphi}{\partial t} = 0.$$

Example   Determine the normal modes of vibration of a fluid in a tube of length l closed at one end and open at the other. (This problem is essentially that of determining the normal modes of vibration of the air in an organ pipe closed at one end.)

We must solve the differential equation

$$\frac{\partial^2 \varphi}{\partial x^2} = \frac{1}{c^2} \frac{\partial^2 \varphi}{\partial t^2}$$

subject to the boundary conditions

$$\frac{\partial \varphi}{\partial x} = 0 \quad \text{at} \quad x = 0$$

and

$$\frac{\partial \varphi}{\partial t} = 0 \quad \text{at} \quad x = l \quad \text{for all } t.$$

Using the method of separation of variables and putting $\varphi = X(x)T(t)$ we obtain

$$\varphi = (A \cos px + B \sin px) \cos (pct + \varepsilon)$$

for arbitrary constants A, B and ε.
   Then

$$\frac{\partial \varphi}{\partial x} = p(-A \sin px + B \cos px) \cos (pct + \varepsilon).$$

Putting $x = 0$, we obtain $B = 0$ and

$$\varphi = A \cos px \cos (pct + \varepsilon).$$

   Therefore

$$\frac{\partial \varphi}{\partial t} = -pcA \cos px \sin (pct + \varepsilon).$$

Putting $x = l$, we obtain

$$\cos pl = 0 \quad \text{(if} \quad A \neq 0).$$

and so $pl = (n + \frac{1}{2})\pi$ for integers n and

$$\varphi = A \cos (n + \tfrac{1}{2}) \frac{\pi x}{l} \cos \left\{(n + \tfrac{1}{2}) \frac{\pi c t}{l} + \varepsilon\right\}$$

for integer values of n. The general solution of this form is

$$\varphi = \sum_n A_n \cos (n + \tfrac{1}{2}) \frac{\pi x}{l} \cos \left\{(n + \tfrac{1}{2}) \frac{\pi c t}{l} + \varepsilon\right\}$$

where the constants ε and A_n ($n = 0,1,2, \ldots$) are determined by the initial conditions of a particular problem.

The normal modes of vibration are represented by

$$\varphi = A_n \cos (n + \tfrac{1}{2}) \frac{\pi x}{l} \cos \left\{(n + \tfrac{1}{2}) \frac{\pi c t}{l} + \varepsilon\right\}$$

Their periods are given by

$$\frac{2l}{(n + \tfrac{1}{2})c}.$$

The period of the fundamental (that is the largest period) is $4l/c$.

Example Prove that for sound waves which are spherically symmetric about the origin, at which there is no source, the velocity potential is given by

$$r\varphi = f(ct - r) - f(ct + r)$$

where c is the velocity of the sound waves.

If, when $t = 0$, $\varphi = 0$ and $\partial \varphi / \partial t = 4c\, e^{-r^2}$ obtain an expression giving φ everywhere at all subsequent times.

The velocity potential φ satisfies the wave equation

$$\frac{1}{c^2} \frac{\partial^2 \varphi}{\partial t^2} = \nabla^2 \varphi = \frac{\partial^2 \varphi}{\partial r^2} + \frac{2}{r} \frac{\partial \varphi}{\partial r}.$$

Therefore

$$\frac{r}{c^2} \frac{\partial^2 \varphi}{\partial t^2} = r \frac{\partial^2 \varphi}{\partial r^2} + 2 \frac{\partial \varphi}{\partial r}.$$

or

$$\frac{1}{c^2} \frac{\partial^2}{\partial t^2} (r\varphi) = \frac{\partial^2}{\partial r^2} (r\varphi).$$

The general solution is

$$r\varphi = f(ct - r) + g(ct + r)$$

for arbitrary functions f and g. Now as $r \to 0$, φ remains finite (since there is no source at $r = 0$) and so

$$f(ct - r) + g(ct + r) \to 0$$

as $r \to 0$ for all t. Therefore $f(ct) + g(ct) = 0$ for all t. Hence f and $-g$ must be the same function.
Therefore

$$r\varphi = f(ct - r) - f(ct + r).$$

Now when $t = 0$ we have

$$f(-r) - f(r) = 0.$$

Also,

$$r\frac{\partial \varphi}{\partial t} = cf'(ct - r) - cf'(ct + r)$$

and so putting $t = 0$ we obtain

$$4cre^{-r^2} = c\{f'(-r) - f'(r)\}.$$

Then, integrating with respect to r, we obtain

$$-2e^{-r^2} = -f(-r) - f(r) + \text{constant}.$$

Hence

$$f(r) = e^{-r^2} + \text{constant}.$$

Therefore

$$r\varphi = e^{-(ct-r)^2} - e^{-(ct+r)^2}.$$

Example   Show that the normal periods of spherically symmetrical air vibrations inside a rigid sphere of radius a are $2\pi/pc$ where p is a root of the equation $\tan ap = ap$.
If the sphere now pulsates radially with a small amplitude so that the velocity on $r = a$ may be taken as $V \cos nt$, show that the pressure at the centre varies between the limits

$$p_0 \pm \frac{\rho_0 V n^2 a^2}{\left(na \cos \dfrac{na}{c} - c \sin \dfrac{na}{c}\right)},$$

where p_0 and ρ_0 are the equilibrium pressure and density and $2\pi/n$ is not a normal period of the system.
Discuss briefly what happens when $2\pi/n$ is close to a normal period of the system.

The velocity potential satisfies the wave equation

$$\frac{1}{c^2}\frac{\partial^2 \varphi}{\partial t^2} = \nabla^2 \varphi = \frac{\partial^2 \varphi}{\partial r^2} + \frac{2}{r}\frac{\partial \varphi}{\partial r}.$$

Therefore

$$\frac{1}{c^2}\frac{\partial^2}{\partial t^2}(r\varphi) = \frac{\partial^2}{\partial r^2}(r\varphi).$$

The boundary conditions are (i) φ remains finite as $r \to 0$, (ii) $\partial\varphi/\partial r = 0$ when $r = a$. The second condition arises because the spherical surface $r = a$ is rigid and so there can be no normal component of velocity across it. Put $r\varphi = R(r)T(t)$. Therefore

$$\frac{1}{c^2 T}\frac{\mathrm{d}^2 T}{\mathrm{d}t^2} = \frac{1}{R}\frac{\mathrm{d}^2 R}{\mathrm{d}r^2}.$$

Put

$$\frac{1}{R}\frac{\mathrm{d}^2 R}{\mathrm{d}r^2} = -p^2.$$

Therefore

$$R = A\cos pr + B\sin pr.$$

Then

$$\frac{1}{c^2 T}\cdot\frac{\mathrm{d}^2 T}{\mathrm{d}t^2} = -p^2.$$

Hence

$$T = C\cos pct + D\sin pct.$$

Therefore

$$\varphi = \frac{1}{r}(A\cos pr + B\sin pr)(C\cos pct + D\sin pct).$$

Since φ remains finite as $r \to 0$, $A = 0$ and so

$$\varphi = B\frac{\sin pr}{r}\cdot T.$$

Therefore

$$\frac{\partial\varphi}{\partial r} = -B\frac{\sin pr}{r^2}T + Bp\frac{\cos pr}{r}T.$$

Then, since $\partial\varphi/\partial r = 0$ when $r = a$, for all t,

$$-\frac{\sin pa}{a^2} + p\frac{\cos pa}{a} = 0 \qquad (\text{if } B \neq 0)$$

and so $\tan ap = ap$.

Hence the periods of the normal modes of vibration are

$$\frac{2\pi}{pc}$$

where p is a root of the above equation. Now

$$\frac{\partial\varphi}{\partial r} = V\cos nt \quad \text{on} \quad r = a \quad \text{for all } t$$

and so

$$B\left(-\frac{\sin pa}{a^2} + \frac{p\cos pa}{a}\right)(C\cos pct + D\sin pct) = V\cos nt \quad \text{for all } t.$$

Therefore

$$pc = n, \quad B\left(-\frac{\sin pa}{a^2} + \frac{p \cos pa}{a}\right)C = V \quad \text{and} \quad D = 0.$$

Then

$$\varphi = \frac{-a^2 V}{\sin \dfrac{na}{c} - \dfrac{na}{c}\cos \dfrac{na}{c}}\left(\frac{\sin \dfrac{nr}{c}}{r}\right)\cos nt.$$

Now

$$\frac{\partial \varphi}{\partial t} = -\int_{p_0}^{p}\frac{\mathrm{d}p}{\rho} \simeq \frac{-(p - p_0)}{\rho_0}.$$

Therefore

$$p = p_0 - \rho_0 \frac{\partial \varphi}{\partial t}$$

to this approximation. When $r = 0$,

$$\frac{\partial \varphi}{\partial t} = \frac{a^2 V n^2}{c \sin \dfrac{na}{c} - na \cos \dfrac{na}{c}}\sin nt$$

since

$$\lim_{r \to 0}\left(\frac{\sin \dfrac{nr}{c}}{r}\right) = \frac{n}{c}.$$

Therefore the pressure at the centre of the sphere is given by

$$p_0 + \frac{\rho_0 V n^2 a^2}{na \cos \dfrac{na}{c} - c \sin \dfrac{na}{c}}\sin nt$$

which varies between the limits

$$p_0 \pm \frac{\rho_0 V n^2 a^2}{na \cos \dfrac{na}{c} - c \sin \dfrac{na}{c}}.$$

As $2\pi/n$ approaches a normal period of the system,

$$\sin (na/c) - (na/c) \cos (na/c)$$

tends to zero so that the amplitude of the potential φ becomes indefinitely large. Thus the assumption of small displacements, used to develop the wave equation, is no longer valid and so the wave equation itself no longer applies. However it is observed that when an oscillation with period approximately equal to a normal period is forced on a system the resulting oscillation has an abnormally large amplitude. This phenomenon is known as *resonance*.

Exercises

1. Determine the normal modes of vibration of a fluid in a tube of length l when
   (a) the tube is closed at both ends and
   (b) the tube is open at both ends.

2. Show that the normal periods of spherically symmetric vibrations of a fluid enclosed between two rigid concentric spheres of radii a and $2a$ are $2\pi/pc$ where p is a root of the equation

$$pa \cos pa = (1 + 2p^2a^2) \sin pa$$

and c is the velocity of wave propagation.

9

The use of complex potential and conformal mappings in the solution of two-dimensional problems

Many two-dimensional problems in hydrodynamics, electrostatics, magnetostatics and steady state heat conduction which involve the solution of Laplace's equation are most easily solved using functions of a complex variable. In this chapter we shall briefly build up the necessary complex variable theory and then consider a few of the practical applications.

The complex quantity $z = x + iy$ in which x and y are real variables and $i^2 = -1$ is called a *complex variable*. The plane in which the variable z is represented by the point (x,y) is known as the z-plane. This representation is also called an Argand diagram. If, in some domain of this plane, with each $z = x + iy$ one and only one complex number $w = u + iv$ for real u and v is associated, then w is said to be a function of z. We can write

$$w = u(x,y) + iv(x,y) = f(z).$$

For example (i) $w = z^2 = (x + iy)^2 = (x^2 - y^2) + i2xy$. Therefore $u = x^2 - y^2$ and $v = 2xy$. (ii) $w = \bar{z} = x - iy$ the complex conjugate of z.

Therefore $u = x$ and $v = -y$. The relation $w^2 = z$ however defines two functions, for if $z = r(\cos \theta + i \sin \theta)$ then w may be either

$$r^{\frac{1}{2}}(\cos \tfrac{1}{2}\theta + i \sin \tfrac{1}{2}\theta)$$

or

$$r^{\frac{1}{2}}(\cos (\tfrac{1}{2}\theta + \pi) + i \sin (\tfrac{1}{2}\theta + \pi)).$$

Thus $w = r^{\frac{1}{2}}(\cos \tfrac{1}{2}\theta + i \sin \tfrac{1}{2}\theta)$ or $w = -r^{\frac{1}{2}}(\cos \tfrac{1}{2}\theta + i \sin \tfrac{1}{2}\theta)$.

Each of the above functions is defined throughout the whole z-plane, but we can have functions which are only defined for certain restricted domains in the z-plane. For example

$$w = \frac{1}{|z| - 1}$$

is not defined when $|z| = 1$, that is when the point representing z lies on the circle of unit radius and centre the origin in the z-plane.

9.1 Continuity, differentiability and the Cauchy–Riemann equations

Let z_0 and l be two complex numbers given respectively by

$$z_0 = x_0 + iy_0, \qquad l = a + ib$$

where x_0, y_0, a and b are real constants, and let the complex function $f(z)$ be equal to $u(x,y) + iv(x,y)$ where u and v are real. Then, if as $z \to z_0$ (that is as $x \to x_0$ and $y \to y_0$) so that the point (x,y) approaches the point (x_0,y_0) in any manner in the z-plane, $f(z)$ approaches l (that is $u \to a$ and $v \to b$), we say that $f(z)$ tends to the limit l as z tends to z_0 and write

$$f(z) \to l \quad \text{as} \quad z \to z_0$$

or

$$\lim_{z \to z_0} f(z) = l.$$

The function $f(z)$ is then said to be *continuous* at $z = z_0$ if $f(z_0)$ is defined and

$$\lim_{z \to z_0} f(z) = f(z_0).$$

Since $f(z) = u(x,y) + iv(x,y)$ and $f(z_0) = u(x_0,y_0) + iv(x_0,y_0)$ this definition implies that the real functions u and v are continuous at the point (x_0,y_0).

A function $f(z)$ which is continuous at all points of a given domain of the z-plane is said to be a continuous function in that domain.

Now let ρ be the complex number $p + iq$ where p and q are real. A function $f(z)$ defined in a domain of the z-plane is said to be *differentiable* at $z = z_0$ in that domain if

$$\frac{f(z_0 + \rho) - f(z_0)}{\rho} \quad \text{tends to a definite finite limit as} \quad \rho \to 0.$$

This limit is then called the *derivative* of $f(z)$ at $z = z_0$ and is denoted by $f'(z_0)$.

Thus

$$f'(z_0) = \lim_{\rho \to 0} \frac{f(z_0 + \rho) - f(z_0)}{\rho} \, .$$

Alternatively, we can write

$$f'(z_0) = \lim_{z \to z_0} \frac{f(z) - f(z_0)}{z - z_0}$$

where now $z = z_0 + \rho$.

Now it is important to emphasize that the above definition of the derivative of $f(z)$ at $z = z_0$ implies that z can approach z_0 in any manner in the z-plane without affecting the value of $f'(z_0)$.

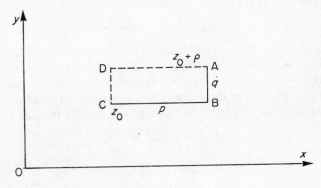

FIGURE 131

Consider the following two special cases:

(i) Let $z \to z_0$ along the curve ABC (figure 131) where BA is parallel to the y-axis and has length q and CB is parallel to the x-axis and has length p. In other words, first keep p fixed and let $q \to 0$ and then let $p \to 0$.

(ii) Let $z \to z_0$ along the curve ADC (figure 131) where DA is parallel to the x-axis and has length p and CD is parallel to the y-axis and has length q. In other words, first keep q fixed and let $p \to 0$ and then let $q \to 0$.

(i) For the curve ABC, we obtain

$$f'(z_0) = \lim_{p \to 0} \left\{ \lim_{q \to 0} \frac{f(z_0 + \rho) - f(z_0)}{\rho} \right\}$$

$$= \lim_{p \to 0} \left\{ \lim_{q \to 0} \frac{u(x_0 + p, y_0 + q) + iv(x_0 + p, y_0 + q) - u(x_0, y_0) - iv(x_0, y_0)}{p + iq} \right\}$$

Now if $f(z)$ is differentiable at z_0 then it must be continuous at z_0 and so $u(x,y)$ and $v(x,y)$ must be continuous. Therefore

$$f'(z_0) = \lim_{p \to 0} \left\{ \frac{u(x_0 + p, y_0) - u(x_0, y_0) + i[v(x_0 + p, y_0) - v(x_0, y_0)]}{p} \right\}$$

$$= \lim_{p \to 0} \left\{ \frac{u(x_0 + p, y_0) - u(x_0, y_0)}{p} \right\} + i \lim_{p \to 0} \left\{ \frac{v(x_0 + p, y_0) - v(x_0, y_0)}{p} \right\}$$

$$= \frac{\partial u}{\partial x} + i \frac{\partial v}{\partial x} \quad \text{evaluated at } z_0.$$

(Since $f(z)$ is differentiable at z_0, the last two limits exist.)

(ii) For the curve ADC we obtain

$$f'(z_0) = \lim_{q \to 0} \left\{ \lim_{p \to 0} \frac{f(z_0 + \rho) - f(z_0)}{\rho} \right\}$$

$$= \lim_{q \to 0} \left\{ \frac{u(x_0, y_0 + q) - u(x_0, y_0)}{iq} \right\} + \lim_{q \to 0} \left\{ \frac{v(x_0, y_0 + q) - v(x_0, y_0)}{q} \right\}.$$

(In a similar manner to the above (i).) Hence

$$f'(z_0) = \frac{1}{i} \frac{\partial u}{\partial y} + \frac{\partial v}{\partial y} = -i \frac{\partial u}{\partial y} + \frac{\partial v}{\partial y}$$

evaluated at z_0.

Hence, if $f(z)$ is differentiable at z_0,

$$\frac{\partial u}{\partial x} + i \frac{\partial v}{\partial x} = -i \frac{\partial u}{\partial y} + \frac{\partial v}{\partial y} \quad \text{at } z_0.$$

Therefore

$$\frac{\partial u}{\partial x} = \frac{\partial v}{\partial y} \quad \text{and} \quad \frac{\partial v}{\partial x} = -\frac{\partial u}{\partial y} \quad \text{at } z_0.$$

These important relations, which are thus *necessary* for differentiability at any point of the function $f(z) = u + iv$, are called the *Cauchy–Riemann equations*.

It can also be shown that if the partial derivatives $\partial u / \partial x$, $\partial u / \partial y$, $\partial v / \partial x$, $\partial v / \partial y$ are continuous and satisfy the Cauchy–Riemann equations in a small area surrounding a given point then the function $f(z) = u + iv$ is differentiable at that point, that is the above conditions are *sufficient* to ensure differentiability.

A function which is defined and differentiable in a neighbourhood of a given point is said to be *analytic* at that point. A function which is analytic at all points in a given domain is said to be *analytic* (or regular or holomorphic) within the domain. For such a function the Cauchy–Riemann equations will of course be satisfied at all points of the domain. A point at

14—(24 pp.)

which the function is not defined or does not have a derivative is called a *singular point* or *singularity* of the function. For example $f(z) = 1/(z - a)$ has a singularity at the point $z = a$ since its value is undefined there.

Example *Show that the derivative of the function* $f(z) = z^2$ *at any point z is 2z.*

$$f(z) = z^2 = (x + iy)^2$$
$$= x^2 - y^2 + 2ixy$$

Therefore $u = x^2 - y^2$ and $v = 2xy$.

$$\frac{\partial u}{\partial x} = 2x = \frac{\partial v}{\partial y} \quad \text{and} \quad \frac{\partial v}{\partial x} = 2y = -\frac{\partial u}{\partial y}.$$

Hence the Cauchy–Riemann equations are satisfied and the partial derivatives are continuous and so the given function is differentiable. Then, from above,

$$f'(z) = \frac{\partial u}{\partial x} + i\frac{\partial v}{\partial x} = -i\frac{\partial u}{\partial y} + \frac{\partial v}{\partial y}$$
$$= 2x + i2y$$
$$= 2(x + iy)$$
$$= 2z.$$

Alternatively,

$$f'(z) = \lim_{\rho \to 0} \frac{f(z + \rho) - f(z)}{\rho}$$
$$= \lim_{\rho \to 0} \frac{(z + \rho)^2 - z^2}{\rho}$$
$$= \lim_{\rho \to 0} (2z + \rho) = 2z.$$

It is easily shown that the familiar rules for differentiating real functions carry over to the differentiation of complex functions. For example

$$\frac{dc}{dz} = 0 \text{ if } c \text{ is a (complex) constant,}$$

$$\frac{dz^n}{dz} = nz^{n-1},$$

$$\frac{d}{dz}(f(z) + g(z)) = \frac{df}{dz} + \frac{dg}{dz} \quad (f, g \text{ both differentiable functions}),$$

$$\frac{d}{dz}(f(z)g(z)) = f\frac{dg}{dz} + g\frac{df}{dz},$$

$$\frac{d}{dz}\left(\frac{f(z)}{g(z)}\right) = \frac{g\frac{df}{dz} - f\frac{dg}{dz}}{g^2}$$

etc.

The chain rule for differentiating a function of a function also holds. For example

$$\frac{d}{dz}(\sin z^2) = 2z \cos z^2.$$

Example *Show from first principles that $f(z) = \bar{z}$ is not differentiable anywhere and verify that the Cauchy–Riemann equations are not satisfied.*

$$\lim_{\rho \to 0} \frac{f(z + \rho) - f(z)}{\rho} = \lim_{\rho \to 0} \frac{\overline{z + \rho} - \bar{z}}{\rho}$$

$$= \lim_{\rho \to 0} \frac{x + p - i(y + q) - x + iy}{p + iq} \quad (\rho = p + iq)$$

$$= \lim_{\rho \to 0} \frac{p - iq}{p + iq}.$$

Now, considering the curve ABC in figure 131, this becomes

$$\lim_{p \to 0} \left\{ \lim_{q \to 0} \frac{p - iq}{p + iq} \right\} = \lim_{p \to 0} 1 = 1.$$

Similarly, considering the curve ADC in figure 131, we obtain

$$\lim_{q \to 0} \left\{ \lim_{p \to 0} \frac{p - iq}{p + iq} \right\} = \lim_{q \to 0} -1 = -1.$$

Since these two values are unequal, the function $f(z)$ is not differentiable. Now $\bar{z} = x - iy$ and so $u = x$ and $v = -y$. Therefore

$$\frac{\partial u}{\partial x} = 1 \neq \frac{\partial v}{\partial y} = -1.$$

Thus the Cauchy–Riemann equations are not both satisfied.

Example *Consider the function $f(z) = 1/(z - a)$ for real constant a and $z \neq a$.*

$$\frac{1}{z - a} = \frac{1}{x - a + iy} = \frac{x - a - iy}{(x - a)^2 + y^2}.$$

Therefore, if $f(z) = u + iv$,

$$u = \frac{x - a}{(x - a)^2 + y^2} \quad \text{and} \quad v = \frac{-y}{(x - a)^2 + y^2}.$$

Hence

$$\frac{\partial u}{\partial x} = \frac{y^2 - (x - a)^2}{[(x - a)^2 + y^2]^2}, \qquad \frac{\partial u}{\partial y} = \frac{-2y(x - a)}{[(x - a)^2 + y^2]^2},$$

$$\frac{\partial v}{\partial x} = \frac{2y(x - a)}{[(x - a)^2 + y^2]^2}, \qquad \frac{\partial v}{\partial y} = \frac{y^2 - (x - a)^2}{[(x - a)^2 + y^2]^2}$$

except when $x = a$ and $y = 0$. Thus it is easily seen that the Cauchy–Riemann equations are satisfied everywhere except where $x = a$ and $y = 0$, that is everywhere except at $z = a$. Also the partial derivatives are continuous. Hence $f(z)$ is analytic except at $z = a$.

$$\frac{df}{dz} = \frac{-1}{(z - a)^2} \qquad (z \neq a).$$

Example *Show that the function $f(z) = (az + b)/(cz + d)$ for real constants a, b, c and d is analytic except where $cz + d = 0$.*

$$\frac{az + b}{cz + d} = \frac{a(x + iy) + b}{c(x + iy) + d} \cdot \frac{c(x - iy) + d}{c(x - iy) + d} \quad \text{(except where } cz + d = 0\text{)}$$

$$= \frac{(ax + b)(cx + d) + acy^2 + i\{ay(cx + d) - cy(ax + b)\}}{(cx + d)^2 + c^2 y^2}.$$

Therefore, if $f(z) = u + iv$, then

$$u = \frac{(ax + b)(cx + d) + acy^2}{(cx + d)^2 + c^2 y^2} \quad \text{and} \quad v = \frac{y(ad - bc)}{(cx + d)^2 + c^2 y^2}.$$

Hence

$$\frac{\partial u}{\partial x} = \frac{(ad - bc)\{(cx + d)^2 - c^2 y^2\}}{\{(cx + d)^2 + c^2 y^2\}^2}$$

$$\text{(after simplification)},$$

$$\frac{\partial u}{\partial y} = \frac{2cy(ad - bc)(cx + d)}{\{(cx + d)^2 + c^2 y^2\}^2},$$

$$\frac{\partial v}{\partial x} = \frac{-2cy(ad - bc)(cx + d)}{\{(cx + d)^2 + c^2 y^2\}^2}$$

and

$$\frac{\partial v}{\partial y} = \frac{(ad - bc)\{(cx + d)^2 - c^2 y^2\}}{\{(cx + d)^2 + c^2 y^2\}^2}.$$

Thus the Cauchy–Riemann equations are satisfied and the partial derivatives are continuous and so $f(z)$ is analytic everywhere except where $cz + d = 0$.

$$\frac{df}{dz} = \frac{(cz + d)a - (az + b)c}{(cz + d)^2} = \frac{ad - bc}{(cz + d)^2} \qquad (cz + d \neq 0).$$

Using the polar form $z = re^{i\theta} = r(\cos \theta + i \sin \theta)$ and putting $f(z) = u(r, \theta) + iv(r, \theta)$, the Cauchy–Riemann equations become

$$\frac{\partial u}{\partial r} = \frac{1}{r} \frac{\partial v}{\partial \theta} \quad \text{and} \quad \frac{\partial v}{\partial r} = -\frac{1}{r} \frac{\partial u}{\partial \theta}.$$

It can also be shown that the derivative of an analytic function is itself analytic (and so is differentiable). Thus a complex function which is once differentiable is differentiable any number of times. (This result of course has no parallel for real functions.) This also implies that if $f(z) = u + iv$ and f is analytic, then u and v have continuous partial derivatives of all orders.

Exercises

1. Verify that the Cauchy–Riemann equations are satisfied for the following functions and evaluate their derivatives

   (i) z^3,   (ii) e^{2z},   (iii) $\cos z^2$,

   (iv) $\ln 3z\,(z \neq 0)$,   (v) $\dfrac{1}{z-1}\,(z \neq 1)$,   (vi) $\dfrac{z}{z-i}\,(z \neq i)$.

2. Show from first principles that the following functions are not differentiable and verify that the Cauchy–Riemann equations are not satisfied.
   (i) $1/\bar{z}$, (ii) $z\bar{z}\,(z \neq 0)$.

9.2 Conjugate functions

If $f(z) = u(x,y) + iv(x,y)$ is analytic, then $u(x,y)$ and $v(x,y)$ are said to form a pair of *conjugate functions*. They of course satisfy the Cauchy–Riemann equations

$$\frac{\partial u}{\partial x} = \frac{\partial v}{\partial y}, \qquad \frac{\partial u}{\partial y} = -\frac{\partial v}{\partial x}.$$

Therefore

$$\frac{\partial^2 u}{\partial x^2} = \frac{\partial^2 v}{\partial x\,\partial y} \quad \text{and} \quad \frac{\partial^2 u}{\partial y^2} = -\frac{\partial^2 v}{\partial x\,\partial y}.$$

Hence

$$\frac{\partial^2 u}{\partial x^2} = -\frac{\partial^2 u}{\partial y^2} \quad \left(= \frac{\partial^2 v}{\partial x\,\partial y}\right)$$

or

$$\frac{\partial^2 u}{\partial x^2} + \frac{\partial^2 u}{\partial y^2} = 0.$$

Similarly, we obtain

$$\frac{\partial^2 v}{\partial x^2} + \frac{\partial^2 v}{\partial y^2} = 0.$$

Thus conjugate functions satisfy Laplace's equation in two dimensions (that is they are also harmonic functions).

Conversely, if a pair of functions u and v satisfy the Cauchy–Riemann equations, and the partial derivatives $\partial u/\partial x$, $\partial u/\partial y$, $\partial v/\partial x$, $\partial v/\partial y$ are continuous, the function $f(z) = u + iv$ is analytic and u and v form a conjugate pair.

Example *Consider the following pairs of functions*

(i) $u = x^3 - 3xy^2$, $v = 3x^2y - y^3$,

(ii) $u = x^2 + y^2$, $v = 2xy$.

(i) $\dfrac{\partial u}{\partial x} = 3x^2 - 3y^2$, $\dfrac{\partial u}{\partial y} = -6xy$, $\dfrac{\partial v}{\partial x} = 6xy$, $\dfrac{\partial v}{\partial y} = 3x^2 - 3y^2$.

Thus the Cauchy–Riemann equations are satisfied and the partial derivatives are continuous. Hence the functions u and v are conjugate and the function $f(z)$ defined by

$$f(z) = u + iv = (x^3 - 3xy^2) + i(3x^2y - y^3) = z^3.$$

is analytic.

(ii) $\dfrac{\partial u}{\partial x} = 2x$, $\dfrac{\partial u}{\partial y} = 2y$, $\dfrac{\partial v}{\partial x} = 2y$, $\dfrac{\partial v}{\partial y} = 2x$.

Thus the Cauchy–Riemann equations are not satisfied (except at the point $x = 0$, $y = 0$) and the functions u and v are not conjugate. Then the function $f(z)$ defined by

$$f(z) = u + iv = (x^2 + y^2) + i2xy$$

is not analytic.

Example *Construct where possible a function conjugate to each of the following functions*

(i) $u = x^2 - y^2$,
(ii) $u = x^2 + y^2$.

For the function u to be one of a conjugate pair, it must satisfy Laplace's equation.

(i) In this case,

$$\frac{\partial^2 u}{\partial x^2} + \frac{\partial^2 u}{\partial y^2} = 2 - 2 = 0$$

and so a function conjugate to u exists.

If this function is v then, from the Cauchy–Riemann equations, we must have

$$\frac{\partial v}{\partial y} = \frac{\partial u}{\partial x} = 2x \quad \text{and} \quad \frac{\partial v}{\partial x} = -\frac{\partial u}{\partial y} = 2y.$$

From the first of these we obtain $v = 2xy + g(x)$ where $g(x)$ is an arbitrary real function of x. Then substituting in the second we obtain

$$\frac{dg}{dx} = 0 \quad \text{and so} \quad g(x) = c$$

where c is a real constant.

Therefore $v = 2xy + c$ for any real constant c is a conjugate function to the function $u = x^2 - y^2$. The function

$$f(z) = u + iv = x^2 - y^2 + i(2xy + c) = z^2 + ic$$

is then analytic for all real values of c.

(ii) Here,

$$\frac{\partial^2 u}{\partial x^2} + \frac{\partial^2 v}{\partial y^2} = 2 + 2 = 4 \neq 0$$

and so the function u is not one of a conjugate pair. That is we cannot construct a function which is conjugate to u.

Now let $u(x,y)$ and $v(x,y)$ be a pair of conjugate functions and suppose that the point $P(x_0, y_0)$ is a point of intersection of the two curves.

$$u(x,y) = c$$

and

$$v(x,y) = k$$

for some constant values c and k.

Then the gradient of the tangent at P to the first curve is

$$-\frac{(\partial u/\partial x)}{(\partial u/\partial y)}$$

evaluated when $x = x_0$ and $y = y_0$. Also, the gradient of the tangent at P to the second curve is

$$-\frac{(\partial v/\partial x)}{(\partial v/\partial y)}$$

evaluated when $x = x_0$ and $y = y_0$. Thus the product of these gradients is

$$\frac{(\partial u/\partial x)(\partial v/\partial x)}{(\partial u/\partial y)(\partial v/\partial y)}$$

which we see, on using the Cauchy–Riemann equations, has the value -1. Hence the two tangents are perpendicular and so the two curves intersect at right angles. Therefore every curve of the family given by

$$u(x,y) = c$$

for different values of the constant c intersects the curves of the family given by

$$v(x,y) = k$$

for different values of the constant k at right angles. The two families of curves are said to be *orthogonal* to each other.

Example   Show that $u = \frac{1}{2} \ln (x^2 + y^2)$ and $v = \tan^{-1} (y/x)$ are conjugate functions and sketch the families of curves $u = $ constant and $v = $ constant.

$$\frac{\partial u}{\partial x} = \frac{x}{x^2 + y^2}, \quad \frac{\partial u}{\partial y} = \frac{y}{x^2 + y^2}, \quad \frac{\partial v}{\partial x} = \frac{-y}{x^2 + y^2}, \quad \frac{\partial v}{\partial y} = \frac{x}{x^2 + y^2}.$$

Therefore the Cauchy–Riemann equations are satisfied and the partial derivatives are continuous. Hence u and v are conjugate functions. $u = $ constant gives $\ln (x^2 + y^2) = $ constant and so $x^2 + y^2 = $ constant. Thus the curves of this family are circles with centre the origin (figure 132). $v = $ constant gives $\tan^{-1} (y/x) = $ constant, and so $y/x = $ constant or $y = $ (constant) x. Thus the curves of this family are straight lines through the origin (figure 132).

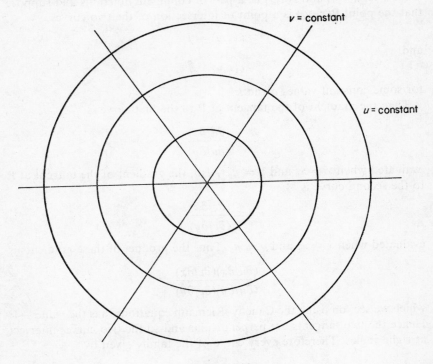

$v = $ constant

$u = $ constant

FIGURE 132

From the above, it is easily seen that the families of curves $u = $ constant and $v = $ constant are orthogonal.

Note the similarity of figure 132 with figure 69 of Chapter 3 which represented the lines of force and equipotentials for an electrostatic point charge.

It is easily seen that the complex function associated with the pair of conjugate functions $u = \frac{1}{2} \ln (x^2 + y^2)$ and $v = \tan^{-1} (y/x)$ is $\ln z$.

Exercises

1. Determine which of the following pairs of functions form conjugate pairs.
   (i) $x/(x^2 + y^2)$, $-y/(x^2 + y^2)$,   (ii) $x^2 - y^2 + x$, $2xy + y$,
   (iii) $x^2 + y^2 + x$, $2xy + y$,   (iv) $e^x(x \cos y - y \sin y)$, $e^x(y \cos y + x \sin y)$.

2. Construct where possible functions conjugate to each of the following functions
   (i) $2y(x + 1)$,   (ii) $x \cos y - y \sin y$,   (iii) $e^x(x \cos y + y \sin y)$,
   (iv) $e^{-x}(x \cos y + y \sin y)$,   (v) $x(x^2 - 3y^2)$.

9.3 Complex potential

9.31 Hydrodynamics

Consider again the two-dimensional, irrotational motion of an inviscid, incompressible fluid (that is an ideal fluid).

Since the flow is irrotational, curl $\mathbf{q} = \mathbf{0}$ if $\mathbf{q}$ is the fluid velocity vector. Thus a velocity potential φ exists for the flow. That is $\mathbf{q} = \mathrm{grad}\ \varphi$ and so

$$u = \frac{\partial \varphi}{\partial x} \quad \text{and} \quad v = \frac{\partial \varphi}{\partial y}$$

where $\mathbf{q} = (u,v,0)$. Then, since the fluid is incompressible, div $\mathbf{q} = 0$ and so $\nabla^2 \varphi = 0$.

Now let ψ be a conjugate function to φ. (This exists, since $\nabla^2 \varphi = 0$, and is unique apart from an arbitrary additive constant.) The curves $\psi = $ constant are then everywhere orthogonal to the curves $\varphi = $ constant, that is to the equipotentials, and so have the direction of grad φ or $\mathbf{q}$ at any point. This means that the curves $\psi = $ constant are thus streamlines for the flow. The function ψ is called the *stream function* of the flow. Since φ and ψ are conjugate functions, the Cauchy–Riemann equations hold.

That is

$$\frac{\partial \varphi}{\partial x} = \frac{\partial \psi}{\partial y} \quad \text{and} \quad \frac{\partial \varphi}{\partial y} = - \frac{\partial \psi}{\partial x}.$$

14A

Therefore

$$u = \frac{\partial \psi}{\partial y} \quad \text{and} \quad v = -\frac{\partial \psi}{\partial x}.$$

Also $\nabla^2 \psi = 0$.

The complex function W defined by

$$W = \varphi + i\psi$$

is called the *complex potential* of the flow.

From the above, we see that the complex potential W for the two-dimensional, irrotational motion of an ideal fluid is analytic and satisfies Laplace's equation $\nabla^2 W = 0$.

If $W = W_1 + W_2$ then W is the complex potential for the flow obtained by adding vectorially the velocity vectors for the flows for which W_1 and W_2 are the complex potentials.

Again, if

$$W = \varphi + i\psi$$
$$\frac{dW}{dz} = \frac{\partial \varphi}{\partial x} + i\frac{\partial \psi}{\partial x}$$
$$= u - iv$$
$$= \bar{q}.$$

Therefore

$$u + iv = \overline{\left(\frac{dW}{dz}\right)}.$$

Also

$$\left| \frac{dW}{dz} \right| = |q|$$

and so stagnation points (that is points at which $q = 0$) occur where $dW/dz = 0$.

Points where dW/dz becomes infinite are called singular points of the flow. It is not possible to determine the direction of a streamline through such a point.

Example   Consider the complex potential $W = Uz$ where U is a constant.

The potential φ is Ux and the stream function ψ is Uy. Therefore the equipotentials are the lines $x = $ constant, that is lines parallel to the y-axis, and the streamlines are the lines $y = $ constant (figure 133), that is lines parallel to the x-axis.

$$\frac{dW}{dz} = U = u - iv.$$

where u and v are the velocity components of the flow.

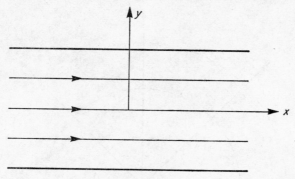

FIGURE 133

Therefore $u = U$ and $v = 0$.

Hence Uz is the complex potential for uniform flow with constant velocity U parallel to the x-axis in the direction of increasing x.

Since, by definition, there is no flow across a streamline, we can replace any of the streamlines for the flow by rigid boundaries without altering the pattern of the flow. Thus Uz may be taken as the complex potential for uniform flow between two parallel rigid boundaries. Again, since the three-dimensional problem of uniform flow between a pair of parallel rigid plane boundaries reduces essentially to the above two-dimensional problem (as the flow is the same in all planes parallel to the x,y-plane and so independent of the third cartesian coordinate) Uz may be taken as the complex potential for this (three-dimensional) problem also.

Similarly $W = -Uz$ represents uniform flow in the direction of decreasing x.

Example   Consider the complex potential $W = C \ln z$ where C is a real constant and $\ln z = \ln |z| + i$ (principal value of arg z).

Using the polar form for z we have

$$W = C(\ln r + i\theta), \qquad -\pi < \theta \leqslant \pi.$$

Hence the equipotentials are given by $C \ln r = $ constant or simply $r = $ constant. That is concentric circles with centre the origin (dashed curves in figure 134.) The streamlines are given by $C\theta = $ constant or simply $\theta = $ constant. That is straight lines through the origin (solid lines in figure 134)

$$\frac{dW}{dz} = \frac{C}{z} = \frac{C}{r} e^{-i\theta} = u - iv.$$

Therefore

$$u = \frac{C \cos \theta}{r} \quad \text{and} \quad v = \frac{C \sin \theta}{r}$$

except at the singular point $z = 0$. Therefore as $r \to \infty$ both u and $v \to 0$. That is the velocity components are very small at large distances from the origin.

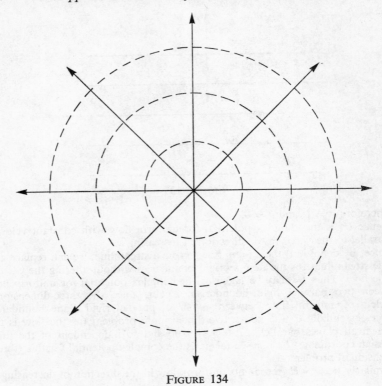

FIGURE 134

If C is positive the flow is radially outwards from the origin and the flow pattern is that for a 'two-dimensional' point source there. In three dimensions this represents the flow due to a uniform line source along the line through the origin perpendicular to the x,y-plane. The rate of volume outflow per unit length of the source is then $2\pi C$ and the source is said to have strength C. Thus the complex potential for a line source of strength m lying along the line through the origin perpendicular to the x,y-plane (or for a two-dimensional point source of strength m at the origin) is $m \ln z$.

By changing the origin, it is easily seen that the complex potential for such a source as the above at the point $z = z_0$ is

$$m \ln (z - z_0).$$

When C is negative the flow is radially inwards to the origin and the flow pattern is that for a 'two-dimensional' sink there or for a uniform line sink along the line through the origin perpendicular to the x,y-plane. Then, as above, the complex potential for a line sink of strength m lying along the line through the point $z = z_0$ perpendicular to the x,y-plane (or for a two-dimensional point sink of strength m at the origin) is $-m \ln (z - z_0)$.

Example Consider the complex potential $W = iC \ln z$ where C is a real constant and $\ln z = \ln |z| + i$ (principal value of arg z).

As above, $W = iC(\ln r + i\theta)$, $-\pi < \theta \leqslant \pi$.

Therefore $\varphi = -C\theta$ and $\psi = C \ln r$. Hence the equipotentials are given by $\theta = $ constant. That is straight lines through the origin (figure 135). The streamlines are given by $r = $ constant. That is concentric circles with centre the origin (figure 135)

$$\frac{\mathrm{d}W}{\mathrm{d}z} = \frac{iC}{z} = \frac{iC}{r}\mathrm{e}^{-i\theta} = u - iv.$$

Therefore

$$u = \frac{C \sin \theta}{r} \quad \text{and} \quad v = -\frac{C \cos \theta}{r}$$

except at the singular point $z = 0$. As $r \to \infty$, both u and $v \to 0$. That is the velocity components are very small at large distances from the origin. If C is positive the flow circulates round the origin in a clockwise sense. The point $z = 0$ is said to be a *vortex*.

Evaluating $\int \mathbf{q} \cdot \mathbf{dl}$ in a clockwise sense round any closed curve enclosing the origin (and in particular round a circle with centre the origin) we obtain the

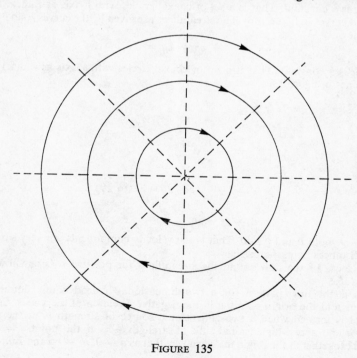

FIGURE 135

value $2\pi C$. This is called the *circulation* of the vortex and C is called its *strength*. Thus the complex potential for a point vortex of strength K (and so circulation $2\pi K$) at the point $z = z_0$ is $iK \ln (z - z_0)$. When C is negative, the flow circulates round the origin in a counterclockwise sense. In this case, the vortex at the origin will have a negative circulation.

Example   Consider the complex potential $W = \mu/z$ where μ is a constant.

$$W = \frac{\mu}{z} = \frac{\mu}{r} e^{-i\theta}.$$

Therefore

$$\varphi = \frac{\mu \cos \theta}{r} \quad \text{and} \quad \psi = -\frac{\mu \sin \theta}{r}.$$

Hence the equipotentials are given by the curves

$$\frac{\mu \cos \theta}{r} = \text{constant} \quad \text{or} \quad \frac{\mu x}{x^2 + y^2} = \text{constant}$$

and so

$$x^2 + y^2 + cx = 0$$

where c is a constant. That is a set of coaxal circles with y-axis as radical axis. (Dashed curves in figure 136). The streamlines are given by the curves $\mu \sin \theta/r = $ constant or

$$x^2 + y^2 + ky = 0$$

where k is a constant. That is a set of coaxal circles with x-axis as radical axis (solid lines in figure 136)

$$\frac{dW}{dz} = -\frac{\mu}{z^2}$$

$$= -\frac{\mu}{r^2} e^{-2i\theta}$$

$$= -\frac{\mu}{r^2} (\cos 2\theta - i \sin 2\theta)$$

$$= u - iv.$$

As $r \to \infty$ both u and $v \to 0$. That is the velocity components are very small at large distances from the origin.

The sense of the flow round the streamlines for positive μ is as shown in figure 136.

This is the flow pattern for a two-dimensional hydrodynamic doublet of moment μ at the origin pointing in the negative direction of the x-axis. Such a doublet is formed when a two-dimensional source of strength m at the point $z = -a$ and a two-dimensional sink of strength $-m$ at the point $z = a$ are brought together at the origin in such a way that as $a \to 0$, $m \to \infty$ and $2am \to \mu$.

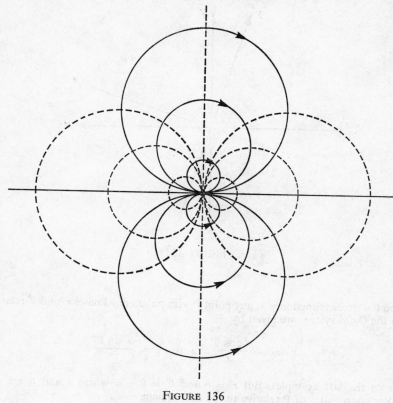

FIGURE 136

The above complex potential can also be looked upon as that for the flow due to a line doublet of moment μ per unit length formed by bringing together in the above fashion a line source of strength m per unit length and a line sink of strength $-m$ per unit length.

If the doublet is at the point $z = z_0$ pointing in the negative direction of the x-axis then the complex potential is

$$W = \frac{\mu}{z - z_0}.$$

Example   Deduce the complex potential for a hydrodynamical doublet of moment μ at the origin when the axis of the doublet makes an angle α with the positive x-axis.

Axes $Ox'y'$ are obtained by rotating the axes Oxy through an angle α (figure 137). The axis of the doublet now lies along the x'-axis and so the potential φ

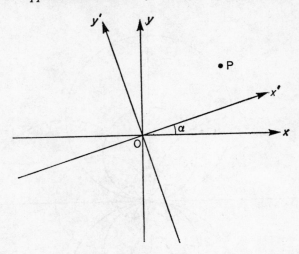

FIGURE 137

and the stream function ψ at any point P with polar coordinates r' and θ' relative to the $Ox'y'$ system are given by

$$\varphi = -\frac{\mu \cos \theta'}{r'} \quad \text{and} \quad \psi = \frac{\mu \sin \theta'}{r'}$$

(from the last example). But $r' = r$ and $\theta' = \theta - \alpha$ where r and θ are the polar coordinates of P relative to the Oxy system.

Therefore

$$\varphi = -\frac{\mu \cos (\theta - \alpha)}{r} \quad \text{and} \quad \psi = \frac{\mu \sin (\theta - \alpha)}{r}.$$

Hence

$$W = \varphi + i\psi$$

$$= -\frac{\mu}{r}\{\cos (\theta - \alpha) - i \sin (\theta - \alpha)\}$$

$$= -\frac{\mu}{r}\,e^{-i(\theta-\alpha)}$$

$$= -\left(\frac{\mu}{r}\,e^{-i\theta}\right)e^{i\alpha}$$

$$= -\frac{\mu}{z}\,e^{i\alpha}.$$

Similarly, of course, when the doublet is not at the origin but at the point z_0 its complex potential is

$$-\frac{\mu}{z - z_0}\,e^{i\alpha}.$$

Example   Consider the complex potential $W = z^2$

$$\varphi = x^2 - y^2 \quad \text{and} \quad \psi = 2xy.$$

Therefore the equipotentials are the curves $x^2 - y^2 = $ constant, that is rectangular hyperbolae with asymptotes $y = \pm x$. (The curves shown by dashed lines in figure 138.) The streamlines are rectangular hyperbolae with asymptotes $x = 0$, $y = 0$. (The curves shown by continuous lines in figure 138.) Since the lines $x = 0$, $y = 0$ are themselves streamlines (corresponding to $\psi = 0$) they can be taken as rigid boundaries in the flow and the flow pattern is then seen to represent the flow in a right-angled bend. Further, one of the other hyperbolae $xy = $ constant (say $xy = 1$) can also be taken as a boundary and the flow interpreted as the flow in a channel bounded by $x = 0$, $y = 0$ and $xy = 1$ (figure 139).

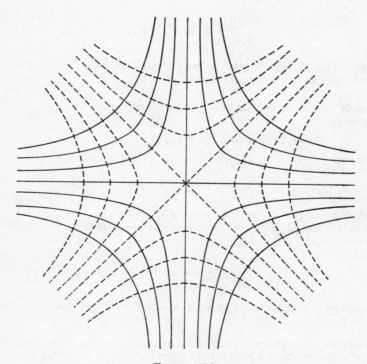

FIGURE 138

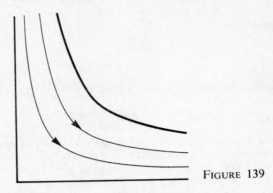

FIGURE 139

$$\frac{\mathrm{d}W}{\mathrm{d}z} = 2z = 2x + i(2y).$$

Therefore the velocity components u and v are $2x$ and $-2y$ respectively. The magnitude of the resultant velocity is

$$\left| \frac{\mathrm{d}W}{\mathrm{d}z} \right| = 2\sqrt{x^2 + y^2}.$$

Example   Consider the complex potential $W = \cosh^{-1}(z/a)$ where a is a real constant and where z is only real when $|z| < a$.*

Since $W = \cosh^{-1}(z/a)$, we have $z = a \cosh W$.
Therefore

$$x + iy = a \cosh(\varphi + i\psi)$$
$$= a(\cosh \varphi \cosh i\psi + \sinh \varphi \sinh i\psi)$$
$$= a(\cosh \varphi \cos \psi + i \sinh \varphi \sin \psi).$$

Therefore $x = a \cosh \varphi \cos \psi$ and $y = a \sinh \varphi \sin \psi$.
   To obtain the equations of the streamlines $\psi = $ constant, first eliminate φ from the above to give

$$\frac{x^2}{a^2 \cos^2 \psi} - \frac{y^2}{a^2 \sin^2 \psi} = \cosh^2 \varphi - \sinh^2 \varphi = 1.$$

Now putting $\psi = $ constant we have the equation of a hyperbola with transverse axis the x-axis.

---

* This (or some similar) restriction on z is necessary to ensure the uniqueness of the point W corresponding to a point z. For further discussion of this rather difficult topic see for example Franklin—Functions of a Complex Variable (Pitman).

Thus the streamlines are a set of confocal hyperbolae with focii $(\pm a, 0)$* and transverse axis the x-axis (figure 140).

For the equipotentials, eliminating ψ gives

$$\frac{x^2}{a^2 \cosh^2 \varphi} + \frac{y^2}{a^2 \sinh^2 \varphi} = \cos^2 \psi + \sin^2 \psi = 1.$$

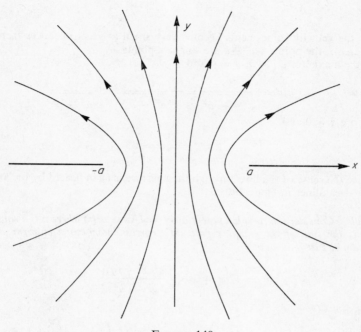

FIGURE 140

Thus the equipotentials are a set of confocal ellipses with focii $(\pm a, 0)$.* When $\psi \to 0$

$$x \to a \cosh \varphi \quad \text{and} \quad y \to 0.$$

That is, as $\psi \to 0$, the points z approach the points on the x-axis for $x \geqslant a$. When $\psi \to \pi$

$$x \to -a \cosh \varphi \quad \text{and} \quad y \to 0.$$

That is, as $\psi \to \pi$, the points z approach the points on the x-axis for $x \leqslant -a$.

---

* The hyperbola $\dfrac{x^2}{a^2} - \dfrac{y^2}{b^2} = 1$ has focii $(\pm \sqrt{a^2 + b^2}, 0)$ and the ellipse $\dfrac{x^2}{a^2} + \dfrac{y^2}{b^2} = 1$ has focii $(\pm \sqrt{a^2 - b^2}, 0)$.

Thus, taking the above two segments of the x-axis as rigid boundaries, we see that the flow pattern can be interpreted as that for flow through an aperture of width $2a$ lying along the x-axis from $x = -a$ to $x = a$.

$$\frac{dW}{dz} = \frac{1}{\sqrt{z^2 - a^2}} = u - iv.$$

Therefore

$$\left|\frac{dW}{dz}\right| \to 0 \quad \text{as} \quad |z| \to \infty,$$

that is the velocity components become very small at great distances from the origin. Also, the points $z = \pm a$ are singular points.

On the x-axis for $|x| < a$, $y = 0$ and

$$u - iv = \frac{1}{\sqrt{x^2 - a^2}} = \frac{1}{i\sqrt{a^2 - x^2}} = \frac{-i}{\sqrt{a^2 - x^2}}.$$

Therefore $u = 0$ and

$$v = \frac{1}{\sqrt{a^2 - x^2}}.$$

Hence the sense of the flow through the aperture is as indicated by the arrows on the streamlines in figure 140.

Example   Consider the complex potential $W = U(z + a^2/z)$ where a is a constant. (That is the sum of the complex potentials corresponding to a uniform stream and a doublet at the origin.)

$$W = U\left\{x + iy + \frac{a^2(x - iy)}{x^2 + y^2}\right\}.$$

Therefore

$$\varphi = Ux\left(1 + \frac{a^2}{x^2 + y^2}\right) \quad \text{and} \quad \psi = Uy\left(1 - \frac{a^2}{x^2 + y^2}\right).$$

Thus the streamlines are given by

$$y\left(1 - \frac{a^2}{x^2 + y^2}\right) = \text{constant}.$$

In particular, the streamline corresponding to $\psi = 0$ is given by

$$y\left(1 - \frac{a^2}{x^2 + y^2}\right) = 0.$$

Therefore this streamline consists of the curves $y = 0$ and $x^2 + y^2 = a^2$, that is the x-axis and the circle centre the origin and radius a.

$$\frac{dW}{dz} = U(1 - a^2/z^2).$$

Therefore, the points $z = \pm a$ are stagnation points for the flow. Also

$$\frac{dW}{dz} \to U \quad \text{as} \quad |z| \to \infty.$$

Thus at large distances from the origin the flow approaches that due to uniform stream with speed U in the positive direction of the x-axis.

The streamlines are as shown in figure 141 and can be interpreted as those for the flow of a uniform parallel stream past a long circular cylinder of radius of

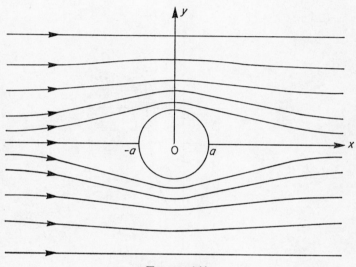

FIGURE 141

cross-section a and axis through the origin perpendicular to the x,y-plane. (The doublet at the origin is sometimes referred to as the image in the cylinder of the uniform stream: compare with Chapter 3.)

Now on the surface of the cylinder $z = ae^{i\theta}$ (where θ is the usual plane polar coordinate) and so we have

$$\frac{dW}{dz} = U(1 - e^{-2i\theta}) = U(1 - \cos 2\theta + i \sin 2\theta)$$

and

$$|q| = \left| \frac{dW}{dz} \right| = U\{(1 - \cos 2\theta)^2 + \sin^2 2\theta\}^{\frac{1}{2}}$$
$$= U(2 - 2\cos 2\theta)^{\frac{1}{2}}$$
$$= 2U |\sin \theta|.$$

Bernoulli's theorem gives for steady flow

$$\frac{p}{\rho} + \tfrac{1}{2}\mathbf{q}^2 = \text{constant}$$

where p is pressure and ρ is density.

Therefore

$$\frac{p}{\rho} + \tfrac{1}{2}\mathbf{q}^2 = \frac{P}{\rho} + \tfrac{1}{2}U^2$$

(where P is the pressure at large distances from the origin).

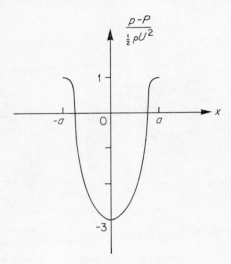

FIGURE 142

Hence, on the surface of the cylinder

$$\frac{p}{\rho} + 2U^2 \sin^2 \theta = \frac{P}{\rho} + \tfrac{1}{2}U^2.$$

Therefore $p - P = \tfrac{1}{2}\rho U^2 (1 - 4 \sin^2 \theta)$.

Hence the pressure distribution over the surface of the cylinder is symmetric about the x-axis and the maximum pressure occurs at the stagnation points. The distribution of pressure over the surface of the cylinder above (or below) the x-axis is shown graphically in figure 142. Because of the symmetry of the pressure distribution about the x-axis there can be no resultant thrust on the cylinder in the y-direction. Similarly, since the pressure distribution is symmetric about the y-axis, there can be no resultant thrust on the cylinder in the x-direction.

Example   Flow with circulation round a circular cylinder.

To obtain the complex potential for this flow we add a simple vortex at the origin to the flow of the last example.

Therefore

$$W = U(z + a^2/z) + iK \ln \frac{z}{a}.$$

Since the wall of the cylinder $|z| = a$ is a streamline for both parts of the above flow, it will also be a streamline in the resultant flow.

$$\varphi = Ux\left(1 + \frac{a^2}{x^2 + y^2}\right) - K \arg\left(\frac{z}{a}\right).$$

$$\psi = Uy\left(1 - \frac{a^2}{x^2 + y^2}\right) + K \ln\left(\frac{\sqrt{x^2 + y^2}}{a}\right).$$

Therefore the streamlines are given by

$$Uy\left(1 - \frac{a^2}{x^2 + y^2}\right) + K \ln\left(\frac{\sqrt{x^2 + y^2}}{a}\right) = \text{constant}.$$

(In particular, on the cylinder $x^2 + y^2 = a^2$ and the above equation is satisfied with the right hand side equal to zero. Hence the cylinder forms part of the streamline for which $\psi = 0$.)

Now

$$\frac{dW}{dz} = U(1 - a^2/z^2) + \frac{iK}{z}$$

and so $dW/dz \to U$ as $|z| \to \infty$. That is at large distances from the cylinder, the flow is nearly uniform and parallel with speed U in the positive direction of the x-axis.

Stagnation points occur where $dW/dz = 0$, that is where

$$U(1 - a^2/z^2) + \frac{iK}{z} = 0$$

or

$$z^2 + \frac{iK}{U}z - a^2 = 0.$$

Therefore

$$z = -\frac{iK}{2U} \pm \sqrt{a^2 - \frac{K^2}{4U^2}}.$$

Thus we must now consider the three cases distinguished by the expression under the square root being positive, zero or negative.

(i) $$\frac{K}{2U} < a.$$

(Note that this includes the case $K = 0$ considered in the previous example.)

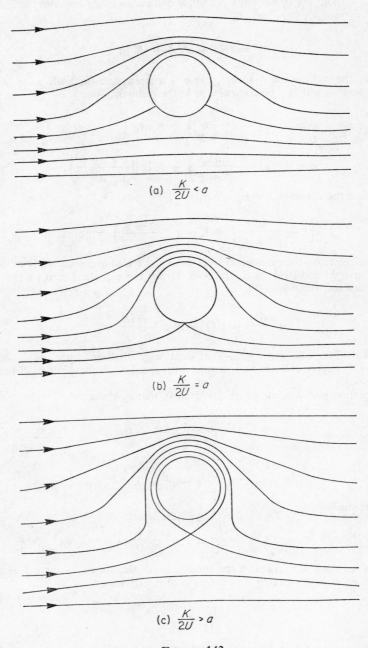

(a) $\dfrac{K}{2U} < a$

(b) $\dfrac{K}{2U} = a$

(c) $\dfrac{K}{2U} > a$

FIGURE 143

426

The modulus of both roots is then

$$\left\{\left(a^2 - \frac{K^2}{4U^2}\right) + \left(\frac{-K}{2U}\right)^2\right\}^{\frac{1}{2}} = a.$$

Hence both stagnation points lie on the surface of the cylinder itself. The streamlines for this flow are sketched in figure 143(a).

(ii)
$$\frac{K}{2U} = a.$$

Both stagnation points coincide in the point $z = -iK/2U = -ia$ on the surface of the cylinder.

The streamlines for this flow are sketched in figure 143(b).

(iii)
$$\frac{K}{2U} > a.$$

Here
$$z = -\frac{iK}{2U} \pm i\sqrt{\frac{K^2}{4U^2} - a^2}$$

where the square root is now real.
Put
$$z_1 = -\frac{iK}{2U} + i\sqrt{\frac{K^2}{4U^2} - a^2}$$

and
$$z_2 = -\frac{iK}{2U} - i\sqrt{\frac{K^2}{4U^2} - a^2}.$$

Now
$$\sqrt{\frac{K^2}{4U^2} - a^2} < \frac{K}{2U}$$

and
$$\frac{K}{2U} - a < \sqrt{\frac{K^2}{4U^2} - a^2}^{\;*}$$

Therefore
$$|z_1| = \frac{K}{2U} - \sqrt{\frac{K^2}{4U^2} - a^2} < \frac{K}{2U} - \left(\frac{K}{2U} - a\right) = a.$$

---

* $(\alpha - \beta)^2 = \alpha^2 - 2\alpha\beta + \beta^2 < \alpha^2 - 2\beta^2 + \beta^2$ if $\alpha > \beta > 0$. Therefore $\alpha - \beta < \sqrt{\alpha^2 - \beta^2}$.

15

That is this stagnation point lies inside the cylinder and so does not have any physical significance.

$$|z_2| = \frac{K}{2U} + \sqrt{\frac{K^2}{4U^2} - a^2} > a + \sqrt{\frac{K^2}{4U^2} - a^2} > a.$$

Thus this stagnation point lies on the imaginary axis (y-axis) and is beneath the cylinder. The streamline through this point crosses over itself at right angles at z_2 and then, because of the symmetry of the whole pattern about the imaginary axis, we see that each branch of this stream-line must be inclined to the imaginary axis at 45°. The streamlines for this flow are sketched in figure 143(c).

Again, putting $z = ae^{i\theta}$, we see that on the surface of the cylinder

$$\frac{dW}{dz} = U(1 - e^{-2i\theta}) + \frac{iK}{a} e^{-i\theta}$$

$$= \left\{ U(1 - \cos 2\theta) + \frac{K}{a} \sin \theta \right\} + i \left\{ U \sin 2\theta + \frac{K}{a} \cos \theta \right\}$$

Therefore

$$|\mathbf{q}| = \left| \frac{dW}{dz} \right| = \left\{ 4U^2 \sin^2 \theta + \frac{K^2}{a^2} + \frac{4KU}{a} \sin \theta \right\}^{\frac{1}{2}}.$$

Bernoulli's theorem gives for steady flow

$$\frac{p}{\rho} + \tfrac{1}{2} q^2 = \text{constant}$$

$$= \frac{P}{\rho} + \tfrac{1}{2} U^2$$

where P is the pressure at very large distances from the cylinder (where the speed is U). Therefore on the surface of the cylinder,

$$\frac{p}{\rho} + \frac{1}{2} \left(4U^2 \sin^2 \theta + \frac{K^2}{a^2} + \frac{4KU}{a} \sin \theta \right) = \frac{P}{\rho} + \tfrac{1}{2} U^2.$$

Therefore

$$p = P + \tfrac{1}{2} \rho U^2 (1 - 4 \sin^2 \theta) - \frac{K^2 \rho}{2a^2} - \frac{2\rho KU}{a} \sin \theta.$$

Since this is symmetric about the y-axis there can be no resultant thrust in the x-direction on the cylinder.

The term $-(2\rho KU/a) \sin \theta$ is not symmetric about the x-axis however and so will produce a resultant thrust in the y-direction on the cylinder. This resultant thrust per unit length of the cylinder is given by the integral of $(2\rho KU/a) \sin^2 \theta$ over the surface of the cylinder (figure 144). That is by

$$\int_0^{2\pi} \frac{2\rho KU}{a} \sin^2 \theta \, ad\theta = 2\pi \rho KU.$$

Thus the cylinder experiences an upthrust of $2\pi \rho KU$ per unit length.

This upthrust is known as the lift force and is essentially what enables an aircraft to remain in the air during flight.

Note that the upthrust becomes zero when K is put equal to zero, thus agreeing with the result obtained in the previous example.

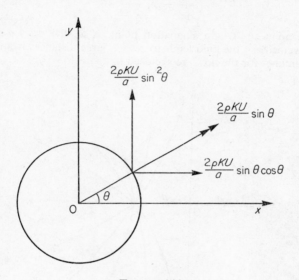

FIGURE 144

Example Deduce the complex potential for a line source of strength m per unit length at a distance a from a rigid plane boundary.

Recalling the method of images discussed in Chapter 3 we consider the complex potential due to the given line source together with its image in the plane boundary, that is another line source of strength m per unit length at a distance a on the other side of the boundary.

Thus, with axes as indicated in figure 145 the complex potential is

$$W = m \ln (z - a) + m \ln (z + a)$$
$$= m \ln (z^2 - a^2)$$
$$= m \ln (x^2 - y^2 - a^2 + 2ixy).$$

Therefore

$$\psi = m \arg (z^2 - a^2)$$
$$= m \tan^{-1} \frac{2xy}{x^2 - y^2 - a^2}.$$

(The principal value of the argument is to be used.)

Hence the streamlines are given by $2xy = C(x^2 - y^2 - a^2)$ for constant C. In particular, when $C = 0$, we have $2xy = 0$ and so $x = 0$ or $y = 0$. Thus the plane boundary is a streamline for the flow.

$$\frac{dW}{dz} = m \frac{2z}{z^2 - a^2}$$

and so the point $z = 0$ is a stagnation point. Also, $dW/dz \to 0$ as $|z| \to \infty$, that is the velocity of the fluid tends to zero at large distances from the source.

The streamlines for the flow are sketched in figure 145.

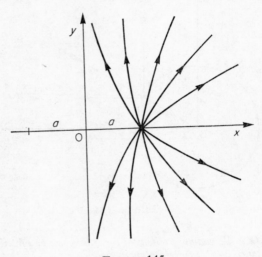

FIGURE 145

Example   Deduce the complex potential for a doublet of moment μ held at a distance a from a plane rigid boundary and inclined at an angle α to the normal to the plane.

The image of the given doublet is another doublet of moment μ at a distance a on the other side of the boundary and inclined at an angle $(\pi - \alpha)$ to the normal to the plane (see chapter 3). Therefore if the rigid boundary is the plane $x = 0$ and the doublet is at the point $(a, 0)$, the complex potential is

$$W = \frac{-\mu e^{i\alpha}}{z - a} - \frac{\mu e^{i(\pi - \alpha)}}{z + a}$$

$$= \frac{\mu e^{-i\alpha}}{z + a} - \frac{\mu e^{i\alpha}}{z - a}.$$

Example   Deduce the complex potential for a point vortex of strength C held at a distance a from a plane rigid boundary.

Choose cartesian coordinate axes such that the rigid boundary coincides with the line $x = 0$ and the given vortex is at the point $(a,0)$.

The complex potential due to the point vortex alone is

$$iC \ln (z - a).$$

Now consider the complex potential W due to the given vortex together with another point vortex of strength $-C$ at the point $(-a,0)$.

$$W = iC \ln (z - a) - iC \ln (z + a)$$

$$= iC \ln \left(\frac{z - a}{z + a} \right).$$

The stream function ψ is then given by

$$\psi = C \ln \left| \frac{z - a}{z + a} \right|.$$

Hence the streamlines are given by the curves

$$C \ln \left| \frac{z - a}{z + a} \right| = \text{constant}$$

or

$$\left| \frac{z - a}{z + a} \right| = k$$

for constants k.

Now, on the line $x = 0$

$$\left| \frac{z - a}{z + a} \right| = \left| \frac{iy - a}{iy + a} \right| = \frac{\sqrt{y^2 + a^2}}{\sqrt{y^2 + a^2}} = 1.$$

Thus the line $x = 0$ is a streamline for the flow and so

$$W = iC \ln \left| \frac{z - a}{z + a} \right|$$

may be regarded as the complex potential due to the given point vortex in the presence of a rigid boundary coinciding with the line $x = 0$.

The vortex of strength $-C$ at the point $(-a,0)$ is, of course, the image in the line $x = 0$ of the vortex of strength C at the point $(a,0)$.

1.311 Milne–Thomson circle theorem. This theorem states that if an irrotational, two-dimensional, incompressible, inviscid fluid flow in the z-plane has complex potential $f(z)$ when there are no rigid boundaries and all the singularities of $f(z)$ lie outside the circle $|z| = a$, then the complex potential for the flow when a rigid boundary coinciding with the circle $|z| = a$ is introduced is $f(z) + \bar{f}(a^2/z)$ where if $f(z) = g(z) + ih(z)$, $\bar{f}(z) = g(z) - ih(z)$. (Note that $\bar{f}(z)$ is not the complex conjugate $\overline{f(z)}$ of $f(z)$.)

To prove this result we must show (i) that the circle $|z| = a$ is in fact a streamline for the flow and (ii) that the singularities in the region outside $|z| = a$ are only those of $f(z)$.

Proof: (i) On the circle $|z| = a$, $z = ae^{i\theta}$ and so $a^2/z = ae^{-i\theta} = \bar{z}$.

Therefore, on the circle,

$$W = f(z) + \bar{f}(a^2/z) = f(z) + \bar{f}(\bar{z}) = f(z) + \overline{f(z)}.$$

Hence W is wholly real and so $\psi = 0$. That is the circle $|z| = a$ is a streamline.

(ii) If

$$f(a^2/z) = u(z) + iv(z),$$

then

$$\bar{f}(a^2/z) = u(z) - iv(z).$$

Hence if a particular value of z is a singularity of $f(a^2/z)$ and so either (or both) of u and $v \to \infty$ as $z \to$ that particular value, then it will also be a singularity of $\bar{f}(a^2/z)$ and vice versa.

Thus the singularities of $\bar{f}(a^2/z)$ are identical with the singularities of $f(a^2/z)$ and these singularities occur when $a^2/z = z_i$ ($i = 1,2,\ldots$) where the z_i are the singularities of $f(z)$. But these singularities z_i lie outside the circle $|z| = a$ and so $|z_i| > a$.

Hence $|a^2/z| > a$ and so $|z| < a$ for the singularities of $f(a^2/z)$. Thus all the singularities of $\bar{f}(a^2/z)$ lie within the circle $|z| = a$ and so the singularities of $W = f(z) + \bar{f}(a^2/z)$ in the region outside $|z| = a$ are only those of $f(z)$.

The extension of this theorem to symmetric three-dimensional motions of the type already considered is obvious. The circle becomes an infinitely long cylinder and any point sources present become line sources etc.

Example   Uniform stream.

The complex potential for uniform stream with speed U at infinity is Uz. Thus $f(z) = Uz$ and the complex potential for uniform stream flowing past a circular cylinder of radius a is then

$$W = f(z) + \bar{f}(a^2/z)$$
$$= Uz + U(a^2/z).$$

The second term on the right hand side is the complex potential for a line doublet of moment Ua^2 per unit length along the axis of the cylinder and with direction opposite to that of the flow of the stream. This doublet is this image in the cylinder of the uniform stream. The flow corresponding to this complex potential has been considered fully in an earlier example on page 422.

Example   A source of strength m at z = b.

The complex potential for the source is $m \ln (z - b)$. Now introduce the rigid boundary $|z| = a$ where $a < b$. The complex potential now is

$$
\begin{aligned}
W &= m \ln (z - b) + m \ln \left(\frac{a^2}{z} - b\right) \\
&= m \ln (z - b) + m \ln (a^2 - bz) - m \ln z \\
&= m \ln (z - b) + m \ln \left(z - \frac{a^2}{b}\right) - m \ln z + m \ln (-b).
\end{aligned}
$$

Since the last term on the right hand side is a constant, this will not affect the flow and so we can take

$$
W = m \ln (z - b) + m \ln \left(z - \frac{a^2}{b}\right) - m \ln z.
$$

The second term on the right hand side is the complex potential for a source of strength m at the point $z = a^2/b$, which is the inverse point of $z = b$ with respect to the circle. The third term at the right-hand side is the complex potential for a sink of strength m at the origin. Thus the image in the circle $|z| = a$ of the source m at $z = b$ is an equal source m at the inverse point $z = a^2/b$ together with a sink of strength m at the origin.

Example   A doublet of moment μ at z = b with axis making an angle α with the real axis.

The complex potential for the doublet is $-[\mu/(z - b)]e^{i\alpha}$. After introducing the rigid boundary $|z| = a$ $(a < b)$, the complex potential becomes

$$
\begin{aligned}
W &= \frac{-\mu}{z - b} e^{i\alpha} - \frac{\mu}{a^2/z - b} e^{-i\alpha} \\
&= \frac{-\mu}{z - b} e^{i\alpha} - \mu e^{-i\alpha}\left(\frac{z}{a^2 - bz}\right) \\
&= \frac{-\mu}{z - b} e^{i\alpha} - \mu e^{-i\alpha}\left(-\frac{1}{b} + \frac{a^2/b}{a^2 - bz}\right) \\
&= \frac{-\mu}{z - b} e^{i\alpha} - \mu e^{-i\alpha}\left(-\frac{1}{b} - \frac{a^2/b^2}{z - a^2/b}\right) \\
&= \frac{-\mu}{z - b} e^{i\alpha} - \frac{\mu(a^2/b^2)}{z - a^2/b} e^{i(\pi-\alpha)} + \frac{\mu}{b} e^{-i\alpha}.
\end{aligned}
$$

Again the third term on the right hand side is a constant and so does not affect the flow. Therefore the complex potential may be taken as

$$
W = \frac{-\mu}{z - b} e^{i\alpha} - \frac{\mu(a^2/b^2)}{z - a^2/b} e^{i(\pi-\alpha)}.
$$

Since the second term on the right hand side is the complex potential for a doublet of moment $\mu a^2/b^2$ and direction making an angle $(\pi - \alpha)$ with the real axis situated at $z = a^2/b$, the inverse point of $z = b$ with respect to the circle $|z| = a$, this doublet is the image in the circle $|z| = a$ of the given doublet at $z = b$.

9.32 Electrostatics

We have seen earlier (Chapter 2) that the electrostatic potential φ satisfies Laplace's equation $\nabla^2 \varphi = 0$ except at points where there are charges. We now consider (in particular) electrostatic fields for which conditions are identical in all planes parallel to a given plane taken to be the x,y-plane. Such fields are said to be two-dimensional and the potential is known everywhere once it has been determined in the x,y-plane.

Since φ satisfies the equation

$$\frac{\partial^2 \varphi}{\partial x^2} + \frac{\partial^2 \varphi}{\partial y^2} = 0$$

a function ψ conjugate to it can be found. The curves $\psi = $ constant are then everywhere orthogonal to the equipotentials ($\varphi = $ constant) and so have the direction of grad φ or $\mathbf{E}$ at any point. This means that the curves $\psi = $ const. are the lines of force for the field. Since φ and ψ are conjugate functions, the Cauchy–Riemann equations hold.

Hence

$$\frac{\partial \varphi}{\partial x} = \frac{\partial \psi}{\partial y} \quad \text{and} \quad \frac{\partial \varphi}{\partial y} = -\frac{\partial \psi}{\partial x}.$$

Therefore

$$\mathbf{E} = -\text{grad } \varphi$$

$$= \left(-\frac{\partial \varphi}{\partial x}, -\frac{\partial \varphi}{\partial y} \right) = \left(-\frac{\partial \psi}{\partial y}, \frac{\partial \psi}{\partial x} \right).$$

The complex function W defined by

$$W = \varphi + i\psi$$

is called the *complex potential* for the field.

$$\frac{dW}{dz} = \frac{\partial \varphi}{\partial x} + i\frac{\partial \psi}{\partial x} = \frac{\partial \varphi}{\partial x} - i\frac{\partial \varphi}{\partial y}.$$

Therefore

$$\left| \frac{dW}{dz} \right| = |\mathbf{E}|$$

and so the equilibrium points (that is points at which $\mathbf{E} = 0$) occur where

$dW/dz = 0$.

Points where dW/dz becomes infinite are again called singular points of the field. It is not possible to determine the direction of a line of force through such a point.

The use of complex potential in electrostatics is very similar to its use in hydrodynamics.

Example   Consider the complex potential $W = Ez$ where E is a constant.

$$\varphi = Ex \quad \text{and} \quad \psi = Ey.$$

Therefore the equipotentials are the lines $x =$ constant and the lines of force are the lines $y =$ constant.

$$\frac{dW}{dz} = E = \frac{\partial\varphi}{\partial x} - i\frac{\partial\varphi}{\partial y}.$$

Therefore

$$\frac{\partial\varphi}{\partial x} = E \quad \text{and} \quad \frac{\partial\varphi}{\partial y} = 0.$$

Hence Ez is the complex potential for a uniform field parallel to the x-axis and in the direction of decreasing x. The magnitude of the field strength is E every-where.

Now since $\varphi =$ constant within a conductor, we can replace any of the equipotentials for the field by conducting plates. Thus Ez may be taken as the complex potential for the field between two parallel conducting plates maintained at constant potentials (figure 146).

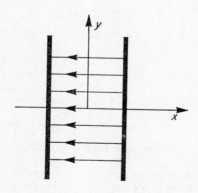

FIGURE 146

Example   Consider the complex potential $W = C \ln z$ where C is a real constant.

By analogy with the hydrodynamic case, the equipotentials in the x,y-plane are concentric circles with centre the origin and the lines of force are straight lines through the origin. If C is negative, the lines of force point radially outwards and so the complex potential is that for a uniform line charge through the origin perpendicular to the x,y-plane. To obtain the charge density e per unit length of the line, apply Gauss' law to the cylinder of radius a and unit length and with axis coincident with the line charge.

This gives $-2\pi C = 4\pi e$ and so the charge density is $-\frac{1}{2}C$.

Thus for a line having charge density e per unit length the complex potential is $-2e \ln z$.

Since any of the surfaces $\varphi = $ constant can be regarded as conducting surfaces maintained at given potentials however we see that the complex potential $C \ln z$ also applies to the field between two conducting cylinders maintained at constant potentials and having the same axis.

By analogy with the hydrodynamic case (and using the result of the above example), we see that the complex potential for a dipole of moment μ at the origin and with axis along the real axis is $2\mu/z$.

Similarly the complex potential for a line charge of density e per unit length at a distance a from a plane conducting boundary is

$$-2e \ln (z - a) + 2e \ln (z + a).$$

Also the complex potential for a line dipole of moment μ per unit length held at a distance a from a plane conducting boundary and with its axis inclined at an angle α to the normal to the plane is

$$\frac{2\mu e^{i\alpha}}{z - a} + \frac{2\mu e^{-i\alpha}}{z + a}.$$

Example   Consider the complex potential $W = \cos^{-1}(z/a)$ where a is a real constant and where z is only real when $|z| < a$.[*]

Since $W = \cos^{-1}(z/a)$, we have $z = a \cos W$. Therefore

$$x + iy = a \cos (\varphi + i\psi)$$
$$= a(\cos \varphi \cos i\psi - \sin \varphi \sin i\psi)$$
$$= a(\cos \varphi \cosh \psi = i \sin \varphi \sinh \psi).$$

[*] This (or some similar) restriction on z is necessary to ensure the uniqueness of the point W corresponding to a point z. For further discussion of this rather difficult topic see for example Franklin—Functions of a Complex Variable (Pitman).

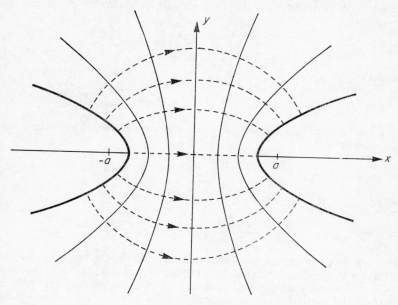

FIGURE 147

Therefore $x = a \cos \varphi \cosh \psi$ and $y = -a \sin \varphi \sinh \psi$. Eliminating φ gives

$$\frac{x^2}{a^2 \cosh^2 \psi} + \frac{y^2}{a^2 \sinh^2 \psi} = \cos^2 \varphi + \sin^2 \varphi = 1.$$

Then putting $\psi = $ constant we have the equation of a line of force. Thus the lines of force are confocal ellipses with focii $(\pm a, 0)$.

Eliminating ψ from the above gives

$$\frac{x^2}{a^2 \cos^2 \varphi} - \frac{y^2}{a^2 \sin^2 \varphi} = \cosh^2 \psi - \sinh^2 \psi = 1.$$

Then putting $\varphi = $ constant we have the equation of an equipotential. Thus the equipotentials are confocal hyperbolae with focii $(\pm a, 0)$.

Taking each of the branches of one of these hyperbolae as conducting surfaces, we see that this complex potential may be interpreted as that for the field between two hyperbolic pole pieces maintained at constant potentials.* In figure 147 the

---

* The values for φ on each of the branches will be different, for if φ is such that $\cos \varphi > 0$ then $x = a \cos \varphi$ is always >0 for positive a whereas $y = -a \sin \varphi \sinh \psi$ can vary over both positive and negative values. Thus such a value of φ gives only the right hand branch of a hyperbola. Similarly a value of φ such that $\cos \varphi < 0$ will give a left hand branch.

field lines (dashed curves) and the equipotentials (solid curves) for this problem are sketched.

$$\frac{dW}{dz} = -\frac{1}{\sqrt{a^2 - z^2}}.$$

Therefore

$$\left|\frac{dW}{dz}\right| \to 0 \quad \text{as} \quad |z| \to \infty.$$

Hence the field strength becomes very small at large distances from the origin. Also, when $x = 0$,

$$\frac{dW}{dz} = -\frac{1}{\sqrt{a^2 + y^2}} = -E_x + iE_y$$

where $\mathbf{E} = (E_x, E_y)$. Therefore

$$E_x = \frac{1}{\sqrt{a^2 + y^2}} \quad \text{and} \quad E_y = 0.$$

Hence the senses of the lines of force are as indicated in the figure.

This same complex potential has another interpretation however, when $\varphi \to 0$, $x \to a \cosh \psi$ and $y \to 0$. That is, as $\varphi \to 0$, the points z approach the points on the x-axis for $x \geqslant a$. Similarly, when $\varphi \to \pi$, $x \to -a \cosh \psi$ and $y \to 0$ so that the points z approach the points on the x-axis for $x \leqslant -a$.

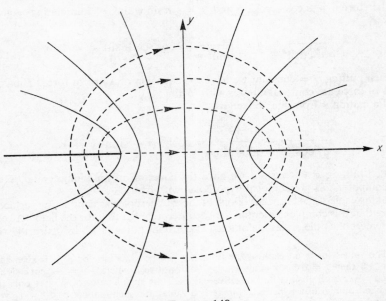

FIGURE 148

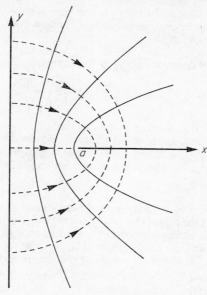

FIGURE 149

Taking both of these segments of the x-axis as conducting surfaces then we see that the above complex potential may be interpreted as that for the field between two semi-infinite conducting plates maintained at constant potentials and separated by a slit of width $2a$ as in figure 148 in which the field lines and equipotentials for this problem are also sketched.

The points $z = \pm a$ are singular points.

Also, when $\varphi = \frac{1}{2}\pi$, $x = 0$ and $y = -a \sinh \psi$ so that the imaginary axis $x = 0$ is an equipotential. Now, taking the x-axis for $x \geqslant a$ and the y-axis as conducting surfaces, we see that the above complex potential may be interpreted as that for the field between two perpendicular conducting plates maintained at constant potentials and situated as indicated in figure 149 in which the field lines and equipotentials are also sketched.

Complex potentials are also used in magnetostatics and in current electricity.

Exercises

1. Sketch the streamlines and equipotentials for the complex potential $W = Ue^{-i\alpha}z$ where U and α are real constants. Show that this complex potential can be interpreted as that for uniform flow with velocity U in a direction making an angle α with the positive direction of the x-axis.

2. Sketch the streamlines and equipotentials for the complex potential

$$W = \cos^{-1}\frac{z}{a}$$

where a is a real constant and where z is only real when $|z| > a$. Show that this complex potential can be interpreted as that for flow round an elliptic cylinder or round a flat plate of length $2a$.

3. Show that for real constants a and U the complex potential

$$W = aU \coth \left(\frac{a}{z}\right)$$

can be interpreted as that for flow in the presence of a rigid boundary coinciding with the plane $y = 0$ and the fixed circular cylinder whose surface has the equation $x^2 + y^2 - 2ay = 0$. Show also that at large distances from the cylinder the flow tends to that for a uniform stream with velocity U in the positive x-direction.

4. Two points vortices each of strength K are situated at the points P$(a,0)$ and Q$(-a,0)$. Show that the origin O is a stagnation point for the resulting flow field A third point vortex is now placed at O. Find its strength if the velocity at Q due to the vortices at P and O is zero and the velocity at P due to the vortices at Q and O is zero. Find the equations of the streamlines for the flow due to the three vortices and show that the points $(0, \pm a/\sqrt{3})$ are stagnation points and that the streamline passing through these stagnation points also passes through the points $(\pm b,0)$ where

$$3\sqrt{3}(b^2 - a^2)^2 = 16a^3b.$$

5. Six hydrodynamical doublets each of moment μ are equally spaced round the circumference of the circle with centre the origin and radius a, one of the doublets being at the point $(a,0)$. If all the doublets are pointing towards the origin, show that the complex potential is

$$W = \frac{-6\mu a^5}{a^6 - z^6}.$$

6. Obtain the complex potential for the flow given by

$$u = \frac{-ax - by}{x^2 + y^2}, \qquad v = \frac{bx - ay}{x^2 + y^2}$$

where a, b are constants and u, v are the velocity components in the x, y directions. Show that the difference in pressure at points distant r_1 and r_2 from the origin is proportional to $1/r_1^2 - 1/r_2^2$.

7. A doublet of moment μ is held at a distance a from a plane rigid boundary. If the axis of the doublet is inclined at an angle $\pi/2$ to the normal to the plane, find the stagnation points for the flow.

8. Obtain the complex potential for the two-dimensional flow due to a point source of strength m at the point (a,a) in the right angle between the two walls $y = 0, x = 0$.

Show that the maximum velocity on the wall $x = 0$ occurs at the point $(0,b)$ where

$$(2a - b)\{a^2 + (a + b)^2\} = (2a + b)\{a^2 + (b - a)^2\}.$$

9. Show that $W = -iVz$, where V is a real, positive constant, is the complex potential for the two-dimensional flow of an ideal fluid parallel to the y-axis and in the direction of increasing y.

An ideal fluid fills the space $y > 0$ on one side of an infinite plane wall $y = 0$ and moves at infinity with velocity V parallel to the y-axis in the direction of increasing y, the motion being completely two-dimensional in the x-y plane. A doublet of strength μ pointing in the negative y-direction is introduced at a distance a from the wall. Show that if $\mu < 4a^2V$ the pressure of the fluid on the wall is a maximum at points distant $a\sqrt{3}$ from O, the foot of the perpendicular from the doublet on to the wall, and is a minimum at O.

If $\mu = 4a^2V$, find the stagnation points for the flow.

10. Obtain the complex potential for the two-dimensional incompressible flow due to a point source of strength m at $z = b$ in the presence of a rigid boundary consisting of the line $x = 0$ together with the semicircular arc $x^2 + y^2 = a^2$ ($0 < a < b$), $x \geqslant 0$.

Determine the velocity of the flow at all points on the y-axis and show that it reaches a maximum at finite distance from the origin.

11. The flow of an ideal fluid is due entirely to the presence of two equal, uniform, parallel line sources of strength m per unit length held a distance $2a$ apart. A rigid cylinder whose radius of cross-section b is less than a is now placed in the fluid with its axis parallel to and midway between the two line sources. Determine the stagnation points for the resulting flow.

12. Sketch the lines of force and equipotentials for the electrostatic complex potential $W = \cosh^{-1}(z/a)$ where a is a real constant and where for real z, $|z| > a$. Show that this complex potential may be interpreted as that for the field between two charged confocal elliptic cylinders.

13. Sketch the lines of force and equipotentials for the complex potential $W = z^{\frac{1}{2}}$ where if z is real, $|z| < 0$ and show that this complex potential may be interpreted as that for the field due to a semi-infinite charged plate of parabolic section.

14. Show that, with suitable choice of coordinate axes, the complex potential due to two parallel line charges of opposite sign, but each having density of magnitude q per unit length and a distance $2a$ apart can be written

$$W = 2q \ln \frac{z + a}{z - a}.$$

Hence show that the equipotential surfaces are a family of circular cylinders and that in particular the cylinder on which the potential has the constant value

$2q\alpha$ has radius $2ae^\alpha/(1 - e^{2\alpha})$. Show also that the axis of this cylinder passes through the point $[-a(1 + e^{2\alpha})/(1 - e^{2\alpha}),0]$.

Indicate how the above complex potential can also be used to give the lines of force and equipotentials due to two eccentric circular cylinders of equal radius and equal and opposite potentials.

9.4 Conformal Mappings

Let $w = f(z) = u(x,y) + iv(x,y)$ be a complex relationship defined for all values of z in a given domain of the z-plane. This relationship then enables us to obtain for each complex number z in the given domain, at least one corresponding complex number w. Values of z can then be represented by points in one Argand diagram called the z-plane and values of w represented in another Argand diagram called the w-plane. (Values of z and of w can of course, if convenient, be represented on the same Argand diagram, but it will be clearer at this stage to consider the two diagrams as separate). The correspondence between points of the z-plane and points of the w-plane set up by the relationship $w = f(z)$ is called a *mapping* (or *transformation*) of points in the z-plane into points in the w-plane. If to each point z in the domain of definition of the relationship there corresponds a *unique* point in the w-plane, the relationship is called a (complex) function. A point in the w-plane corresponding to a particular point P in the z-plane is known as an *image* of P with respect to the mapping. If every point in the domain of the z-plane for which a mapping is defined has a unique image point in the w-plane and if this image point corresponds to only the one point in the z domain, then the mapping is said to be *one-to-one*.

Now let $z = x(t) + iy(t)$ define in terms of a parameter t, a smooth curve C lying in the domain of the z-plane for which the function $f(z)$ is defined. Let $f(z)$ be analytic in this domain and consider the mapping $w = f(z) = u(x,y) + iv(x,y)$. Under this mapping the curve C in the z-plane will correspond to a curve C′ in the w-plane having equation $w = f[z(t)]$. The tangent to the curve C at the point $z_0 = z(t_0)$ makes with the positive direction of the x-axis the angle

$$\left[\tan^{-1}\left(\frac{dy}{dt}\middle/\frac{dx}{dt}\right)\right]_{z=z_0} = \arg(\dot{z}_0)$$

where $\dot{z}_0 = (dz/dt)_{z=z_0}$ and the tangent to the curve C′ at the corresponding point $w_0 = f(z_0) = f[z(t_0)]$ makes with the positive direction of the

u-axis the angle

$$\left[\tan^{-1}\left(\frac{dv}{dt}\Big/\frac{du}{dt}\right)\right]_{w=w_0} = \arg \dot{w}_0$$

where $\dot{w}_0 = (dw/dt)_{w=w_0}$.

But

$$\frac{dw}{dt} = \frac{df}{dz}\frac{dz}{dt}$$

and so

$$\arg \dot{w} = \arg f' + \arg \dot{z}$$

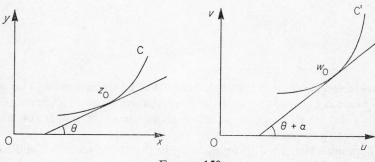

FIGURE 150

where $f' = df/dz$ is not zero or infinite. Then, when $t = t_0$,

$$\arg \dot{w}_0 = \arg f_0' + \arg \dot{z}_0.$$

Hence the angle between the tangents to C and C' at the corresponding points z_0 and w_0 is

$$\arg \dot{w}_0 - \arg \dot{z}_0 = \arg f_0' = \alpha \text{ (say)}$$

which is independent of the curve C (figure 150).

Now suppose that two curves C and C_1 pass through the point z_0 in the z-plane and that the tangents to C and C_1 at z_0 make angles θ and φ respectively with the positive direction of the x-axis (figure 151(a)). The corresponding curves C' and C_1' in the w-plane will of course pass through the point w_0 the image point of z_0 with respect to the mapping. Tangents to C' and C_1' at w_0 will make angles $(\theta + \alpha)$ and $(\varphi + \alpha)$ respectively with the positive direction of the u-axis (figure 151(b)).

Hence the angle between the tangents to C and C_1 at z_0 is equal to the angle between the tangents to C' and C_1' at w_0 (that is $\varphi - \theta$). Thus we have shown that the mapping associated with an analytic function $f(z)$ is

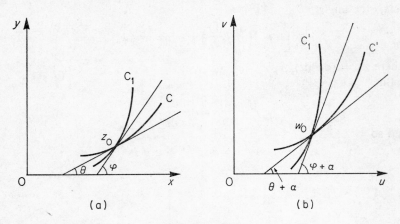

FIGURE 151

'angle-preserving' in that the angle between any two intersecting curves in the z-plane is equal to the angle between the corresponding curves at the image point of intersection in the w-plane except at points for which df/dz is zero or infinite.

Mappings which are angle-preserving in the above sense are said to be *conformal*. Hence a mapping which is defined by the analytic function $f(z)$ is conformal except at points for which df/dz is zero or infinite.

In particular, orthogonal curves in the z-plane will be mapped into orthogonal curves in the w-plane on application of a conformal mapping.

Example $W = z + a$ *for constant a.*

$f(z) = z + a$ is analytic and $f'(z) = 1$. Hence the given mapping is conformal. The lines $x = $ constant are mapped onto the lines $u = $ constant and the lines $y = $ constant are mapped onto the lines $v = $ constant.

This mapping represents a simple change of origin.

Example *Show that the mapping* $W = e^{i\alpha}z$, *where* α *is a real constant, is conformal and represents a rotation through* α *radians.*

$$f(z) = e^{i\alpha}z$$
$$= (\cos \alpha + i \sin \alpha)(x + iy)$$
$$= (x \cos \alpha - y \sin \alpha) + i(x \sin \alpha + y \cos \alpha).$$

Therefore $f(z)$ is analytic and $f'(z) = e^{i\alpha}$.

Hence the mapping is conformal.
Now

$$|w| = |e^{i\alpha}z| = |e^{i\alpha}| \, |z| = |z|$$

since $|e^{i\alpha}| = 1$. Also

$$\arg w = \arg (e^{i\alpha}z) = \arg e^{i\alpha} + \arg z = \alpha + \arg z.$$

Hence if any point P has polar coordinates (r,θ) in the z-plane, then its coordinates in the w-plane will be $(r, \theta + \alpha)$ (figure 152).

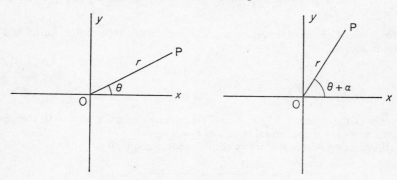

FIGURE 152

Thus the mapping represents a rotation through α radians in an anticlockwise sense.

In particular, when $\alpha = \frac{1}{2}\pi$, $w = e^{i\frac{1}{2}\pi}z = iz$ represents a rotation through $\frac{1}{2}\pi$ radians in an anticlockwise sense.

Also, when $\alpha = \pi$, $w = e^{i\pi}z = -z$ represents a rotation through π radians and when $\alpha = \frac{3}{2}\pi$, $w = e^{i\frac{3}{2}\pi}z = -iz$ represents a rotation through $\frac{3}{2}\pi$ radians.

Example $w = z + a^2/z$ where a is a real constant (*Joukowski's transformation*).

$$f(z) = z + \frac{a^2}{z}$$

$$= re^{i\theta} + \frac{a^2}{r}e^{-i\theta}$$

$$= r(\cos \theta + i \sin \theta) + \frac{a^2}{r}(\cos \theta - i \sin \theta)$$

$$= \left(r + \frac{a^2}{r}\right)\cos \theta + i\left(r - \frac{a^2}{r}\right)\sin \theta$$

$$= u + iv.$$

Therefore

$$u = \left(r + \frac{a^2}{r}\right)\cos \theta$$

and

$$v = \left(r - \frac{a^2}{r}\right) \sin \theta.$$

Then

$$\frac{\partial u}{\partial r} = \left(1 - \frac{a^2}{r^2}\right) \cos \theta, \qquad \frac{\partial u}{\partial \theta} = -\left(r + \frac{a^2}{r}\right) \sin \theta,$$

$$\frac{\partial v}{\partial r} = \left(1 + \frac{a^2}{r^2}\right) \sin \theta \quad \text{and} \quad \frac{\partial v}{\partial \theta} = \left(r - \frac{a^2}{r}\right) \cos \theta.$$

Hence the Cauchy–Riemann equations are satisfied (in polar form) and so $f(z)$ is analytic.

$$f'(z) = 1 - \frac{a^2}{z^2}$$

$$\neq 0$$

except at the points $z = \pm a$. Also $f'(z)$ remains finite for all $z \neq 0$. Hence the given mapping is conformal except at $z = \pm a, 0$.

Eliminating θ from the above relations for u and v gives

$$\frac{u^2}{(r + a^2/r)^2} + \frac{v^2}{(r - a^2/r)^2} = 1$$

and so we see that the circles $r = $ constant in the z-plane are mapped onto confocal ellipses in the w-plane.* In particular, for the circle $r = a$, we have

$$u = 2a \cos \theta \quad \text{and} \quad v = 0.$$

Thus this circle is mapped onto the real axis in the w-plane between $u = -2a$ and $u = 2a$. Note too that both the circles with radii r_1 and a^2/r_1 are mapped onto the same ellipse.

Also, eliminating r from the above relations for u and v gives

$$\frac{u^2}{\cos^2 \theta} - \frac{v^2}{\sin^2 \theta} = (r + a^2/r)^2 - (r - a^2/r)^2 = 4a^2$$

or

$$\frac{u^2}{(2a \cos \theta)^2} - \frac{v^2}{(2a \sin \theta)^2} = 1.$$

Therefore the radial lines $\theta = $ constant in the z-plane are mapped onto confocal hyperbolae in the w-plane. In particular, for the positive x-axis $\theta = 0$ and then

$$u = (r + a^2/r) \quad \text{and} \quad v = 0$$

Thus the positive real axis in the z-plane is mapped onto the real axis for $u \geqslant 2a$ in the w-plane.

* Since $(r + a^2/r)^2 - (r - a^2/r)^2 = 4a^2$, the focii are $(\pm 2a, 0)$.

Similarly, the negative real axis in the z-plane is mapped onto the real axis for $u \leqslant -2a$ in the w-plane.

Hence this mapping may be considered as mapping either the exterior or the interior of the circle $|z| = a$ onto the whole w-plane.

Conversely, we have

$$z = \frac{w \pm \sqrt{w^2 - 4a^2}}{2}$$

and it can be shown that $z = (w + \sqrt{w^2 - 4a^2})/2$ maps the whole of the w-plane onto the region $|z| \geqslant a$ while $z = (w - \sqrt{w^2 - 4a^2})/2$ maps the whole of the w-plane onto the region $|z| \leqslant a$.

Now consider the general *bilinear transformation* or *bilinear mapping* given by

$$w = \frac{az + b}{cz + d}$$

where the constants a, b, c and d are real or complex.

It was shown in an earlier example in section 9.2 that $f(z) = (az + b)/(cz + d)$ is analytic for real constants a, b, c and d except where $cz + d = 0$. This result also holds when a, b, c and d are complex. Thus the given function is analytic except where $cz + d = 0$.

$$\frac{dw}{dz} = \frac{a(cz + d) - c(az + b)}{(cz + d)^2} = \frac{ad - bc}{(cz + d)^2} \neq 0 \quad \text{unless} \quad ad - bc = 0.$$

Hence, if $ad - bc \neq 0$, the given mapping is conformal except where $cz + d = 0$.

Now it is easily seen that to each point z such that $cz + d \neq 0$ there corresponds one and only one point w. If in addition we let the point z such that $cz + d = 0$ correspond to the 'point at infinity' $w = \infty$, then we can say that to every point z there corresponds one and only one point w.

Similarly, on considering the inverse mapping $z = (-dw + b)(cw - a)$ obtained by solving for z in terms of w, we see that to each point w such that $cw - a \neq 0$ there corresponds one and only one point z. Again, letting the point w such that $cw - a = 0$ correspond to the 'point at infinity' $z = \infty$, we can say that to every point w there corresponds one and only one point z.

Thus we have shown that the given mapping is one-to-one. Conversely, it can be shown that every one-to-one mapping by an analytic function with a finite number of isolated singularities† of the plane onto itself (regarding the z-plane and w-plane as coincident) is a bilinear mapping.

A *fixed point* of a mapping is a point which is mapped onto itself. Thus the fixed points of a mapping $w = f(z)$ are the points z which satisfy the equation

$$z = f(z).$$

For the above bilinear mapping, the fixed points are the roots of

$$z = \frac{az + b}{cz + d}$$

or

$$cz^2 + (d - a)z - b = 0.$$

Since this is a quadratic in z, we see that in general the bilinear mapping will have two fixed points. When $c = 0$ one of the fixed points of the mapping is the point at infinity.

Example   Consider the mapping

$$w = i\left(\frac{z - 1}{z + 1}\right).$$

Since this is a bilinear mapping it is conformal and one-to-one.

$$w = u + iv$$
$$= i\left(\frac{x + iy - 1}{x + iy + 1}\right)$$
$$= i\left(\frac{x^2 - 1 + y^2 + 2iy}{x^2 + y^2 + 2x + 1}\right).$$

Therefore

$$u = \frac{-2y}{x^2 + y^2 + 2x + 1}$$

and

$$v = \frac{x^2 + y^2 - 1}{x^2 + y^2 + 2x + 1}.$$

On the imaginary axis, $x = 0$ and so

$$u = \frac{-2y}{y^2 + 1} \quad \text{and} \quad v = \frac{y^2 - 1}{y^2 + 1}.$$

Then, eliminating y gives

$$u^2 + v^2 = \frac{(y^2 - 1)^2 + 4y^2}{(y^2 + 1)^2} = 1.$$

† An *isolated* singularity is a singularity for which there exists a small area surrounding it which does not include any other singularities.

Thus the imaginary axis in the z-plane is mapped onto the circle with unit radius and centre the origin in the w-plane, that is the circle given by $|w| = 1$.

Now the two regions into which the z-plane is divided by the line $x = 0$ must correspond to the two regions into which the w-plane is divided by the curve $|w| = 1$.

Since the image of $z = 2$ ($x = 2 > 0$) is $w = \frac{1}{3}i$ ($|w| = \frac{1}{3} < 1$) we see that the half-plane $x > 0$ is mapped onto the interior of the circle $|w| = 1$ that is onto the region $|w| < 1$ (figure 153).

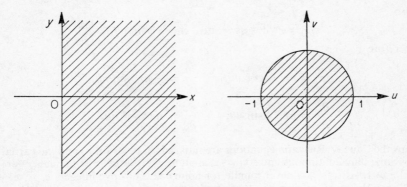

FIGURE 153

The half-plane $x < 0$ is then mapped onto the exterior of the circle $|w| = 1$, that is onto the region $|w| > 1$. (As a check, note that in particular $z = -2$ is mapped onto $w = 3i$). Alternatively,

$$z = -\frac{(i + w)}{w - i}$$

$$= -\frac{(i + w)(\bar{w} + i)}{(w - i)(\bar{w} + i)}$$

$$= -\frac{(i\bar{w} - 1 + w\bar{w} + iw)}{|w - i|^2}.$$

Hence

$$\mathrm{Re}(z) = \frac{1 - u^2 - v^2}{|w - i|^2}.$$

Therefore if $\mathrm{Re}(z) > 0$, $u^2 + v^2 < 1$.

That is the half-plane $x > 0$ is mapped onto the interior of the circle $|w| = 1$.

Similarly, if $\mathrm{Re}(z) < 0$, $u^2 + v^2 > 1$ and so the half-plane $x < 0$ is mapped onto the exterior of the circle $|w| = 1$.

The following examples illustrate mappings which are obtained using other well known functions and which have interesting applications which will be discussed later.

Example $w = z^n$ *for positive integers* n.

$$f(z) = z^n = r^n e^{in\theta} = r^n(\cos n\theta + i \sin n\theta) = u + iv.$$

Therefore

$$|z^n| = r^n = |z|^n$$

and

$$\arg z^n = n\theta = n \arg z.$$

Also

$$u = r^n \cos n\theta \quad \text{and} \quad v = r^n \sin n\theta.$$

Therefore

$$\frac{\partial u}{\partial r} = nr^{n-1} \cos n\theta, \qquad \frac{\partial u}{\partial \theta} = -nr^n \sin n\theta,$$

$$\frac{\partial v}{\partial r} = nr^{n-1} \sin n\theta, \qquad \frac{\partial v}{\partial \theta} = nr^n \cos n\theta.$$

Thus the Cauchy–Riemann equations are satisfied (in polar form) and the partial derivatives are continuous and so $f(z)$ is analytic. $f'(z) = nz^{n-1} \neq 0$ unless $z = 0$ and $n \neq 1$. Also $f'(z)$ is never infinite for finite z and positive integers n.

Hence the given mapping is conformal except at $z = 0$ when $n \neq 1$. (When $n = 1$, the mapping is the identity mapping $w = z$ which is also conformal at $z = 0$).

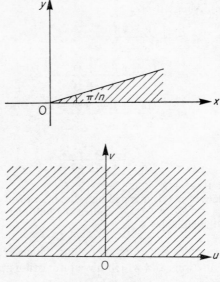

FIGURE 154

Now $|w| = |z^n| = |z|^n = r^n$. Hence circles with centre the origin in the z-plane, that is the curves given by $r =$ constant are mapped onto the curves given by $|w| =$ constant that is onto circles with centre the origin in the w-plane.

Also $\arg w = \arg z^n = n \arg z = n\theta$. Hence radial lines through the origin in the z-plane, that is the lines given by $\theta =$ constant are mapped onto the curves given by $\arg w =$ constant that is onto radial lines in the w-plane. In particular, the line for which $\theta = \alpha$ is mapped onto the line making an angle $n\alpha$ with the positive u-axis.

Thus the angular region in the z-plane defined by $0 \leqslant \arg z \leqslant \pi/n$ is mapped onto the upper half-plane in the w-plane (figure 154).

Example $w = e^z$.

$$f(z) = e^z = e^{x+iy} = e^x(\cos y + i \sin y) = u + iv.$$

Therefore $u = e^x \cos y$ and $v = e^x \sin y$. Then

$$\frac{\partial u}{\partial x} = e^x \cos y, \qquad \frac{\partial u}{\partial y} = -e^x \sin y,$$

$$\frac{\partial v}{\partial x} = e^x \sin y \quad \text{and} \quad \frac{\partial v}{\partial y} = e^x \cos y.$$

Hence the Cauchy–Riemann equations are satisfied and so $f(z)$ is analytic.

$$f'(z) = e^z \neq 0.$$

Also $f'(z)$ remains finite for all finite z. Therefore the mapping is conformal everywhere except at the point at infinity. Eliminating y from the above relations for u and v gives

$$u^2 + v^2 = e^{2x}.$$

Hence the lines $x =$ constant in the z-plane are mapped onto circles with centre the origin in the w-plane. In particular, the line $x = 0$ is mapped onto the circle $u^2 + v^2 = 1$. Since $|e^z| = e^x$ it is easily seen that the half-plane $x > 0$ is mapped onto the exterior of the circle $|w| = 1$ while the half-plane $x < 0$ is mapped onto the interior.

Eliminating x from the relations for u and v gives

$$\frac{v}{u} = \tan y \quad \text{or} \quad v = u \tan y.$$

Hence the lines $y =$ constant in the z-plane are mapped onto lines through the origin in the w-plane (figure 155).

In particular, when $y = 0$, $u = e^x$ and $v = 0$ so that the line $y = 0$ is mapped onto the positive real axis in the w-plane.

Also, when $y = \pi$, $u = -e^x$ and $v = 0$ so that the line $y = \pi$ is mapped onto the negative real axis in the w-plane.

It is then easily seen that the region in the z-plane between the lines $y = 0$ and $y = \pi$ is mapped onto the half-plane $v > 0$ in the w-plane (figure 156).

Similarly, the region in the z-plane between the lines $y = -\pi$ and $y = 0$ is mapped onto the half-plane $v < 0$ in the w-plane.

Further, the region in the z-plane between (and including) the lines $y = c$ and $y = c + 2\pi$ for any real constant c is mapped onto the whole w-plane.

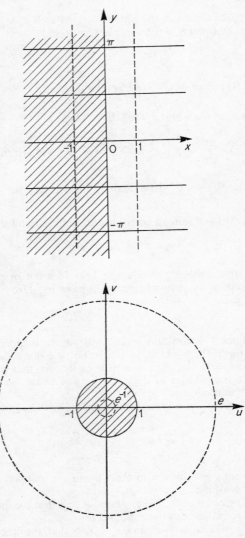

FIGURE 155

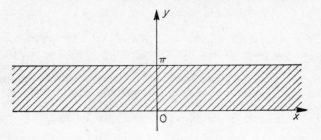

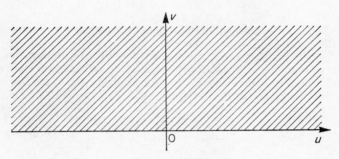

FIGURE 156

Example $w = \ln z$, *where* $\ln z = \ln |z| + i$ (*principal value of* arg z).

If $w = \ln z$ then $z = e^w$. This mapping is conformal for all finite z except $z = 0$ and its properties can be deduced immediately from the last example. For example the whole z-plane will be mapped onto the region in the w-plane between (and including) the lines $v = c$ and $v = c + 2\pi$ for any constant c.

Circles centre the origin in the z-plane are mapped onto the lines $u =$ constant in the w-plane etc.

Example $w = \sin z$.

$$f(z) = \sin (x + iy)$$
$$= \sin x \cosh y + i \cos x \sinh y$$
$$= u + iv.$$

Therefore $u = \sin x \cosh y$ and $v = \cos x \sinh y$. Hence

$$\frac{\partial u}{\partial x} = \cos x \cosh y, \qquad \frac{\partial u}{\partial y} = \sin x \sinh y,$$

$$\frac{\partial v}{\partial x} = -\sin x \sinh y \quad \text{and} \quad \frac{\partial v}{\partial y} = \cos x \cosh y.$$

Therefore the Cauchy–Riemann equations are satisfied everywhere and so $f(z)$ is analytic.

$f'(z) = \cos z \neq 0$ except where $z = \pm[(2n-1)\pi]/2$ for positive integers n. Thus the given mapping is conformal except at the points $z = \pm[(2n-1)\pi]/2$ for positive integers n.

Now it is clear from the above relations for u and v that they (and so w also) are periodic in x. Thus we need only consider the strip $-\pi < x \leqslant \pi$ in the z-plane.

Eliminating y from the expressions for u and v we obtain

$$\frac{u^2}{\sin^2 x} - \frac{v^2}{\cos^2 x} = 1.$$

Hence the lines $x = \text{constant}$ in the z-plane are mapped onto confocal hyperbolae (with focii $(\pm1,0)$) in the w-plane.

Note however that for $-\pi < x < 0$ the values of $\sin x$ are negative and so the values of u are negative and for $0 < x < \pi$ the values of $\sin x$ are positive and so the values of u are positive. Thus for positive c ($< \pi$) the line $x = c$ is mapped onto the branch of the hyperbola

$$\frac{u^2}{\sin^2 c} - \frac{v^2}{\cos^2 c} = 1$$

lying in the right-half plane of the w-plane. Similarly, the line $x = -c$ is mapped onto the left hand branch of the above hyperbola.

Eliminating x from the expressions for u and v gives

$$\frac{u^2}{\cosh^2 y} + \frac{v^2}{\sinh^2 y} = 1.$$

Hence the lines $y = \text{constant}$ in the z-plane are mapped onto confocal ellipses (with focii $(\pm1,0)$) in the w-plane (figure 157).

Consider now the region in the z-plane defined by $y \geqslant 0$, $-\frac{1}{2}\pi \leqslant x \leqslant \frac{1}{2}\pi$ (figure 158). When $y = 0$, $u = \sin x$, $v = 0$. Therefore the real axis between $x = -\frac{1}{2}\pi$ and $x = \frac{1}{2}\pi$ in the z-plane is mapped onto the real axis between $u = -1$ and $u = 1$ in the w-plane.

When $x = \frac{1}{2}\pi$, $u = \cosh y$, $v = 0$. Therefore the line $x = \frac{1}{2}\pi$, $y \geqslant 0$ in the z-plane is mapped onto the real axis for $u \geqslant 1$ in the w-plane.

Similarly, putting $x = -\frac{1}{2}\pi$ we see that the line $x = -\frac{1}{2}\pi$, $y \geqslant 0$ in the z-plane is mapped onto the real axis for $u \leqslant -1$ in the w-plane.

It is then easily seen that the region $y \geqslant 0$, $-\frac{1}{2}\pi \leqslant x \leqslant \frac{1}{2}\pi$ is mapped onto the upper half-plane in the w-plane (figure 158).

Similarly the region $y \leqslant 0$, $-\frac{1}{2}\pi \leqslant x \leqslant \frac{1}{2}\pi$ is mapped onto the lower half-plane in the w-plane.

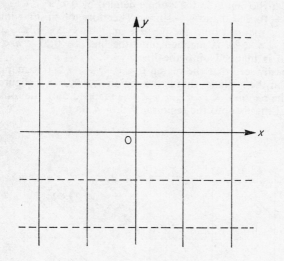

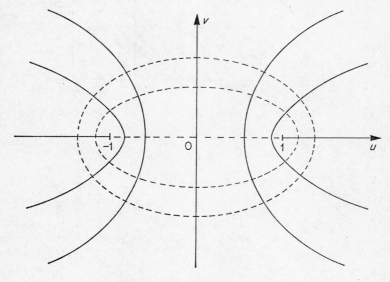

Figure 157

Now consider the region in the z-plane defined by $0 \leqslant x \leqslant \frac{1}{2}\pi$, $y \geqslant 0$ (figure 159). When $x = 0$, $u = 0$ and $v = \sinh y$. Therefore the positive imaginary axis in the z-plane is mapped onto the positive imaginary axis in the w-plane. The line $y = 0$, $0 \leqslant x \leqslant \frac{1}{2}\pi$ is mapped onto the line $v = 0$, $0 \leqslant u \leqslant 1$. The line $x = \frac{1}{2}\pi$, $y \geqslant 0$ is mapped onto the line $v = 0$, $u \geqslant 1$.

It is then easily seen that the region $0 \leqslant x \leqslant \frac{1}{2}\pi$, $y \geqslant 0$ is mapped onto the first quadrant in the w-plane, that is the region $u \geqslant 0$, $v \geqslant 0$ (figure 159).

Similarly, the region $0 \leqslant x \leqslant \frac{1}{2}\pi$, $y \leqslant 0$ is mapped onto the fourth quadrant in the w-plane, that is onto the region $u \geqslant 0$, $v \leqslant 0$.

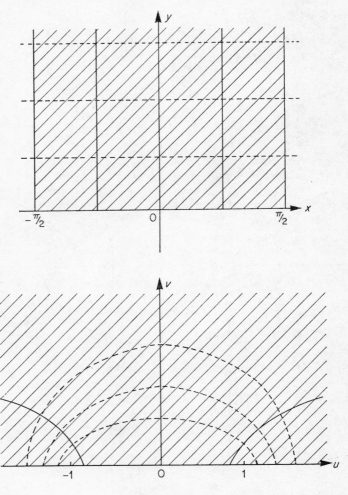

FIGURE 158

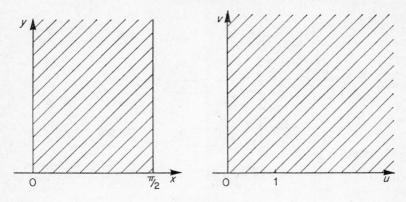

FIGURE 159

Example $w = \sinh z$.

This can be tackled from first principles as in the previous example, but it is perhaps quicker to proceed as follows.

$$w = \sinh z = -\mathrm{i} \sin (\mathrm{i} z).$$

Thus the mapping is obtained by firstly rotating through $\frac{1}{2}\pi$ radians, applying the sine function mapping of the previous example and then rotating through $\frac{3}{2}\pi$ (or $-\frac{1}{2}\pi$) radians.

Hence the lines $x = \text{constant}$ are mapped onto confocal ellipses and the lines $y = \text{constant}$ are mapped onto confocal hyperbolae as indicated in figure 160 (cf. figure 157).

The strip $x \geqslant 0$, $-\frac{1}{2}\pi \leqslant y \leqslant \frac{1}{2}\pi$ is mapped onto the right-half plane $u \geqslant 0$ as in figure 161 (cf. figure 158).

Also the strip $x \geqslant 0$, $-\frac{1}{2}\pi \leqslant y \leqslant 0$ is mapped onto the fourth quadrant $u \geqslant 0$, $v \leqslant 0$ in the w-plane as in figure 162 (cf. figure 159).

9.5 Applications

The previous examples on the use of complex potential are of course applications of conformal mappings.

We put $W = \varphi + \mathrm{i}\psi$ then the mutually orthogonal systems of straight lines $\varphi = \text{constant}$ and $\psi = \text{constant}$ are mapped onto the equipotentials and streamlines for a particular hydrodynamical problem. (In electrostatics the lines $\psi = \text{constant}$ are mapped onto the field lines.) For example $z = \mathrm{e}^{W/m}$ maps the lines $\varphi = \text{constant}$ in the W-plane onto

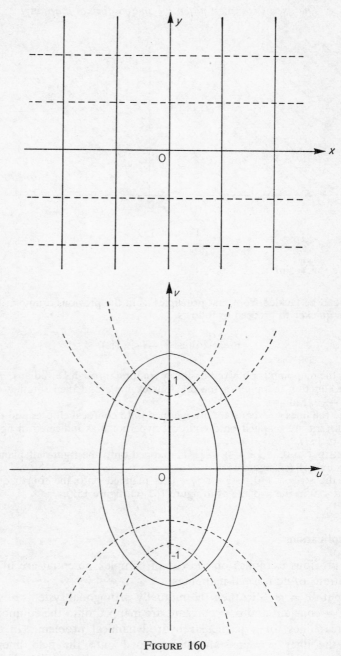

Figure 160

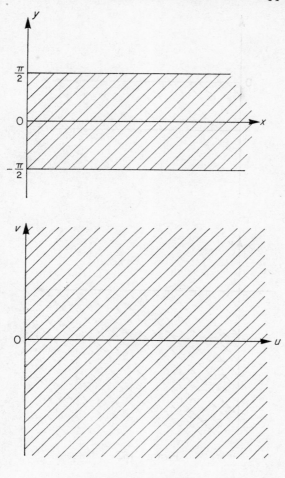

FIGURE 161

concentric circles with centre the origin in the z-plane and maps the lines $\psi =$ constant in the W-plane onto lines through the origin in the z-plane. Thus this mapping gives the streamlines and equipotentials for a two-dimensional point source (if m is positive) at the origin in the z-plane. The complex potential for a two-dimensional point source of strength m at the origin is $W = m \ln z$.

Also, if the streamlines and equipotentials for a particular flow are known, then a suitable conformal mapping may transform these into the

16

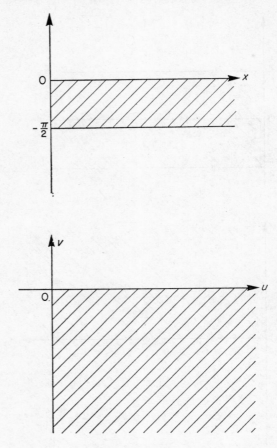

FIGURE 162

streamlines and equipotentials for some other problem of interest. These applications are illustrated in the following examples.

Example *Since $w = z^3$ maps the angular region $0 \leqslant \arg z \leqslant \frac{1}{3}\pi$ in the z-plane onto the upper half-plane in the w-plane (see a previous example in section 9.4, p. 450) this mapping can be used to give the streamlines for the two-dimensional, irrotational, incompressible flow of an inviscid fluid in an angular region of angle $\frac{1}{3}\pi$*

$$\varphi + i\psi = W = z^3 = (x + iy)^3.$$

Hence the streamlines $\psi = c$ for constant c ($\geqslant 0$) in the W-plane are mapped onto the curves

$$3x^2y - y^3 = c \qquad \text{in the } z\text{-plane}$$

Thus the streamlines for the flow in the angular region $0 \leqslant \arg z \leqslant \frac{1}{3}\pi$ in the z-plane have equations

$$3x^2y - y^3 = c \qquad \text{for constants } c \geqslant 0.$$

Flow in the angular region is mapped into uniform flow parallel to the real axis in the upper half-plane of the W-plane (figure 163).

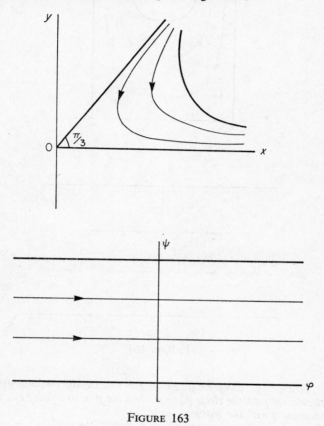

FIGURE 163

Example   *Obtain the streamlines for the flow round a double right-angled bend.*

We have seen that the mapping $w = \sin z$ maps the strip $y \geqslant 0$, $-\frac{1}{2}\pi \leqslant x \leqslant \frac{1}{2}\pi$ in the z-plane onto the upper half-plane in the W-plane and that the boundary of the strip in the z-plane is mapped onto the real axis in the W-plane. This mapping can then be used to give the equations of the streamlines for the required flow in the z-plane.

$$\varphi + i\psi = W = \sin z = \sin (x + iy).$$

Hence the lines $\psi = c \ (\geqslant 0)$ are mapped onto the curves $\cos x \sinh y = c$ which are the required streamlines (figure 164).

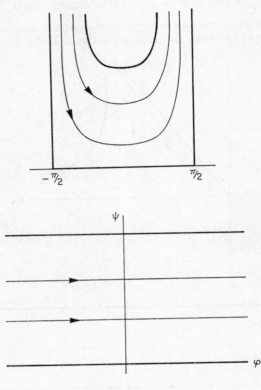

$-\pi/2$ $\pi/2$

FIGURE 164

Example   Deduce the complex potential for the steady two-dimensional flow without circulation past the circle $|z| = a$ when the flow at infinity has velocity U making an angle α with the positive x-axis.

Show that under the transformation $Z = z' + a^2/z'$ the circle $|z'| = a$ becomes a line of length $4a$. Hence obtain the complex potential for the two-dimensional flow without circulation past a flat plate at an angle α to the direction of the flow at infinity. In particular, when the plate is at right angles to the stream determine the points of the plate at which the velocity is equal in magnitude to the velocity at infinity.

The complex potential for uniform stream with velocity U in the positive x-direction is

$$W = Uz.$$

Hence, using the Milne–Thomson circle theorem, the complex potential for the uniform stream in the presence of the circle $|z| = a$ is

$$W = U\left(z + \frac{a^2}{z}\right).$$

Now putting $z' = e^{i\alpha}z$ rotates the axis through an angle α so that

$$W = U\left(z'e^{-i\alpha} + \frac{a^2}{z'e^{-i\alpha}}\right)$$

gives the complex potential for the steady two-dimensional flow without circulation past the circle $|z'| = a$ when the flow at infinity has velocity U and makes an angle α with the positive x-axis.

On the circle $|z'| = a$, $z' = ae^{i\theta}$ and so

$$Z = z' + \frac{a^2}{z'} = ae^{i\theta} + \frac{a^2}{ae^{i\theta}} = a(e^{i\theta} + e^{-i\theta}) = 2a\cos\theta.$$

Thus as θ takes all values between 0 and 2π, Z takes all values between $-2a$ and $+2a$.

Hence under the given transformation the circle $|z'| = a$ becomes the line of length $4a$ between $X = -2a$, $Y = 0$ and $X = 2a$, $Y = 0$ where $Z = X + iY$.

Thus under this transformation the streamline $|z'| = a$ for the flow represented by the above complex potential becomes the line $-2a \leqslant X \leqslant 2a$, $Y = 0$ and so the complex potential becomes that for flow past a flat plate of length $4a$. Now $Z = z' + a^2/z'$ and so

$$z' = \frac{Z \pm \sqrt{Z^2 - 4a^2}}{2}.$$

We use the positive square root since, as was seen in an earlier example on p. 447, this maps the *outside* of the circle $|z'| = a$ onto the z-plane.

Therefore

$$W = U\left(\frac{Z + \sqrt{Z^2 - 4a^2}}{2}e^{-i\alpha} + \frac{2}{Z + \sqrt{Z^2 - 4a^2}}a^2e^{i\alpha}\right)$$

$$= \frac{U}{2}\{(Z + \sqrt{Z^2 - 4a^2})e^{-i\alpha} + (Z - \sqrt{Z^2 - 4a^2})e^{i\alpha}\}$$

$$= U(Z\cos\alpha - i\sqrt{Z^2 - 4a^2}\sin\alpha)$$

is the required complex potential.

In particular, when $\alpha = \frac{1}{2}\pi$,

$$W = -iU\sqrt{Z^2 - 4a^2}$$

Therefore

$$\frac{dW}{dZ} = \frac{-iUZ}{\sqrt{Z^2 - 4a^2}}.$$

On the plate $Z = X$ where $-2a \leqslant X \leqslant 2a$ and so

$$\frac{dW}{dZ} = \frac{-iUX}{\sqrt{X^2 - 4a^2}} = \frac{-UX}{\sqrt{4a^2 - X^2}}.$$

Now the magnitude of the velocity is $|dW/dZ|$ and so this will be equal to U when

$$\frac{X}{\sqrt{4a^2 - X^2}} = 1.$$

Therefore $X^2 = 4a^2 - X^2$ and so $X^2 = 2a^2$. Therefore $X = \pm \sqrt{2}a$.

Example   Seepage flow under a gravity dam.

Let the base of the dam have width $2a$ and let the depth of water held back by the dam be h. The pressure due to this head of water is then $\rho g h = p_0$ where ρ is the density of the water.

Choose cartesian coordinates x and y with origin at the midpoint of the base of the dam and x-axis along the surface of the ground pointing downstream (figure 165).

The mapping $W = C \cos^{-1} z/a$ for a real constant C and for z real only when $|z| < a$ then gives the equations of the streamlines and equipotentials for the flow.

$$x = a \cos\left(\frac{\varphi}{C}\right) \cosh\left(\frac{\psi}{C}\right) \quad \text{and} \quad y = -a \sin\left(\frac{\varphi}{C}\right) \sinh\left(\frac{\psi}{C}\right).$$

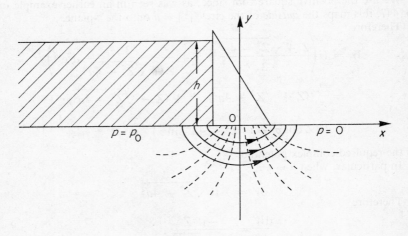

FIGURE 165

The streamlines are confocal ellipses with focii at the points $(\pm a, 0)$, that is at the toe and heel of the dam and the equipotentials are confocal hyperbolae with focii at the points $(\pm a, 0)$ (figure 165).

Now in Chapter 2 we saw that the potential φ for seepage flow is given by

$$\varphi = -\frac{K}{\eta} p$$

where p is pressure, K is the coefficient of permeability of the soil and η is the coefficient of viscosity of the liquid (water in this case).

Thus, since $p = 0$ on the ground downstream from the toe of the dam, that is for $x \geqslant a$ and $y = 0$ and $p = p_0$ on the ground upstream from the heel of the dam, that is for $x \leqslant -a$ and $y = 0$, we have

(i) $\varphi = 0$ for $x \geqslant a$ and $y = 0$ and
(ii) $\varphi = -(K/\eta)p_0 = \varphi_0$ for $x \leqslant -a$ and $y = 0$.

(Note that to allow for the constant atmospheric pressure P merely involves adding the constant $(K/\eta)P$ to the solution which will be obtained for φ.)

Putting $\varphi = 0$ in $x = a \cos(\varphi/C) \cosh(\psi/C)$ and $y = -a \sin(\varphi/C) \sinh(\psi/C)$, gives $x = a \cosh(\psi/C) \geqslant a$ and $y = 0$.

Thus the above condition (i) is satisfied. Putting $\varphi = \varphi_0$, gives $x = a \cos(\varphi_0/C) \cosh(\psi/C)$ and $y = -a \sin(\varphi_0/C) \sinh(\psi/C)$. If then we put $x \leqslant -a$ and $y = 0$ (for all ψ) we must have

$$\cos\left(\frac{\varphi_0}{C}\right) = -1 \quad \text{and} \quad \sin\left(\frac{\varphi_0}{C}\right) = 0.$$

These are both satisfied if $C = \varphi_0/\pi$. Thus the pressure distribution in the soil under the dam is given by

$$p = -\frac{\eta}{K} \varphi$$

where φ is the real part of $W = (\varphi_0/\pi)\cos^{-1}(z/a)$. Now the bottom surface of the dam itself is the streamline $\psi = 0$ and so on this surface,

$$x = a \cos\left(\frac{\varphi}{C}\right) \qquad (\text{and } y = 0).$$

Therefore

$$\varphi = C \cos^{-1}\frac{x}{a} = \frac{\varphi_0}{\pi} \cos^{-1}\frac{x}{a}.$$

Therefore the pressure distribution over the bottom of the dam is given by

$$p = -\frac{\eta \varphi_0}{\pi K} \cos^{-1}\frac{x}{a}.$$

This distribution is sketched in figure 166. Hence
The total thrust upwards on the bottom of the dam is

$$\int_{-a}^{a} p\, dx = -\frac{\eta \varphi_0}{\pi K} \int_{-a}^{a} \cos^{-1} \frac{x}{a}\, dx$$

$$= \frac{p_0}{\pi} \left\{ \left[x \cos^{-1} \frac{x}{a} \right]_{-a}^{a} - \int_{-a}^{a} \frac{x}{a} \frac{1}{\sqrt{1 - x^2/a^2}}\, dx \right\}$$

$$= \frac{p_0}{\pi} \left\{ a\pi + \left[\sqrt{a^2 - x^2} \right]_{-a}^{a} \right\} = p_0 a.$$

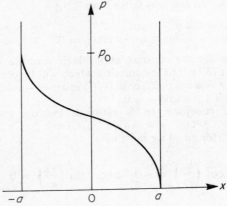

FIGURE 166

This uplift is what it would be if the entire base of the dam were subjected to a pressure equal to one half of the pressure due to the head of water above the dam or if the pressure decreased linearly from the static pressure p_0 at the heel to zero at the toe.

Exercises

1. Use the Joukowski transformation

$$Z = z + \frac{a^2}{z}$$

to obtain the complex potential for the flow with circulation $2\pi K$ past an elliptic cylinder of cross-section

$$\frac{X^2}{25} + \frac{Y^2}{9} = \frac{a^2}{4}$$

when the flow at infinity has magnitude U and is parallel to the X-axis.

2. Sketch the streamlines in the region $x > 0$ for the flow whose complex potential is

$$W = (z - \tfrac{1}{2})^2.$$

Show that the transformation $Z = z^2 - \tfrac{1}{4}$ maps the curve $X = -Y^2$ onto the line $x = \tfrac{1}{2}$, $y = 0$ and that the regions on the right hand sides of these two curves correspond.

Hence show that the complex potential

$$W = (\sqrt{z^2 + \tfrac{1}{4}} - \tfrac{1}{2})^2$$

represents uniform flow from $X = -\infty$ parallel to the x-axis past a parabolic cylinder.

Appendix 1

The gamma (or factorial) function

Consider the integral

$$\int_0^\infty t^{x-1} e^{-t} \, dt.$$

To show that this infinite integral is meaningful, consider firstly

$$\int_\varepsilon^1 t^{x-1} e^{-t} \, dt \quad \text{and} \quad \int_1^T t^{x-1} e^{-t} \, dt \quad \text{where} \quad \varepsilon > 0 \quad \text{and} \quad T > 1.$$

In the first of the above two integrals, when t is small the integrand behaves like t^{x-1} and provided $x - 1 > -1$ the integral does not become infinite as the lower limit $\varepsilon \to 0$. Now consider the second of the above two integrals when $x > 0$.

$$e^t > \frac{t^n}{n!} > \frac{t^{x+1}}{n!} \quad \text{for} \quad n > x + 1 \quad \text{and} \quad t > 1.$$

Therefore

$$t^{x-1} e^{-t} < \frac{n!}{t^2}.$$

Hence the integral does not exceed a constant multiple of $\int_1^T \frac{dt}{t^2}$. But $\int_1^T \frac{dt}{t^2} = \left[-\frac{1}{t} \right]_1^T = 1 - \frac{1}{T} \to 1$ as $T \to \infty$ and so $\int_1^T t^{x-1} e^{-t} \, dt$ remains finite as $T \to \infty$.

Thus the integral $\int_0^\infty t^{(x-1)} e^{-t} \, dt$ is convergent when $x > 0$.

Define then the function $\Gamma(x)$ by

$$\Gamma(x) = \int_0^\infty t^{x-1} e^{-t} \, dt \quad \text{for} \quad x > 0.$$

468

If $x > 1$, we can integrate by parts as follows:

$$\Gamma(x) = \left[-t^{x-1}e^{-t} \right]_0^\infty + (x-1)\int_0^\infty t^{x-2}e^{-t}\,dt$$

$$= (x-1)\int_0^\infty t^{x-2}e^{-t}\,dt$$

$$= (x-1)\Gamma(x-1).$$

Continuing in this way, we have, for the case when x is a positive integer n (say)

$$\Gamma(n) = (n-1)\Gamma(n-1)$$

$$= (n-1)(n-2)\Gamma(n-2)$$

$$= \ldots\ldots$$

$$= (n-1)!\Gamma(1)$$

$$= (n-1)!\int_0^\infty e^{-t}\,dt$$

$$= (n-1)!\left[-e^{-t} \right]_0^\infty$$

$$= (n-1)!$$

Thus $\Gamma(x)$ is seen to be a generalization of the factorial.

The definition of $\Gamma(x)$ can now be extended into the domain $x < 0$, for non-integral values of x, in the following way:

Define $\Gamma(x)$ for non-integral $x < 0$ by

$$\Gamma(x) = \frac{\Gamma(x+n)}{(x+n-1)(x+n-2)\ldots(x+1)x}$$

where n is an integer such that $x + n > 0$.

In order that this may be a valid definition however it is necessary to show that the above expression for $\Gamma(x)$ is independent of the choice of n provided only that $x + n > 0$. Let m be any integer $> n$, then

$$\frac{\Gamma(x+m)}{(x+m-1)(x+m-2)\ldots(x+1)x}$$

$$= \frac{(x+m-1)(x+m-2)\ldots(x+n)\Gamma(x+n)}{(x+m-1)(x+m-2)\ldots(x+1)x}$$

$$= \frac{\Gamma(x+n)}{(x+n-1)(x+n-2)\ldots(x+1)x}$$

and so the above definition is unique. Note that as $x \to$ a negative integer $\Gamma(x) \to \pm\infty$.

Appendix 2

The Bessel function $Y_\nu(x)$

Consider firstly Bessel's equation of order 0.
$$x^2 y'' + x^2 y = 0.$$

Putting $y(x) = \sum_{n=0}^{\infty} a_n x^{n+c}$ with $a_0 \neq 0$ into this equation we obtain, on equating coefficients of powers of x,
$$c^2 a_0 = 0 \qquad \text{(Indicial equation)}$$

and so
$$c = 0 \qquad \text{(twice)}.$$

Also
$$(c + 1)^2 a_1 = 0$$

and
$$(m + c)^2 a_m + a_{m-2} = 0 \qquad (m \geqslant 2).$$

Hence
$$a_1 = a_3 = a_5 = \ldots = 0$$

and
$$\{y_1(x,c)\}_{c=0} = \{x^c(a_0 + a_2 x^2 + a_4 x^4 + \ldots)\}_{c=0} = k J_0(x),$$

for arbitrary constants k, is a solution of the differential equation.

The second independent solution is given by

$$\left\{ \frac{\partial}{\partial c} y_1(x,c) \right\}_{c=0} = \left\{ x^c \ln x \, (a_0 + a_2 x^2 + a_4 x^4 + \ldots) \right.$$

$$\left. + \, x^c \frac{\partial}{\partial c} (a_0 + a_2 x^2 + \ldots) \right\}_{c=0}$$

$$= \ln x \{x^c (a_0 + a_2 x^2 + \ldots)\}_{c=0}$$

$$+ \left\{ x^c \frac{\partial}{\partial c} (a_0 + a_2 x^2 + \ldots) \right\}_{c=0}$$

$$= k \ln x \, J_0(x) + \left\{ \frac{\partial}{\partial c} (a_0 + a_2 x^2 + \ldots) \right\}_{c=0}$$

$$= K Y_0(x).$$

for some constant K.

Hence $Y_0(x)$ contains the term $J_0(x) \ln x$.

Now consider Bessel's equation of order v where v is a positive integer,

$$x^2 y'' + xy' + (x^2 - v^2)y = 0.$$

Again putting $y(x) = \sum_{n=0}^{\infty} a_n x^n$ with $a_0 \neq 0$ we obtain

$$(c^2 - v^2)a_0 = 0 \qquad \text{(Indicial equation)}$$

and so

$$c = \pm v.$$

Also

$$((c+1)^2 - v^2)a_1 = 0$$

and

$$((m+c)^2 - v^2)a_m + a_{m-2} = 0 \qquad (m \geqslant 2).$$

Hence

$$a_1 = a_3 = a_5 = \ldots = 0.$$

However, when $c = -v$ and $m = 2v$ the above recurrence relation in the a's cannot be solved for finite a_{2v}.

Therefore, put $a_0 = k(c + v)$ with $k \neq 0$ and it is now possible to solve the recurrence relation for a finite value of a_{2v}.

We have

$$a_2 = \frac{-k(c+v)}{(2+c)^2 - v^2},$$

$$a_4 = -\frac{a_2}{(4+c)^2 - v^2} = \frac{k(c+v)}{\{(2+c)^2 - v^2\}\{(4+c)^2 - v^2\}},$$

etc.

Then

$$y_1(x,c) = x^c\{k(c+v) + a_2 x^2 + a_4 x^4 + \ldots\}$$

and

$$\{y_1(x,c)\}_{c=-v} = \alpha J_{-v}(x)$$

for arbitrary constant α is a solution of the differential equation.

The second independent solution is given by

$$\left\{\frac{\partial}{\partial c} y_1(x,c)\right\}_{c=-v} = \ln x \left\{x^c(k(c+v) + a_2 x^2 + \ldots)\right.$$

$$+ x^c \frac{\partial}{\partial c}\left. (k(c+v) + a_2 x^2 + \ldots)\right\}_{c=-v}$$

$$= \alpha \ln x \, J_{-v}(x) + \{x^c(k(c+v) + a_2 x^2 + \ldots))\}_{c=-v}$$

$$= \beta Y_v(x)$$

for some constant β.

Hence $Y_v(x)$ contains the term $J_{-v}(x) \ln x$.

Appendix 3

Leibnitz theorem

If u and v are two functions of x each having nth derivatives, then

$$\frac{d^n}{dx^n}(uv) = \frac{d^n u}{dx^n}v + \binom{n}{1}\frac{d^{n-1}u}{dx^{n-1}}\frac{dv}{dx} + \binom{n}{2}\frac{d^{n-2}u}{dx^{n-2}}\cdot\frac{d^2 v}{dx^2} + \cdots + u\frac{d^n v}{dx^n}$$

where

$$\binom{n}{r} = \frac{n!}{(r-1)!(n-r)!},$$

Proof: Suppose the theorem is true for $n = m$. Then

$$\frac{d^m}{dx^m}(uv) = \frac{d^m u}{dx^m}v + \binom{m}{1}\frac{d^{m-1}u}{dx^{m-1}}\frac{dv}{dx} + \cdots + u\frac{d^m v}{dx^m}$$

and

$$\frac{d^{m+1}}{dx^{m+1}}(uv) = \frac{d}{dx}\left(\frac{d^m(uv)}{dx^m}\right).$$

The coefficient of the term $\dfrac{d^{m+1-r}u}{dx^{m+1-r}}\cdot\dfrac{d^r v}{dx^r}$ is then

$$\binom{m}{r} + \binom{m}{r-1} = \frac{m!}{r!(m-r)!} + \frac{m!}{(r-1)!(m-r+1)!}$$

$$= \frac{m!}{r!(m-r+1)!}(m+1)$$

$$= \frac{(m+1)!}{r!(m+1-r)!} = \binom{m+1}{r}.$$

472

Thus

$$\frac{d^{m+1}}{dx^{m+1}}(uv) = \frac{d^{m+1}u}{dx^{m+1}}v + \binom{m+1}{1}\frac{d^m u}{dx^m}\frac{dv}{dx} + \cdots$$

$$+ \binom{m+1}{r}\frac{d^{m+1-r}u}{dx^{m+1-r}}\frac{d^r v}{dx^r} + \cdots + u\frac{d^{m+1}v}{dx^{m+1}}.$$

Hence the theorem is true for $n = m + 1$. But

$$\frac{d}{dx}(uv) = \frac{du}{dx}v + u\frac{dv}{dx}$$

and so the theorem is true for $n = 1$.

Hence by induction the theorem is true for all positive integers n.

Answers

Chapter 1

§ *1.1*

1. $(a^2/4)(\pi a - 4)$.
2. $\frac{9}{4}$.
3. (i) $\frac{5}{4}$; (ii) $\frac{3}{4}$.
4. (i) $\frac{9}{2}$; (ii) $\frac{71}{16}$; (iii) $\frac{59}{16}$; (iv) $23\frac{19}{144}$.
5. (a) 1; (b) $\frac{5}{6}$.
6. (a) 1; (b) 1; (c) 1.
7. (a) $\frac{61}{5}$; (b) 16.

§ *1.2*

1. $\frac{4}{3}$.
2. 3.
4. πa^3.
5. $47\frac{1}{3}$.
6. $\pi a^2(c_1 + c_2 + c_3)$ where $\mathbf{c} = (c_1, c_2, c_3)$ in terms of cartesian coordinates.

§ *1.3*

1. $\frac{1}{2}$.
3. $\frac{1}{24}$.
4. $\frac{5}{9}h^4$.

§ *1.5*

1. $(4, -3, 12)$.
2. -1, $(7, -5, 3)$.
3. $(\frac{3}{5}, \frac{4}{5}, 0)$.
4. $\frac{10}{3}$.

§ *1.6*

1. $3x^2 + y^2/x^2 + 2 + 3xz$.
3. $a = 2$, $b = -1$, $c = -2$.
4. $\mathbf{a} = (3x^2y^4z^2, 4x^3y^3z^2, 2x^3y^4z)$, div $\mathbf{a} = 2xy^2(3y^2z^2 + 6x^2z^2 + x^2y^2)$.

§ *1.7*

1. curl $\mathbf{F} = (-xy - 2zy, -y^2 + 2z, 3yz - 2y)$.

§ *1.8*

1. $\nabla^2\varphi = 6xy + 2y^2z + 2x^2z + 6z^3y + 6zy^3$.
6. (i) $(8xyz(2x^2 - z^2), 4x^2z^2(x^2 - z^2), 8x^2yz(x^2 - 2z^2))$; (ii) 0.

§ *1.94*

1. $\frac{1}{2}\pi a^5$.
2. a^2.

474

3. $8\sqrt{3}\pi$.

4. $\frac{1}{6}\pi h a^3$.

5. $\frac{3}{2}\pi a^4$.

6. $-(1/r^3)\mathbf{r}$.

7. grad $\varphi = (\cos\theta - \sin\theta, -\sin\theta - \cos\theta, 0)$.
   Unit vector is $(1/\sqrt{2})(\cos\alpha - \sin\alpha, -\sin\alpha - \cos\alpha, 0)$.

8. div $\mathbf{a} = (1/\sin\theta)(\frac{5}{2}\sin 2\theta + 1)$.

10. div $\mathbf{a} = (e^\rho \sin\psi)/\rho$.

11. $\nabla^2\varphi = -1/r\sin\theta$.

12. 2.

13. $\nabla^2\varphi = \cos 2\theta/r^2\sin\theta$.

14. curl $\mathbf{a} = (\psi\cot\theta, -2\psi, 3\sin\theta)$.

§ 1.10

2. $\frac{4}{5}\pi c^5$.

§ 1.11

1. The value of the integral is $-\pi a^2$.

§ 1.12

1. $\varphi = x^2 + xz^2 + yz + c$ where c is an arbitrary constant.

2. $f = 3xy^2$, $\varphi = x^4/4 + 3xy^2z + c$ where c is an arbitrary constant.

3. (i) No scalar potential; (ii) Scalar potential $\frac{1}{2}(\mathbf{c}\cdot\mathbf{r})^2$; (iii) Scalar potential $r(\mathbf{c}\cdot\mathbf{r})$; (iv) No scalar potential.

4. (i) Scalar potential $c(x^2 + y^2)(z^3 + y^3)$; (ii) Scalar potential $-c\rho^3\cos\psi + \frac{2}{3}cz^3$; (iii) Scalar potential $-cr^2\sin\theta\cos\psi$.

§ 1.13

1. $f = 3z^2 - 3x^2 + 7x$; $g(x,y,z) = z^3 - 3x^2z + 7xz$; $h(x,y,z) = -3x^2z - 7yz + 6xyz + z^3$.

2. Both $(z/\rho,0,0)$ and $(0,0,-\ln\rho)$ are vector potentials for $\nabla\psi$. Both $(0,0,\psi)$ and $(0,-z,0)$ are vector potentials for $\nabla\ln\rho$.

Chapter 2

§ 2.173

1. Streamlines are the curves $yx^2 = $ constant, $xz = $ constant.

2. Streamlines are the two-dimensional curves $(x + t)^2 - (y + t)^2 = $ constant.

3. Streamlines are the two-dimensional curves $(x - t)(y + t) = $ constant. Parametric equations of the trajectories are $x = Ae^{2t} + t + \frac{1}{2}$, $y = Be^{-2t} - t + \frac{1}{2}$.

4. Free surfaces have the form $\alpha^2/\rho^2 + 2gz = $ constant.

§ 2.29

1. Potential $\dfrac{1}{4\pi\varepsilon_0}\left\{\dfrac{e}{2a}(4 + \sqrt{2}) + \dfrac{\sqrt{2}e'}{a}\right\}$. Charge $(-e/4)(2\sqrt{2} + 1)$.

2. Force at origin is $(e/48\pi\varepsilon_0 a^2)(3(4 - \sqrt{2}), -(4 + 3\sqrt{2}), 0)$. Required force is $(\sqrt{5}e^2/20\pi\varepsilon_0 a^2)(-1,1,0)$.

3. Electrostatic field is, in terms of cylindrical polar coordinates with axis along the wire, $(\lambda/2\pi\varepsilon_0\rho, 0, 0)$.
4. Electrostatic field is $\rho\mathbf{r}/3\varepsilon_0$, $0 \leqslant r \leqslant a$.
5. $\cos^{-1}\frac{1}{3}$.
7. Force field is $-((2 - \sqrt{2})e/16\pi\varepsilon_0 a^2)(1,1,0)$.
8. $\sqrt{2}m/r^3$ where r is distance from the dipole.
9. $V = \begin{cases} (V_1/3 \ln 2) \ln \rho + 2V_1/3, & 2 \leqslant \rho \leqslant 4 \\ (2V_1/3 \ln 2) \ln \rho, & 4 \leqslant \rho \leqslant 8. \end{cases}$

§ 2.3

3. Take origin at the centre of the sphere and x-axis parallel to $\mathbf{I}$ and x increasing in the positive sense of $\mathbf{I}$.
Then the potential is $Ia^3/3\mu_0 x^2$ for $x > a$ and $-Ia^3/3\mu_0 x^2$ for $x < -a$.
The field intensity vector is $2I/3\mu_0 |x|^3$ for $|x| > a$.
4. Take origin at the centre of the cylinder and z-axis along the axis of the cylinder with z increasing in the positive sense of $\mathbf{I}$.
Then the potential is

$$(I/2\mu_0)\{\sqrt{a^2 + (z - h)^2} - \sqrt{a^2 + (z + h)^2} + 2h\} \text{ for } z > h$$

and

$$\{(I/2\mu_0)\{\sqrt{a^2 + (z - h)^2} - \sqrt{a^2 + (z + h)^2} - 2h\} \text{ for } z < -h.$$

The field intensity vector is $\dfrac{I}{2\mu_0}\left\{\dfrac{z + h}{\sqrt{a^2 + (z + h)^2}} - \dfrac{z - h}{\sqrt{a^2 + (z - h)^2}}\right\}$ for $|z| > h$.

§ 2.4

1. In cylindrical polar coordinates with z-axis parallel to, and midway between the two wires and with z increasing in the positive sense of the current, the magnetic field at P is $(0, Id/\pi(d^2 - a^2), 0)$. The field at Q has magnitude $Ia/\pi d^2$, direction normal to the common perpendicular to the two wires which lies in the plane of Q and sense from Q towards this common perpendicular.
2. If a plane perpendicular to the wires intersects the wires in the points A and B, then the equipotentials in this plane are the circles with AB as a chord. Hence the equipotential surfaces are the corresponding cylinders with axes parallel to the wires.

§ 2.5

1. Potential $2\gamma m/a$. Force field $(-\gamma m/4a^2, \gamma m/3a^2, 0)$.
2. Potential $(\gamma M/a) \ln [(1 + \sqrt{2})/(1 - \sqrt{2})]$. Force $\gamma M/a^2$ towards B.
4. Force is $(\pi\gamma\rho_0 r/3)(4 + 3r[(k - 1)/a])$ towards the centre of the sphere, where r is distance from the centre.
5. $(\gamma M/a^2)(3 - \sqrt{2})$, $(\gamma M/a^2)(\sqrt{2} - 1)$. For the last part, consider the attraction of a complete sphere of mass $2M$ and subtract the attraction due to a hemisphere.

6. $\varphi = \begin{cases} \frac{2}{3}\pi\gamma\rho(21a^2 - r^2), & r < a \\ \frac{4}{3}\pi\gamma\rho(12a^2 - r^2 - a^3/r), & a < r < 2a \\ 20\pi\gamma\rho a^3/r, & r > 2a \end{cases}$

where r is distance from the centre. Attraction is $\frac{4}{5}\pi\gamma\rho a$.

§ 2.63

1. Surface distribution of density $15a^2 \cos^2 \theta/4\pi$ on the surface of the sphere $r = a$ and a volume distribution of density $3r/\pi$ inside the sphere $r = a$.
2. Surface distribution of density $3\varepsilon e \cos^2 \theta$ on the surface of the sphere $r = a$ and a volume distribution of density $6\varepsilon e/5a$ inside the sphere $r = a$. Self-energy $78\varepsilon\pi e^2 a^3/25$.
3. Mean density is k.
4. Surface distribution of density $3\varepsilon A \cos \theta$ on the surface of the sphere $r = a$. Self-energy $2\pi A^2 a^3 \varepsilon$.
8. $H > 10mr^{-3}$.

Chapter 3

§ 3.24

1. $-3ea/2\pi(a^2 + b^2)^{\frac{3}{2}}$.
2. The given charge e at the point $(a,b,0)$ together with a charge e at each of the points $(-b,a,0)$, $(-a,-b,0)$, $(b,-a,0)$ and a charge $-e$ at each of the points $(b,a,0)$, $(a,-b,0)$, $(-b,-a,0)$, $(-a,b,0)$.
3. An attraction $3(3\sqrt{2} - 2)M^2/16\pi\varepsilon_0 r^4$ towards the intersection of the planes. r is distance from the intersection.
4. $11e/18\pi a^2$, $-37e/100\pi a^2$.
7. $(e^2/4\pi a^2\varepsilon_0)(5/4 - 3/5\sqrt{5})$ in the direction of the axis of the boss and towards the plate. $(e^2/4\pi a^2\varepsilon_0)(1/5\sqrt{5} - 1)$ normal to the axis of the boss.
9. For points outside the dielectric and on the same side of the conductor as the given charge—

e at $(a,0,a)$, $\qquad\qquad$ $-e$ at $(-a,0,a)$

$-e/\sqrt{2}$ at $(\frac{1}{2}a,0,\frac{1}{2}a)$, $\qquad$ $e/\sqrt{2}$ at $(-\frac{1}{2}a,0,\frac{1}{2}a)$

$\left(\dfrac{1 - K}{1 + K}\right)e$ at $(a,0,-a)$, $\qquad$ $-\left(\dfrac{1 - K}{1 + K}\right)e$ at $(-a,0,-a)$

$-\left(\dfrac{1 - K}{1 + K}\right)\dfrac{e}{\sqrt{2}}$ at $\left(\dfrac{a}{2},0,-\dfrac{a}{2}\right)$, $\quad$ $\left(\dfrac{1 - K}{1 + K}\right)\dfrac{e}{\sqrt{2}}$ at $\left(-\dfrac{a}{2},0,-\dfrac{a}{2}\right)$.

For points inside the dielectric and on the same side of the conductor as the given charge—

$2e/(1 + K)$ at $(a,0,a)$, $\quad -2e/(1 + K)$ at $(-a,0,a)$

$-\sqrt{2}e/(1 + K)$ at $(\frac{1}{2}a,0,\frac{1}{2}a)$, $\quad \sqrt{2}e/(1 + K)$ at $(-\frac{1}{2}a,0,\frac{1}{2}a)$.

10. $3m^2(K - 1)/64\pi\varepsilon_0 d^4(K + 1)$.

Chapter 4

§ 4.2

1. $\dfrac{2}{\pi}\left(\sin x + \dfrac{\sin 3x}{3} + \dfrac{\sin 5x}{5} + \ldots\right)$, 0, 0, 0.

2. $\frac{1}{4}(\pi + 2) - \dfrac{2}{\pi}\sum_1^\infty \dfrac{1}{(2m-1)^2}\cos(2m-1)x - \dfrac{1}{\pi}\sum_1^\infty \dfrac{(\pi-1)(-1)^n + 1}{n}\sin nx.$

   Sum of series in $\pi^2/8$.

3. $\dfrac{1}{\pi} - \dfrac{2}{\pi}\sum_1^\infty \dfrac{1}{4m^2-1}\cos 2mx - \frac{1}{2}\sin x.$

4. $\dfrac{\sinh \pi}{\pi} + \dfrac{2\sinh \pi}{\pi}\left\{\sum_1^\infty \dfrac{(-1)^n}{1+n^2}(\cos nx - n\sin nx)\right\}.$

5. $\dfrac{1}{2\pi}(1 - \cos \pi) + \sum_{n=1}^\infty \dfrac{1}{\pi(1+n^2)}$
   $\times \{(1 - (-1)^n \cosh \pi)\cos nx - (-1)^n n \sinh \pi \sin nx\}.$

6. $\dfrac{\pi^2}{6} + \sum_{n=1}^\infty \left\{(-1)^n \dfrac{2}{n^2}\cos nx + \left[(-1)^n\left(\dfrac{2}{n^3\pi} - \dfrac{\pi}{n}\right) - \dfrac{2}{n^3\pi}\right]\sin nx\right\}.$

§ 4.3

1. (i) even; (ii) neither; (iii) neither; (iv) odd; (v) even; (vi) odd.

3. $\dfrac{1}{\pi} + \dfrac{1}{2}\cos x - \dfrac{2}{\pi}\sum_{m=1}^\infty \dfrac{(-1)^m}{4m^2-1}\cos 2mx.$

4. $\dfrac{2\sinh \pi}{\pi}\sum_1^\infty (-1)^{n+1}\left(\dfrac{n}{1+n^2}\right)\sin nx.$

5. Sum of series is $(1 - 2\pi^2/27)$; obtained by putting $x = 2\pi/3$.

§ 4.4

1. $\dfrac{2}{\pi} - \dfrac{4}{\pi}\sum_{n=1}^\infty \dfrac{1}{4n^2-1}\cos 2nx.$

2. $\sum_{n=1}^\infty \dfrac{2}{n}\sin nx.$

3. $\dfrac{h}{\pi} + \dfrac{2}{\pi h}\sum_{n=1}^\infty \dfrac{\sin^2 nh}{n^2}\cos nx.$ Sum of series is $\dfrac{1}{4h}(\pi - 2h).$

§ 4.5

1. $1 + \dfrac{2}{\pi}\sum_{n=1}^\infty \dfrac{1}{n}(1 - 2(-1)^n)\sin n\pi x.$ Sums of series are $\pi/4$ and $\sqrt{2}\pi/4$.

2. $\sum_{n=1}^\infty b_n \sin \dfrac{n\pi x}{2}$ where $b_n = \begin{cases} -2/n\pi \text{ for } n \text{ even,} \\ \dfrac{2}{n\pi}\left(1 + \dfrac{1}{n\pi}(-1)^{(n-1)/2}\right) \text{ for } n \text{ odd.} \end{cases}$

3. $\dfrac{2}{3} + \dfrac{2}{\pi} \displaystyle\sum_{n=1}^{\infty} \dfrac{1}{n} \sin \dfrac{2n\pi}{3} \cos n\pi x.$ Sum of series is $\sqrt{3}\pi/9$.

§4.6

$$\dfrac{\sinh \pi}{\pi} + \dfrac{2 \sinh \pi}{\pi} \sum_{n=1}^{\infty} \dfrac{(-1)^n}{1 + n^2} \cos nx.$$

§4.8

$$-\tfrac{1}{2} \cos x + \dfrac{4}{\pi} \sum_{m=1}^{\infty} \dfrac{m}{4m^2 - 1} \sin 2mx.$$

Chapter 5

§5.1

1. $A \cos \sqrt{x} + B \sin \sqrt{x}.$

2. $\dfrac{1}{(1 - x)^2} [A + B(\ln x - x)].$

3. $A(1 + 2x^2) + Bx(1 + x^2)^{\frac{1}{2}}.$

4. $A/x + Be^{-x}(1 + 1/x).$

5. $A \displaystyle\sum_{m=0}^{\infty} \dfrac{1}{(m!)^2} \left(\dfrac{2x^3}{3^2}\right)^m + B\left\{\ln x \displaystyle\sum_{m=0}^{\infty} \dfrac{1}{(m!)^2} \left(\dfrac{2x^3}{3^2}\right)^m - \dfrac{2}{3} \displaystyle\sum_{m=0}^{\infty} \dfrac{S_m}{(m!)^2} \left(\dfrac{2x^3}{3^2}\right)^m\right\}$

   where $S_m = 1 + \tfrac{1}{2} + \tfrac{1}{3} + \ldots + \dfrac{1}{m}.$

6. $A(1 + x + x^2) + Bx^3(1 - x)^{-1}.$

7. $A(1 - 3x) + Bx^{\frac{1}{2}} \displaystyle\sum_{n=0}^{\infty} \dfrac{n + 1}{(2n + 1)(2n - 1)} x^n.$

8. $x(A + Bx)(1 - x^2)^{-1}.$

9. $Ax^2(e^x + e^{-x}) + Bx^2(e^x - e^{-x})$ or $x^2(A^1 \cosh x + B^1 \sinh x).$

10. $x^2(1 - x)^{-2}(A + B \ln x) + Bx(1 - x)^{-1}.$

§5.32

1. $y = Ax^{\nu} J_{\nu}(x) + \begin{cases} Bx^{\nu} Y_{\nu}(x) & \text{if } \nu \text{ is an integer or zero,} \\ Bx^{\nu} J_{-\nu}(x) & \text{if } \nu \text{ is not an integer or zero.} \end{cases}$

3. $x^2(xJ_1(x) - J_2(x)).$

§5.44

1. $\displaystyle\int_{-1}^{1} xP_n(x)\, dx = \begin{cases} 0, & n \neq 1 \\ \tfrac{2}{3}, & n = 1. \end{cases}$ $\displaystyle\int_{-1}^{1} xP_n(x)P_{n-1}(x)\, dx = \dfrac{2n}{4n^2 - 1}.$

4. $\tfrac{2}{3} \tfrac{2}{3} P_0(x) - 2P_1(x) + \tfrac{8}{3} \tfrac{0}{3} P_2(x) + 8P_4(x).$

Chapter 6

§ 6.2

1. (a) $\frac{3}{4} \sin \pi x \cdot e^{-\pi y} - \frac{1}{4} \sin 3\pi x \cdot e^{-3\pi y}$.

   (b) $\frac{1}{2} + \frac{4}{\pi^2} \sum_{n=1}^{\infty} \frac{1}{(2n-1)^2} \cos(2n-1)\pi x \cdot e^{-(2n-1)\pi y}$

2. $-\frac{8a^2}{\pi^3} \sum_1^{\infty} \frac{1}{(2n-1)^3} \sin \frac{(2n-1)\pi x}{a} \cdot \frac{\sinh[(2n-1)\pi y/a]}{\sinh[(2n-1)\pi b/a]}$.

3. $-\frac{16}{\pi^2} \sum_1^{\infty} \frac{n}{(4n^2-1)^2} \left\{ a \sin \frac{2n\pi x}{a} \cdot \frac{\sinh(2n\pi y/a)}{\sinh(2n\pi b/a)} + b \sin \frac{2n\pi y}{b} \frac{\sinh(2n\pi x/b)}{\sinh(2n\pi a/b)} \right\}$.

§ 6.32

1. $\left(\frac{\varphi_a - \varphi_b}{\ln a/b} \right) \ln \rho + \frac{\varphi_b \ln a - \varphi_a \ln b}{\ln a/b}$ $(b \leqslant \rho \leqslant a)$.

3. $-\frac{2E}{1+K} \rho \cos \psi, \quad -E\rho \cos' \psi - \frac{a^2 E}{\rho} \frac{(1-K)}{(1+K)} \cos \psi$.

4. $-E\rho \cos \psi + \frac{Eb^2(b^2 - a^2)(\varepsilon^2 - \varepsilon_0^2)}{b^2(\varepsilon + \varepsilon_0)^2 - a^2(\varepsilon - \varepsilon_0)^2} \cdot \frac{\cos \psi}{\rho}$.

§ 6.4

1. $\frac{e}{4\pi\varepsilon_0 r}, \quad \frac{e}{4\pi K \varepsilon_0 r} + \frac{e}{4\pi\varepsilon_0 b} \left(\frac{K-1}{K} \right)$.

3. $\frac{ka}{3\varepsilon_0} + \frac{kr^2}{15a\varepsilon_0} (3\cos^2\theta - 1)$ for $r < a$. $\frac{ka^2}{3\varepsilon_0 r} + \frac{ka^4}{15\varepsilon_0 r^3} (3\cos^2\theta - 1)$ for $r > a$.

5. $\frac{e^2(\varepsilon_1 - \varepsilon_2)}{4\pi b^2 \varepsilon_2} \sum_{n=1}^{\infty} \frac{n(n+1)}{\{(n+1)\varepsilon_1 + n\varepsilon_2\}} \left(\frac{b}{a} \right)^{2n+1}$.

7. $2\gamma m \left\{ \frac{1}{2}\frac{1}{r} - \frac{1}{8}\frac{a^2}{r^3} P_2(\cos\theta) + \frac{1}{16}\frac{a^4}{r^5} P_4(\cos\theta) - \frac{5}{128}\frac{a^6}{r^7} P_6(\cos\theta) + \ldots \right\}$

$$\text{for } r > a$$

$\frac{2\gamma m}{a} \left\{ 1 - \frac{r}{a} P_1(\cos\theta) + \frac{1}{2}\frac{r^2}{a^2} P_2(\cos\theta) \right.$

$$\left. - \frac{1}{8}\frac{r^4}{a^4} P_4(\cos\theta) + \frac{1}{16}\frac{r^6}{a^6} P_6(\cos\theta) + \ldots \right\}$$

$$r < a$$

8. $-\frac{3E_0 r \cos\theta}{2+K}$ for $r < a$, $-E_0 r \cos\theta \left(1 + \frac{(1-K)}{2+K}\frac{a^3}{r^3} \right)$ for $r > a$.

9. $\frac{m}{4\pi\mu_0}\frac{\cos\theta}{r^2} + \frac{(\mu_0 - \mu)r \cos\theta}{\pi a^3 \mu_0 (2\mu + \mu_0)}$ $r < a$, $\frac{3m \cos\theta}{4\pi r^2 (2\mu + \mu_0)}$ $r > a$.

10. $\frac{9\mu m b^3 \cos\theta}{4\pi r^2 \{b^3(\mu_0 + 2\mu)(\mu + 2\mu_0) - 2a^3(\mu - \mu_0)^2\}}$.

11. $\dfrac{3b^3 E_0 \cos\theta}{2a^3(1 - K) - b^3(2 + K)}\left(r - \dfrac{a^3}{r^2}\right).$

Chapter 7

§ 7.23

1. (i) $\dfrac{4l}{\pi^2}\displaystyle\sum_{n=1}^{\infty}\dfrac{(-1)^{n-1}}{(2n-1)^2}\sin\dfrac{(2n-1)\pi x}{l}\exp\dfrac{-(2n-1)\pi^2 Kt}{l^2}$

   (ii) $\dfrac{l}{4} - \dfrac{2l}{\pi^2}\displaystyle\sum_{1}^{\infty}\dfrac{1}{(2n-1)^2}\cos\dfrac{2(2n-1)\pi x}{l}\exp\dfrac{-4(2n-1)^2\pi^2 Kt}{l^2}\,.$

2. $\dfrac{2a}{\pi}\displaystyle\sum_{1}^{\infty}\dfrac{(-1)^{n+1}}{n}\sin\dfrac{n\pi x}{a}\exp\dfrac{-n^2\pi^2}{a^2}t.$

3. $\dfrac{\varphi_0 x}{l} + \dfrac{2\varphi_0}{\pi}\displaystyle\sum_{n=1}^{\infty}\dfrac{(-1)^n}{n}\sin\dfrac{n\pi x}{l}\exp\dfrac{-n^2\pi^2 Kt}{l^2}$

   $\varphi(\tfrac{1}{2},2) = 26{\cdot}28,\ \varphi(\tfrac{1}{2},4) = 41{\cdot}16,\ \varphi(\tfrac{1}{2},6) = 46{\cdot}70,\ \varphi(\tfrac{1}{2},8) = 48{\cdot}77,$
   $\varphi(\tfrac{1}{2},10) = 49{\cdot}54.$

4. $\dfrac{4\varphi_0}{\pi}\displaystyle\sum_{n=1}^{\infty}\dfrac{1}{(2n+1)}\sin\dfrac{(2n-1)\pi x}{a}\cdot\dfrac{\sinh\left[(2n-1)\pi y/a\right]}{\sinh\left[(2n-1)\pi b/a\right]}.$

5. With $y = 0$ insulated,

   $$\varphi = \sum_{n,m} A_{n,m}\sin\frac{n\pi x}{a}\cos\frac{(2m+1)\pi y}{2a}\exp\left\{-K\pi^2\left[\frac{n^2}{a^2} + \frac{(2m+1)^2}{4a^2}\right]t\right\}$$

   where $A_{n,m} = \dfrac{4}{ab}\left\{\dfrac{-a^2}{n\pi}(-1)^n + \dfrac{2a^2}{(2m+1)\pi}(-1)^m - \dfrac{2^2 a^2}{(2m+1)^2\pi^2}\right\}.$

6. $2\varphi_0\displaystyle\sum^{\infty}\dfrac{\cosh\left(\alpha_n z/a\right)J_0(\alpha_n\rho/a)}{\alpha_n J_1(\alpha_n)\cosh\left(\alpha_n h/a\right)}.$

8. $\varphi_0\left\{\dfrac{8}{15} - \dfrac{8}{21}\dfrac{r^2}{a^2}(3\cos^2\theta - 1) + \dfrac{1}{35}\dfrac{r^4}{a^4}(35\cos^4\theta - 30\cos^2\theta + 3)\right\}.$

9. $\dfrac{10x}{l} + \dfrac{10}{\pi}\displaystyle\sum_{1}^{\infty}\dfrac{1}{n}\sin\dfrac{n\pi x}{l}\exp\left(\dfrac{-n^2\pi^2 t}{l^2 RC}\right)$ (B corresponds to $x = 0$).

10. $\dfrac{4}{5}10^{-3} + \dfrac{33}{\pi^2}10^8\displaystyle\sum_{1}^{\infty}\dfrac{1}{n}\sin\left(\dfrac{n\pi x}{1{\cdot}5\times 10^6}\right)\exp\left(\dfrac{-n^2\pi^2 t}{5{\cdot}625\times 10^9}\right).$

11. $V = a\exp\left(-x\sqrt{\omega\lambda\mu/2}\right)\sin\left(\omega t - x\sqrt{\omega\lambda\mu/2}\right).$

Chapter 8

§ 8.3

1. Displacement is $\dfrac{8h}{\pi^2}\displaystyle\sum_{n=1}^{\infty}\dfrac{1}{n^2}\sin\dfrac{n\pi}{4}(2 + \cos n\pi)\sin\dfrac{n\pi x}{l}\cos\dfrac{n\pi ct}{l}\,.$

§ 8.4

2.

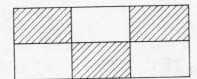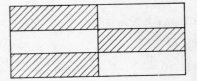

FIGURE 167

3. $x = 0,a; y = 0,a; y = x; y = -x + a.$

4. $\dfrac{64k}{\pi^6} \sum_{n,m} \dfrac{1}{(2n-1)^3} \dfrac{1}{(2m-1)^3} \sin n\pi x \sin m\pi y \cos rct$

$$\text{where } r^2 = \pi^2\{(2n-1)^2 + (2m-1)^2\}$$

5.

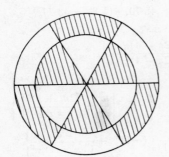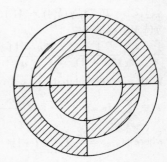

FIGURE 168

6. Displacement is $\dfrac{16k}{\pi} \sum_i \dfrac{J_0(p_i r)\cos p_i ct}{p_i^3 J_1(p_i)}$ where p_i is the ith zero of $J_0(x)$.

§ 8.5

4. Frequencies are $[(2n-1)/4l]\sqrt{\lambda/\rho}$ for positive integer values of n.

§ 8.6

1. (a) Normal modes are $A_n \cos (n\pi x/l) \cos \{(n\pi ct/l) + \varepsilon\}$ for positive integer values of n.
   (b) Normal modes are $A_n \sin (n\pi x/l) \cos \{(n\pi ct/l) + \varepsilon\}$ for positive integer values of n.

Chapter 9

§ 9.1

1. (i) $3z^2$, (ii) $2e^{2z}$, (iii) $-2z \sin z^2$
   (iv) $1/z$, (v) $-1/(z-1)^2$, (vi) $-i/(z-i)^2$.

§ 9.2

1. (i) Conjugate pair. (ii) Conjugate pair. (iii) Not a conjugate pair. (iv) Conjugate pair.
2. (i) $v = y^2 - x^2 - 2x + c$. (ii) No conjugate function exists. (iii) No conjugate function exists. (iv) $v = e^{-x}(y \cos y - x \sin y) + c$. (v) $v = y(3x^2 - y^2) + c$.

§ 9.32

4. Strength is $-\frac{1}{2}K$. Streamlines are $|z^2 - a^2|^2 = c |z|$ for arbitrary constants c.
6. Complex potential $W = -(a \ln r + b(\frac{1}{2}\pi - \theta)) - i(a\theta + b \ln r)$ where $r^2 = x^2 + y^2$ and $\theta = \tan^{-1}(y/x)$.
7. Stagnation points are on the plane boundary a distance a on either side of the normal from the doublet to the plane.
8. Complex potential $W = m \ln (z^4 + 4a^4)$.
10. Complex potential $W = m \ln (z^2 - b^2) + m \ln (z^2 - a^4/b^2) - m \ln z^2$

   Velocity on the y-axis is $v = 2my\left(\dfrac{1}{y^2 + b^2} + \dfrac{1}{y^2 + a^4/b^2} - \dfrac{1}{y^2}\right)$.

   Now $v = 0$ when $y = a$ and $v \to 0$ as $y \to \infty$. Hence v must have a maximum value at some finite distance $(>a)$ from the origin.
11. If the axis of the cylinder is the line $x = 0$, $y = 0$ and the line sources are along the lines $x = \pm a$, $y = 0$, then the stagnation points lie along the lines $x = 0$, $y = \pm b$ and $x = \pm b$, $y = 0$.

§ 9.5

1. Complex potential $W = \dfrac{u}{2}\left\{5Z - 3\sqrt{\dfrac{Z^2}{a^2} - 4}\right\} + i\, K \ln \dfrac{1}{4}\left\{\dfrac{Z}{a} + \sqrt{\dfrac{Z^2}{a^2} - 4}\right\}$.

Index